D1029105

CHEMISTRY AND PHYSICS OF CARBON

Volume 9

CHEMISTRY AND PHYSICS OF CARBON

A SERIES OF ADVANCES

Volume 9

Edited by

P. L. Walker, Jr. and Peter A. Thrower

DEPARTMENT OF MATERIAL SCIENCES
THE PENNSYLVANIA STATE UNIVERSITY
UNIVERSITY PARK, PENNSYLVANIA

1973

MARCEL DEKKER, INC. New York

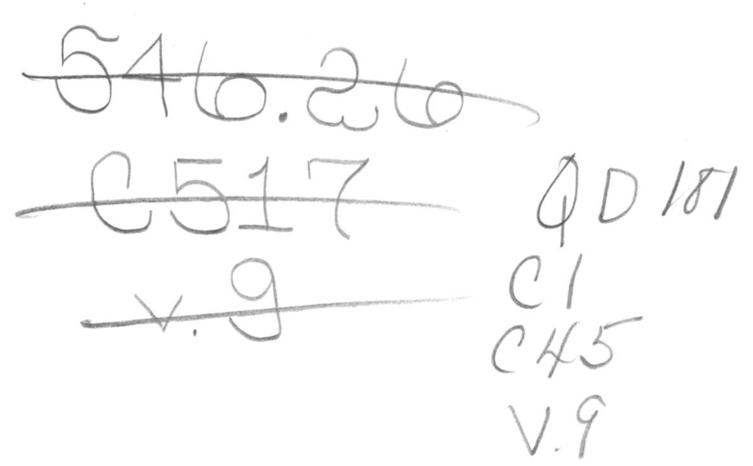

MARCEL DEKKER, INC.

95 Madison Avenue, New York, New York 10016

LIBRARY OF CONGRESS CATALOG CARD NUMBER: 65-27434

ISBN: 0-8247-6019-0

PRINTED IN THE UNITED STATES OF AMERICA

PREFACE TO VOLUME 9

While the previous volume in this series was of more theoretical and fundamental interest as far as the properties of carbon materials are concerned, Volume 9 turns our attention to practical applications. Of those uses which have been found for carbon materials in recent years the most important are probably in the realms of strong materials and of biomaterials. These two topics form the basis of the three chapters which make up this volume.

The subject of carbon fibers is one aspect of modern technology that has formed the basis of many recent articles, both in the scientific literature and in the popular press. The high specific modulus and strength of these materials have opened up a new dimension in engineering applications of composite materials, which is only just beginning to be exploited. In the manufacture of the individual fibers there have been two major starting points, one using cellulose fibers and the other using polyacrylonitrile (PAN) fibers. It is the former which is the subject of the first chapter by Roger Bacon, who has been involved in the manufacture of carbon fibers from cellulosic precursors for many years. The author has prepared an invaluable contribution to those seeking to find their way through the large number of publications that have appeared in recent years on this topic. It is later hoped to include a chapter of similar scope on fibers produced from PAN precursors.

The second chapter of this volume is concerned with pyrocarbon but in an application which was only beginning to be explored when this series began to appear. Turning from the application of pyrocarbons for the containment of nuclear fuels, Jack Bokros became increasingly involved with the use of these materials in prosthetic heart valves. In the last few years he has been much involved in the use of pyrocarbons in the whole realm of bioengineering. Together with two of his co-workers he has prepared a most fascinating and informative contribution on this subject, which perhaps allows us a glimpse into new possibilities for the material that is the subject of this series of volumes.

In the third chapter we return to the subject of carbon fibers, one of whose most spectacular uses is in the field of composite materials, where they may be incorporated into matrices of various types. One possible matrix is carbon itself, which is deposited around the fibers by the technique of chemical vapor deposition in the form of pyrolytic carbon. This approach to the production of composite materials is naturally not restricted to carbon fibers and may be applied to any other filler. The deposition of pyrolytic carbon in porous solids is the subject of the final chapter in this volume. W. V. Kotlensky reviews the deposition mechanisms and the structure and properties of pyrolytic carbon so formed. The porous materials considered are themselves mainly carbonaceous but are not so well oriented as the fibers discussed in the first chapter. The potential applications for this new family of carbon materials are discussed and these highlight the growing importance of carbon in the modern world.

Peter A. Thrower
P. L. Walker, Jr.

CONTRIBUTORS TO VOLUME 9

ROGER BACON, Union Carbide Corporation, Carbon Products Division, Parma Technical Center, Parma, Ohio

J. C. BOKROS, Gulf Oil Corporation, Gulf Energy and Environmental Systems Division, San Diego, California

W. V. KOTLENSKY, Super-Temp Company, Santa Fe Springs, California

L. D. LaGRANGE, Gulf Oil Corporation, Gulf Energy and Environmental Systems Division, San Diego, California

F. J. SCHOEN, Gulf Oil Corporation, Gulf Energy and Environmental Systems Division, San Diego, California

CONTENTS OF VOLUME 9

CONTENTS OF OTHER VOLUMES

VOLUME 8

VOLUME 10

VOLUME 11

CHEMISTRY AND PHYSICS OF CARBON

Volume 9

CARBON FIBERS FROM RAYON PRECURSORS

Roger Bacon

Union Carbide Corporation
Carbon Products Division
Parma Technical Center
Parma, Ohio

I. INTRODUCTION

High-performance carbon fibers for use as reinforcing agents in resin matrices have been known since 1965. They have formed the basis for a new class of lightweight, stiff, and strong materials that promise to find extensive use, first in aerospace structures and later in a variety of applications, such as sporting and recreational equipment. This chapter is concerned solely with the fibers themselves, their preparation, their internal structure, and their properties.

Commercial carbon fibers and textiles were made first from rayon precursors during the late 1950s and later from polyacrylonitrile, lignin, and pitch fibers. Processing refinements led to the first laboratory preparations of highly oriented high-modulus fibers: in the United States from rayon in 1964 [1], in England from polyacrylonitrile in 1964 [2, 3], and in Canada and Japan from pitch in 1969 [4, 5]. High-modulus carbon fibers prepared from pitch will probably be in commercial production in the near future.

Carbon fibers derived from rayon precursors are the subject of this chapter. However, many similarities in structure between carbon fibers made from rayon and those made from polyacrylonitrile are apparent, especially if the fibers are "graphitized" to temperatures higher than 2000 °C. These similarities are frequently emphasized in Section IV. Available evidence shows that these structural similarities also extend to graphitized fibers from many other precursors (provided they are first rendered infusible by crosslinking), such as polyvinyl alcohol, polybenzimidazole, polyamide, polyimide, resins, and pitches.

The situation is different for fibers "carbonized" by raising them to a temperature no higher than approximately 1500 °C: important differences in structure occur, depending on the precursor. For example, carbonized polyacrylonitrile (PAN) fibers possess a structure that is well oriented, yet highly nongraphitic, a combination leading to very high strain-to-failure tensile properties. Exploitation of the structural differences at low heat-treatment temperatures among the various possible precursors may be the key to future developments in carbon-fiber technology.

II. MECHANICAL PROPERTIES OF SEVERAL FORMS OF GRAPHITE

A. Single-Crystal Graphite

The highly anisotropic nature of the graphite crystal is well known. Young's modulus measured parallel to the basal plane ($1/s_{11}$) is 148×10^6 psi (five times that of steel), whereas that measured normal to the basal plane ($1/s_{33}$) is 5.3×10^6 psi (one-half that of aluminum) [6]. The shear modulus for basal-plane shear ($c_{44} = 1/s_{44}$) is 0.6×10^6 psi for dislocation-free graphite and more than an order of magnitude lower than this value (equivalent to that of polyethylene) for graphite containing dislocations [6, 7]. The tensile strength measured parallel to the basal plane is, in theory, at least 15×10^6 psi [8], and strengths as high as 3×10^6 psi have been measured on graphite whiskers [9]. By contrast, the shear strength between basal planes for dislocation-containing, but otherwise perfect, single-crystal graphite is only approximately 100 psi [10]. Crystals containing delamination voids are even weaker.

B. Conventional Polycrystalline Graphites

The mechanical properties of conventional "two-phase" polycrystalline graphites are strongly affected by the low basal-plane shear stiffness and the low shear strength of the graphite crystal [11]. A typical value of Young's modulus for an extruded graphite is 1.5×10^6 psi parallel to the extrusion axis and perhaps half that perpendicular to the axis; the tensile strength is typically 1500 psi. When the internal flaws (cracks parallel to the layer structure) are smaller — that is, when finer grained filler or grist materials are used — the resulting graphites are much stronger; however, a flexural strength of approximately 10,000 psi is considered to be the upper limit for graphites made by conventional processing methods [11].

C. Polymer Carbons

Very fine-grained "polymer carbons" can be made by unconventional processing methods [12]. These are single-phase materials made by carbonizing various resins of the nongraphitizing variety. Unless the starting material is in fiber form, these materials are always isotropic — that is, they exhibit no preferred orientation. The flexural strengths of polymer carbons are typically near 15,000 psi (30,000 psi maximum) for the 1000 to 2000 °C forms; Young's modulus is 4×10^6 to 4.5×10^6 psi. When these materials are heat-treated to 3000 °C, both strength and stiffness decrease.

D. Pyrolytic Graphite

The attainment of significantly higher strengths and stiffnesses in carbon materials requires that they possess a high degree of preferred orientation. Pyrolytic graphite [13], formed by cracking hydrocarbon gases onto a hot substrate, exhibits the molded-graphite type of preferred orientation (graphite c axes preferentially oriented along a single direction). Since Young's modulus and, to a lesser degree, tensile strength parallel to the basal-plane direction are very sensitive to the degree of orientation (which in turn is highly dependent on processing variables), typical values cannot be given. However, the extreme values that have been attained in highly oriented pyrolytic graphite illustrate the significance of preferred orientation in polycrystalline graphite. In the best case pyrolytic graphite that has been compression-annealed [14] exhibits elastic constants considered to be equivalent to those of the single crystal [6]; for example, Young's modulus is as high as 148×10^6 psi. Pyrolytic graphite that has been hot-worked in tension at elevated temperatures exhibits a room-temperature strength as high as 45,000 psi (higher than 100,000 psi at elevated temperatures) [15].

Although these pyrolytic graphites are turbostratic in the as-deposited state, they are fully graphitized after high-temperature annealing or stress annealing. They exhibit, therefore, the low basal-plane shear strength and shear modulus [10] and the low c-axis tensile strength and tensile modulus that are characteristic of the graphite crystal. The low basal-plane shear strength of highly oriented pyrolytic graphite is partly responsible for the fact that the highest tensile strengths are only approximately 3 times those of polymer carbons, although the modulus is 20 to 30 times greater. An interesting intermediate case is that of laminar pyrolytic carbon deposited at approximately 1400 °C; the tensile strength is approximately 7×10^4 psi, whereas the corresponding modulus is only 7×10^6 psi, indicating that this material is not very highly oriented [16]. Although this poorly oriented material must possess many structural faults, it has a high tensile strength, probably the result of a relatively high basal shear strength.

E. Carbon Fibers

High-modulus carbon fibers exhibit the extruded-graphite type of preferred orientation in which the graphite c axes are preferentially confined to the plane transverse to the fiber axis. Thus the high-strength and high-modulus properties are found only along the fiber-axis direction. Very highly oriented graphitized fibers possess a Young's modulus (after correction for porosity) as high as 120×10^6 psi and a corresponding tensile strength of 5.5×10^5 psi. As in the case of pyrolytic carbon, however, the highest tensile strengths are not necessarily exhibited by fibers possessing the highest Young's moduli. Carbon fibers from polyacrylonitrile that have been heat-treated to about 1400 °C exhibit strengths sometimes exceeding 6×10^5 psi, although their Young's modulus is only approximately 40×10^6 psi.

Because they are essentially polymer carbons insofar as their chemical origins are concerned, carbon fibers are very fine grained in addition to being highly oriented. Although the graphitic layers may extend nearly straight for thousands of angstroms parallel to the fiber axis, they are very narrow, or exhibit high tortuosity, in the transverse direction. These structural characteristics are responsible for the relatively high longitudinal shear strengths (greater than 25,000 psi) of carbon fibers, a characteristic believed to be necessary for the achievement of very high tensile strengths.

III. PREPARATION OF HIGH-MODULUS CARBON FIBERS FROM RAYON

Every process for converting rayon into high-modulus carbon fibers includes three basic steps: heat treatment, carbonization, and graphitization. Application of appreciable degrees of stretch during graphitization (stress graphitization) is essential to the achievement of a truly high-modulus fiber from rayon. However, from the point of view of process control, the most critical part of the process is the heat treatment. This step involves the complete destruction of the cellulose-polymer structure and the formation of a primary char; it is accompanied by a weight loss of roughly 50% after the yarn has been heated to approximately 300 to 400 °C. The use of chemical impregnants and/or reactive atmospheres during heating has permitted a reduction in heat-treatment times from many hours to a few minutes in presently used continuous processes. After the yarn has been heat-treated, carbonization is usually carried out in an inert atmosphere to a final temperature of 1000 to 1500° C; processing time may be well under 1 min. Stress graphitization is then carried out at temperatures of 2500 to 3000° C; the residence time in the furnace is only a few seconds. The overall weight yield for processing rayon into graphitized yarn is usually in the range 20 to 30%.

A. History

Public disclosures of practical processes for converting cellulose textiles into useful carbon fibers have been confined largely to the patent literature. Edison [17] is credited with the first invention of this kind: an incandescent-lamp filament made by carbonizing a filament of natural or regenerated cellulosic material. After the advent of tungsten lamp filaments in the early 1900s, filamentary carbon was not used significantly by industry until the middle 1950s, when Abbott [18] developed a process for converting rayon into a fibrous carbon material for such uses as insulation, filtration, and adsorption. This material, produced by the Carbon Wool Corporation, was carbonized at temperatures of up to 1000° C; it possessed tensile strengths as high as 40,000 psi [19]. In 1959 the Union Carbide Corporation began commercial production of "graphite" cloth as well as other fibrous forms [20, 21] made by a slow batch process [22] consisting of baking the fibers in an inert atmosphere to approximately 900° C, followed by graphitizing to temperatures usually higher than 2500° C. Following Union Carbide into this field were such companies as HITCO [23], the 3M Company [24], the Great Lakes Carbon Corporation [25], and the Carborundum Company [26]. In addition to graphite fibers, lower temperature carbon fibers (usually heat-treated to 1200 to 1800° C) were also produced, primarily in cloth and felt forms. The single-filament tensile strengths of these fibers ranged from 50,000 to over 100,000 psi, while their Young's moduli were only 4×10^6 to 7×10^6 psi. The most important application was for the reinforcement of phenolic resins to form ablative materials for rocket and missile components.

Schmidt and Hawkins [27] have reviewed the early history and applications of carbon fibers.

In the first years of this new carbon-fiber industry several companies experimented with chemical pretreatments of the rayon raw material. These treatments proved capable of increasing carbon yields and speeding up processing rates, especially during the early "heat-treatment" stage. The first commercially produced carbon textile made according to this type of process was 3M Company's Pluton, introduced in 1961 [24]. An advantage of the chemical pretreatments is the ability to use rapid continuous-processing methods. The first continuously processed fibers [28] were Union Carbide's carbon and graphite yarns, introduced to the market in 1963. Similar types of yarn were subsequently produced by HITCO and the Carborundum Company.

The availability of strong and uniform continuously processed carbon yarns permitted, for the first time, very high degrees of stretch to be applied to these yarns during graphitization to temperatures in excess of 2500°C [29, 30]. These experiments led to the commercial production of high-strength and high-modulus graphite fibers (Thornel 25) late in 1965 by the Union Carbide Corporation. That same year an Air Force Materials Laboratory contract [31] was awarded to Union Carbide to further develop this stress-graphitization process and to discover new processes for producing high-performance carbon fibers. Higher modulus yarns were subsequently produced commercially, the most recent (1970) being Thornel 75, a yarn with a modulus of 75×10^6 psi. The only other major producer of high-modulus carbon yarns from a rayon precursor is HITCO, which entered the market in 1967 with yarns produced by similar methods [32].

B. Raw Material

The most commonly employed cellulosic raw material for making high-modulus graphite yarn is the textile rayon Villwyte (IRC Fibers Co., subsidiary of American Cyanamid Co.). Figure 1 shows a photomicrograph of several filament cross sections and an X-ray fiber pattern of a Villwyte yarn. Several other types of regenerated-cellulose fiber have been tried as raw materials, including saponified acetate rayon yarn (Fig. 2), tire cord (Fig. 3), and cuprammonium rayon (Fig. 4). However, each of these fibers suffers from one or more defects as a carbon-fiber raw material [33]: tire yarns possess modifiers that tend to cause interfilament sticking during

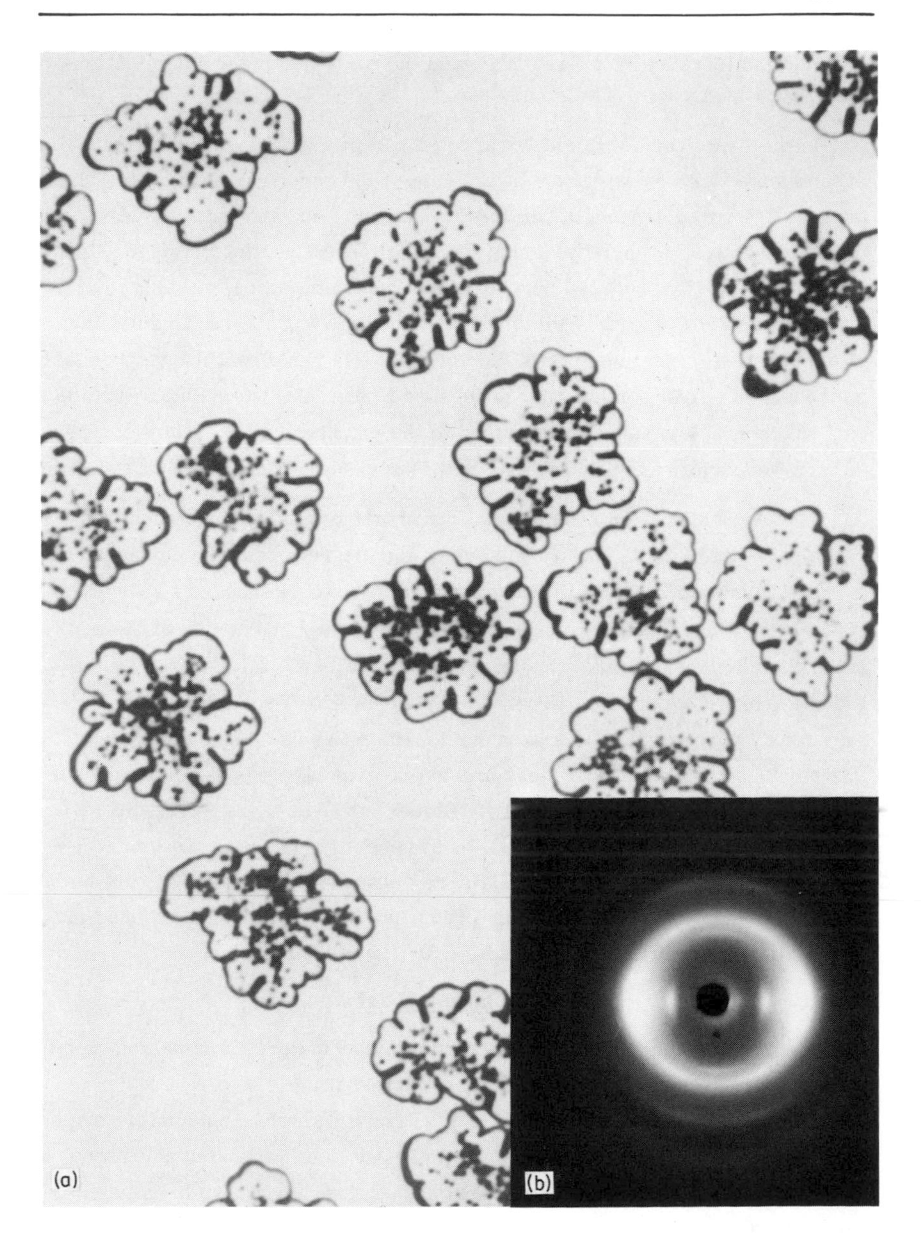

FIG. 1. (a) Photomicrograph of cross sections of Villwyte rayon. (b) X-Ray fiber diffraction pattern of Villwyte rayon. From Ref. 33.

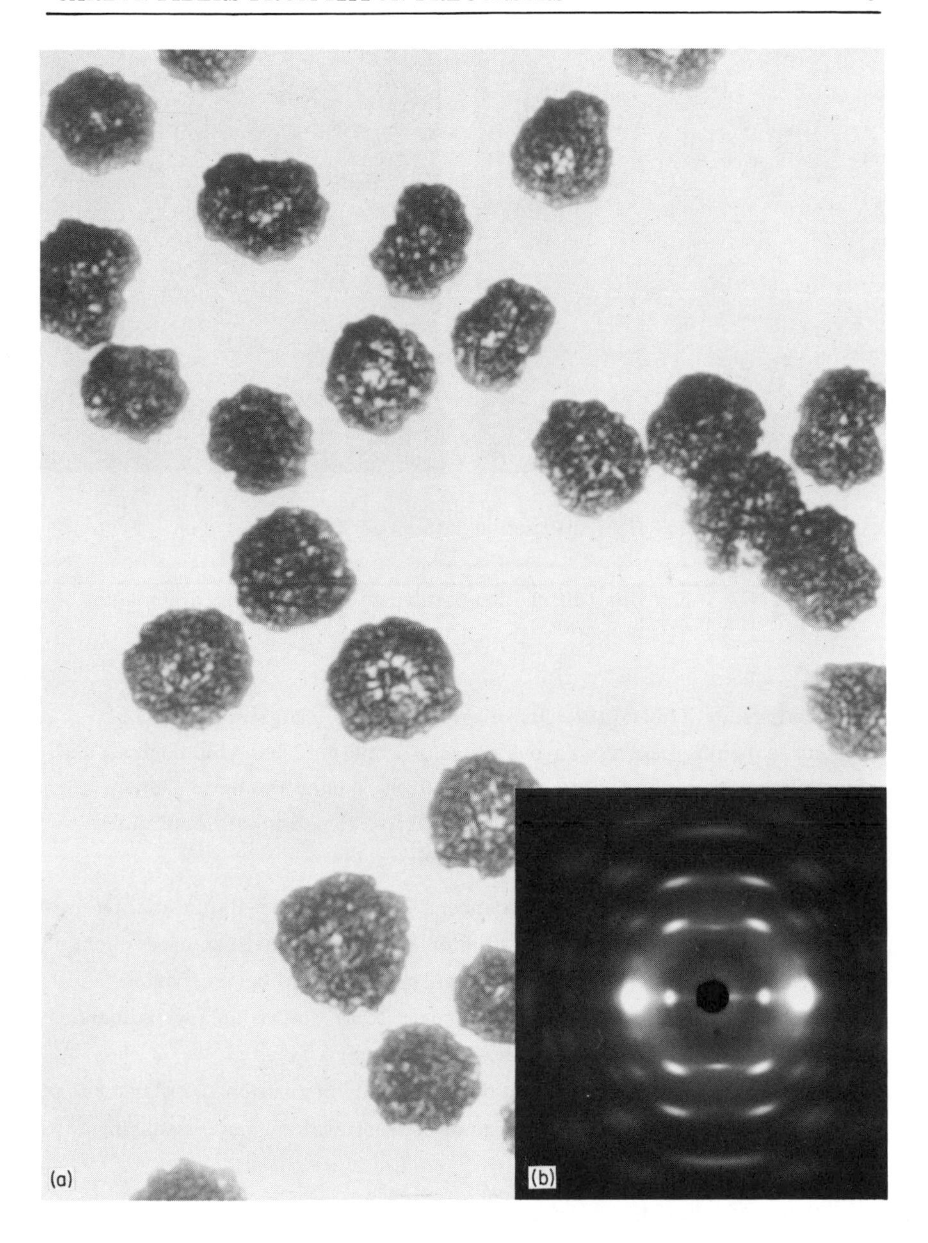

FIG. 2. (a) Photomicrograph of cross sections of Fortisan saponified acetate rayon. (b) X-Ray fiber diffraction pattern of Fortisan. From Ref. 33.

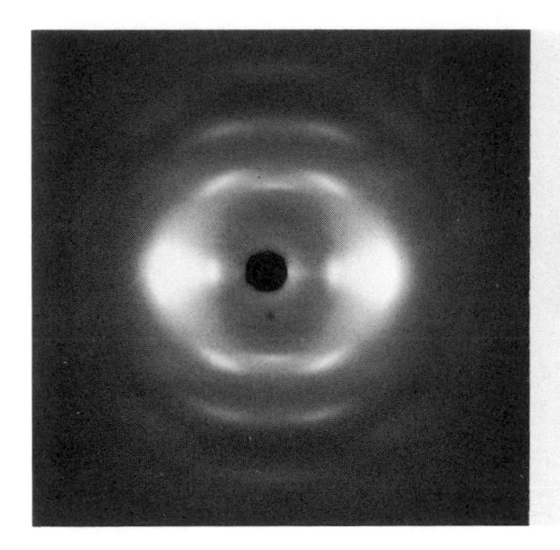

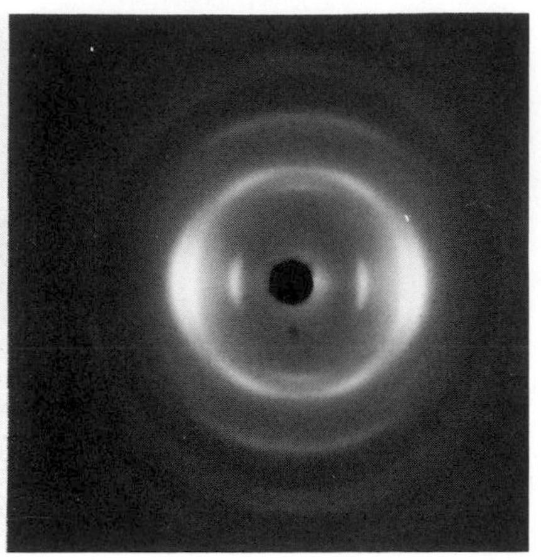

FIG. 3. X-Ray fiber diffraction pattern of Tyrex tire rayon. From Ref. 33.

FIG. 4. X-Ray fiber diffraction pattern of Bemberg cuprammonium rayon. From Ref. 33.

processing, and, in addition, these yarns possess a high internal-void content; saponified acetate rayons possess a high internal-void content; and cuprammonium rayons possess a high degree of interfilament bonding. High void content results in weak filaments after carbonization; interfilament sticking or bonding results in yarn brittleness.

A patent issued to Société le Carbone Lorraine [34] claims that the new polynosic rayons are superior raw materials for carbon-fiber production, but no high-modulus yarns made from such fibers have been reported. A patent issued to the Nippon Carbon Company [35] claims an approximately threefold improvement in the strength of yarn carbonized at 800°C when viscose-rayon fibers possessing a degree of polymerization (DP) higher than 450 are used as raw material instead of conventional rayons possessing a DP of 250.

Natural cellulosic fibers, such as cotton, can be processed without tension (e.g., in cloth form) to produce flexible carbon textiles, but they have generally proved to be inferior to the synthetic cellulosic fibers [18, 24]. They are unsuitable for the production of high-modulus yarns; since the fibers are discontinuous, they cannot be stretched. Structurally they are quite different from rayons: their crystal structure is cellulose I, whereas that of synthetic rayons is cellulose II [36]. The degree of crystallinity is very high, a fact that seems to be detrimental to obtaining good carbon-fiber properties: low carbon yields and poor carbon-fiber strengths are obtained from cotton and, especially, from the highly crystalline ramie [37, 38]. A very important cause of poor strength with these fibers may be the existence of structural flaws, which can initiate brittle fracture in the carbonized fiber.

C. Heat Treatment

The purpose of the heat-treatment step is to completely convert the cellulose structure into a char suitable for rapid carbonization to temperatures exceeding 1000°C. Although water is the principal product evolved during heat treatment, many other substances, including complex tars, are also given off.

Avoidance of excessive tar formation and, especially, of tar redeposition onto the filaments is probably the most important single consideration for the successful heat treatment of rayon. Tarry deposits cause interfilament cementation, leading to eventual brittleness and friability of the yarn. A somewhat more obvious pitfall in heat-treating rayon is the use of excessive heating rates, which can cause disruption of the filament structure. The result is weak, often fuzzy, yarn.

Some of the principal methods that have been used to heat-treat rayon are discussed in this section. They may be classified according to whether they involve pyrolysis in an inert atmosphere, in reactive atmospheres, or after chemical preimpregnation.

1. Inert Atmosphere

A basic process for heat-treating rayon in an inert atmosphere is that described in the patent issued to Ford and Mitchell [22]. After the rayon is dried at about 100°C to remove adsorbed moisture, it is heated very slowly (e.g., 10°C/h) to 400°C, after which it is ready for more rapid carbonization. Although such very slow heating rates might not be necessary in a continuous process for yarn production (where a good inert-gas flow can be maintained over the yarn to remove tarry vapors), this process nonetheless requires many hours.

Removal of tarry substances is effectively accomplished if the heat treatment is carried out in an oil bath [26]. This "French-frying" technique has the additional advantage of providing a better heat-transfer medium; however, the process apparently requires more than 10 h.

The complex reactions that occur during the carbonization of rayon and other forms of cellulose have been considered previously [39, 40] and are not discussed here. Due primarily to the production of tars, the weight yield of graphitized fibers that have been initially heat-treated in an inert atmosphere is only approximately 15%.

2. Reactive Atmospheres

Processing rates and overall weight yields can be improved markedly if heat treatment is performed in the presence of reactive atmospheres, such as air or oxygen [33, 41, 42], chlorine [43, 44], and hydrochloric acid vapor [45, 46]. Although the pyrolysis reactions obviously vary with the atmosphere used, they all seem to promote dehydration of the cellulose structure and to inhibit the formation of tars [45, 47, 48]. The dehydration reactions begin to occur much earlier than they do when heat treatment is performed in an inert atmosphere, and the major part of the pyrolysis takes place over a wider range of temperatures (Figs. 5 and 6). Shindo [45] found that when HCl is used, pyrolysis begins as early as 110°C (Fig. 5) and proceeds almost entirely through the elimination of water. After pyrolysis is completed, only one carbon atom out of six is lost; tar formation is obviously suppressed. The influence of an oxygen atmosphere on the low-temperature pyrolysis (Fig. 6) is similar to that of HCl, but not nearly so severe. A better weight yield than that indicated in Fig. 6 is obtained if the heating rate is reduced through the critical region between 240 and 290°C.

Structural and mechanical property changes during the heat treatment of rayon in HCl [45] and in oxygen [41, 42] atmospheres are shown in Figs. 7 through 10. These properties are plotted as functions of weight loss rather than of temperature, since they tend to change exceedingly rapidly with temperature in the critical pyrolysis range. Tensile strength (Figs. 7 and 8) falls very rapidly during the initial stage of decomposition, which seems to occur exclusively in the amorphous regions: no loss of X-ray diffraction intensity from the crystalline cellulose is detectable until a 15 or 20% weight loss has occurred. The tensile strength reaches a minimum near the point at

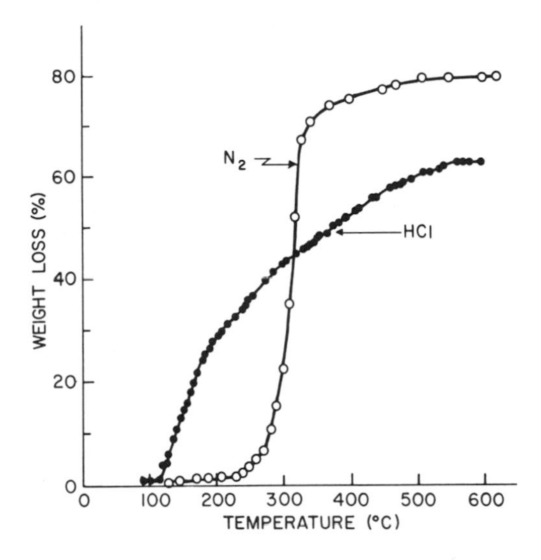

FIG. 5. Effect of HCl vapor on weight loss during pyrolysis of rayon. Heating rate under HCl was 60° C/h; that under nitrogen was 120° C/h. From Ref. 45.

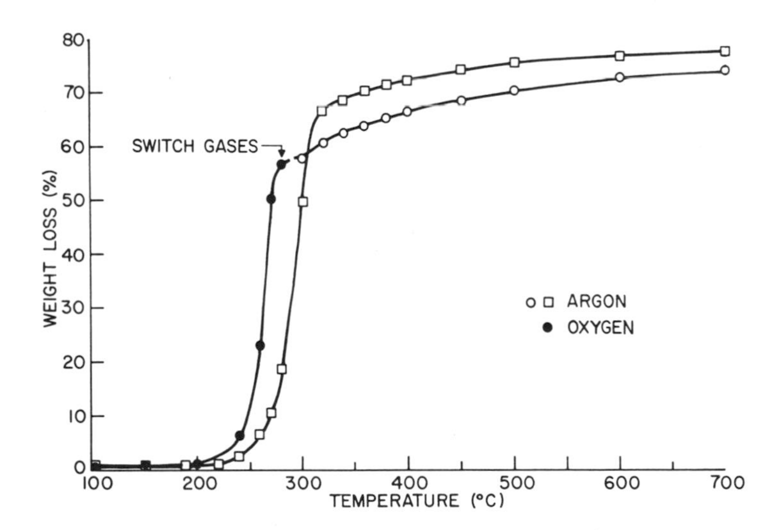

FIG. 6. Effect of an oxygen atmosphere on weight loss during pyrolysis of Villwyte rayon. Heating rate was 60° C/h.

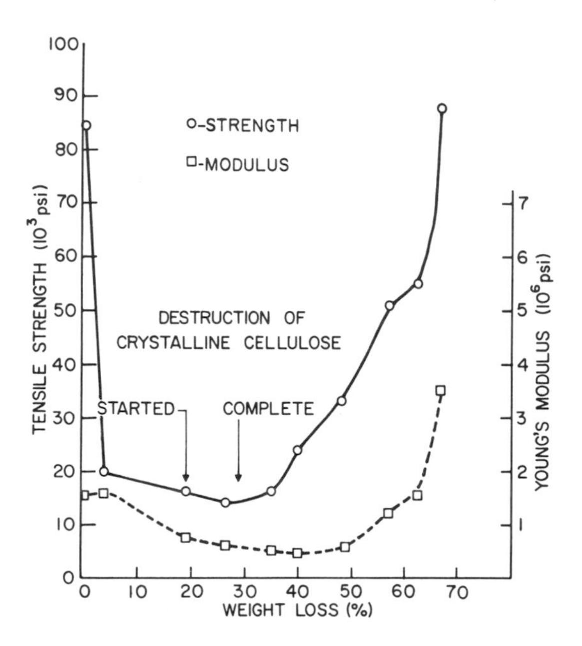

FIG. 7. Tensile strength and modulus of rayon heat-treated under HCl vapor, at various stages of heat treatment and carbonization to 1000° C. Heating rate was 60° C/h. From Ref. 45.

which the destruction of the crystalline cellulose is complete and then begins its remarkable recovery. Contrary to the initial behavior of the tensile strength, the fiber modulus increases during the initial pyrolysis. Thereafter tensile strength and modulus both fall together, but the modulus continues to fall long after the tensile strength has begun its recovery. Finally, the modulus begins to increase markedly as the graphitic layer structure forms and the fiber becomes a brittle carbon material. Additional insight into the changes in physical nature of the fiber during pyrolysis is provided by plots of the fiber elongation at break (Figs. 9 and 10). The initial drop in elongation reflects the degradation of the cellulose structure, at first in the amorphous regions and then in the crystalline regions; as soon as the cellulose degradation is complete, a rapid improvement in elongation takes place until the modulus reaches its minimum, after which the fiber starts to become brittle and the elongation drops again.

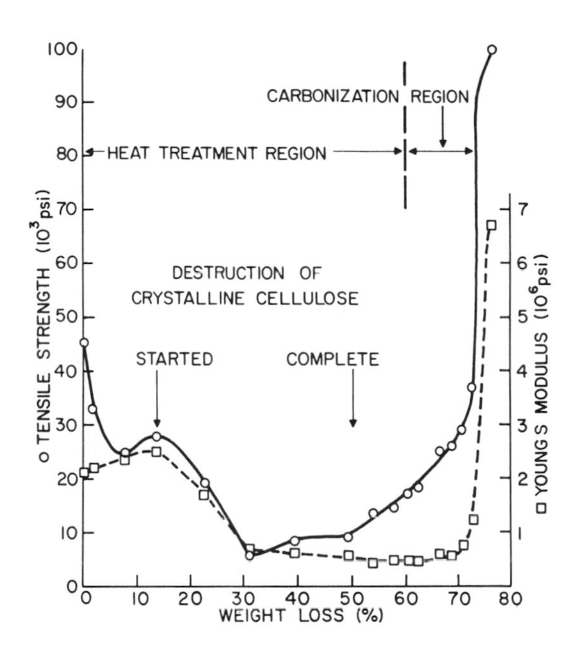

FIG. 8. Tensile strength and modulus of Villwyte rayon heat-treated under oxygen, at various stages of heat treatment and carbonization to 1300°C. Heating schedule: 18 min under oxygen to 40% weight loss; 12 min under nitrogen to 60% weight loss; 16 min to 73% weight loss. From Ref. 41.

Total processing times for fibers heat-treated in reactive atmospheres are reported to lie generally in the range of approximately 20 min to 10 h [33, 44, 46]. Carbon weight yields for fibers carbonized to 1000°C are approximately 23% for an air or oxygen atmosphere [33], 27% for a chlorine atmosphere [44], and 35% for an HCl atmosphere [45]. (Some additional weight loss occurs during graphitization to 2800 to 3000°C.) The theoretical carbon yield for pure cellulose is 44%.

3. Chemical Preimpregnation

The effectiveness of reactive atmospheres in increasing processing rates and improving weight yields is limited by the diffusion rates of the reactive gas species into the interior of the fiber. Although the outside skin readily reacts with the atmosphere, the core material may not be reached by the reacting species before elapsed times of several hours. Diffusion

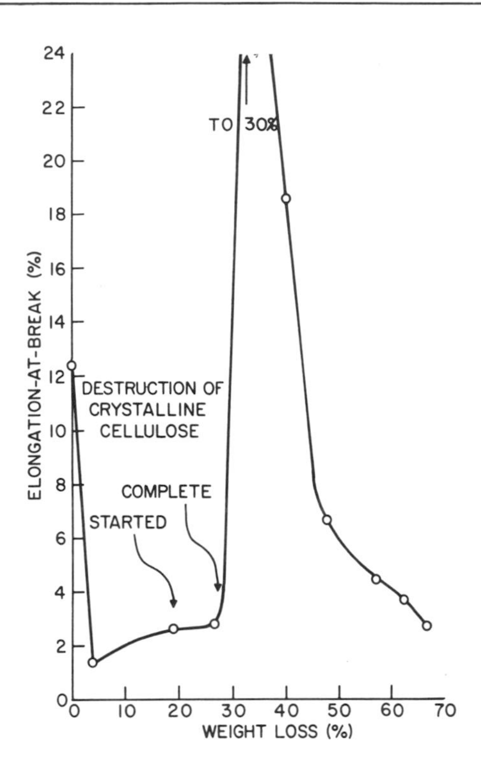

FIG. 9. Elongation at break of rayon heat-treated under HCl vapor, at various stages of heat treatment and carbonization (see also Fig. 7). From Ref. 45.

rates can be increased by increasing the heat-treatment temperature; however, this approach is limited by the fact that degradation of the fiber structure occurs due to excessive pyrolysis rates of unreacted cellulose at temperatures approaching 290° C. This diffusion limitation is overcome if the fiber is impregnated with the reactive chemical prior to heat treatment. Rayon is well suited to this approach because it is easily swelled in water and can be quickly impregnated with salts, acids, etc., in aqueous solution.

A variety of carbonization promoters have been used to impregnate rayon prior to heat treatment. They include nitrogenous salts of strong acids [24], acids or acidic salts [28], metal halides [49], and various derivatives of phosphoric acid [50]. All of these substances, as well as others reported in the literature [47, 48, 51-56], are effective flame retardants for cellulose.

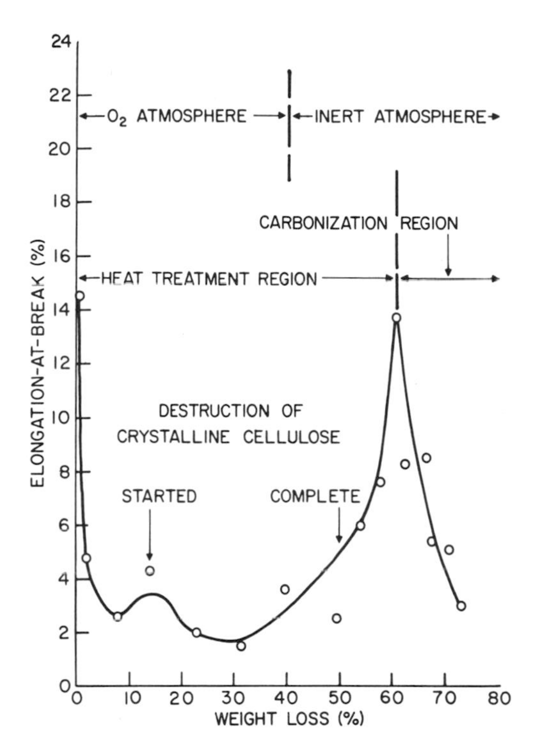

FIG. 10. Elongation at break of Villwyte rayon heat-treated under oxygen, at various stages of heat treatment and carbonization (see also Fig. 8). From Ref. 41.

They promote the dehydration of the cellulose structure at low temperatures (150 to 200°C) and stabilize the molecule against excessive formation of volatile tars, the substances responsible for the flammability of cellulose [48]. Thus flame retardants serve also as carbonization promoters since they speed reaction rates, spread the reactions out over a wider temperature range, and improve carbonization yields. The applications of these types of treatment to the carbonization of rayon textiles have been reviewed by Ross [57], and the mechanisms by which flame retardant cellulosic fabrics are made by various nitrogen- and phosphorus-containing compounds have been recently discussed by Reeves et al. [55].

The heat treatment of rayon impregnated with carbonization promoters as described here can be accomplished in a few minutes. Carbon weight yields after graphitization are approximately 30% of the dry starting-material weight.

4. Use of Tension during Heat Treatment

The use of tension during the heat treatment of rayon has been investigated as a possible means of increasing the preferred orientation of the final carbon fibers (a procedure known to be effective with PAN precursors [58]). The conclusion of the rayon investigation [41] was that tensioning or stretching is essentially ineffective during the heat-treatment region as defined, for example, in Fig. 10 (usually up to about 300°C). The reason for this ineffectiveness is that the polymer structure is being severely degraded throughout most of this period, due to thermal and hydrolytic scission of the ether linkages connecting cellobiose units. Consequently no amount of tension or stretching is capable of maintaining or increasing molecular preferred orientation. When thermal treatment is carried beyond the conventional heat-treatment stage into the carbonization stage, stretching is effective in improving the modulus and strength of the resulting final carbon or graphite fibers. In this latter case a polymerized structure (incipient graphite structure) is beginning to form, and stretching orients this structure to some degree. This process has been called stress-carbonization and is further described in the next section.

D. Carbonization

Properly heat-treated rayon can be carbonized to temperatures usually ranging from 1000 to 1500°C in times considerably less than 1 min. Carbonization is usually carried out in an inert atmosphere. The use of various reactive atmospheres, including hydrochloric acid [45, 46], chlorine [47], and ammonia and alkylamines [44], is reported in the scientific and patent literature.

The use of tension during the carbonization of heat-treated rayon yarn has been found to increase the degree of preferred orientation and hence to improve fiber mechanical properties [59, 60]. Although stretching during carbonization raised the Young's modulus of a 900°C carbon fiber from 6×10^6 to 10×10^6 psi, the effect was much more striking after the fiber was graphitized (stress-free). Figure 11 shows that the Young's modulus of the fiber after graphitization can be increased from 4×10^6 to 30×10^6 psi by

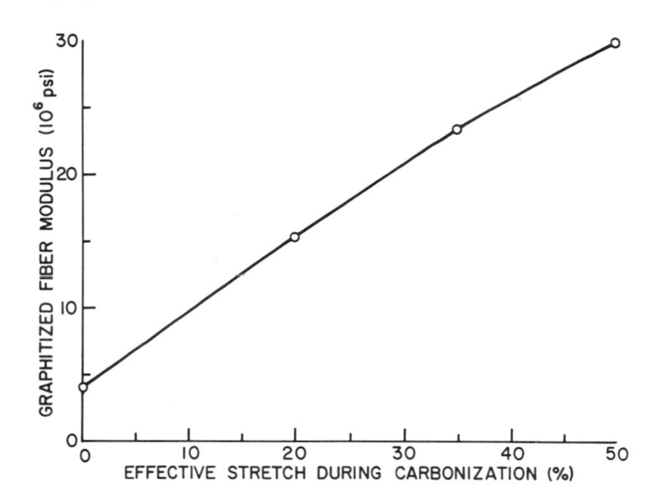

FIG. 11. Effect of stretching during carbonization to 900° C on graphite-fiber Young's modulus. From Ref. 59.

stretching by an amount that is effectively 50% during carbonization. (Effective stretch is the increase in length of a stress-carbonized fiber compared with a fiber that has been carbonized without stress; the latter actually shrinks approximately 15%.) The data of Fig. 11 were obtained in static loading experiments. The tensile stress in the fiber after 50% effective stretch was approximately 10,000 psi.

In stress-carbonization experiments the most effective stretching occurs early in the carbonization process, that is, when the fiber is rapidly developing the incipient graphite structure, but is still quite plastic. A much more effective way to attain a high modulus in rayon-base carbon fibers is by stress graphitization. The fibers go through another highly plastic stage during the graphitization process beginning at approximately 2500° C, and at these temperatures stretching is extremely effective in inducing preferred orientation. This process will be described in the next section.

E. Graphitization

The carbonization step increases the strength of the heat-treated fiber from about 20,000 to about 100,000 psi while at the same time removing a large part of the residual volatiles. The carbon-yarn strength does not depend very much on the final carbonization temperature between 1000 and 1500°C or even as high as 2000°C [33]. The residual volatilization remaining after carbonization declines sharply between 1000 and 2000°C (very roughly, from 20 to 5%). At the same time the percentage of total fiber porosity that is inaccessible ("closed pores") increases from near zero to more than 90% [61]. Regardless of the carbonization temperature, the yarn is capable of being graphitized very rapidly. The graphitization temperature is usually above 2800°C. Residence times in continuous graphitization as short as 0.1 sec have been reported [32]. Good-quality carbon yarn must also be capable of withstanding the high tensile stresses used in the stress-graphitization process to produce high-modulus fibers.

The preparation of truly high-modulus (50×10^6 psi or greater) graphite yarn from a rayon precursor is accomplished only if stress is applied to the yarn during graphitization [29, 30, 32]. The results of early static stress-graphitization experiments in which static loads were affixed to carbon yarn while it was heated to graphitization temperatures are shown in Fig. 12. In these experiments the sample began to stretch at approximately 1600°C and

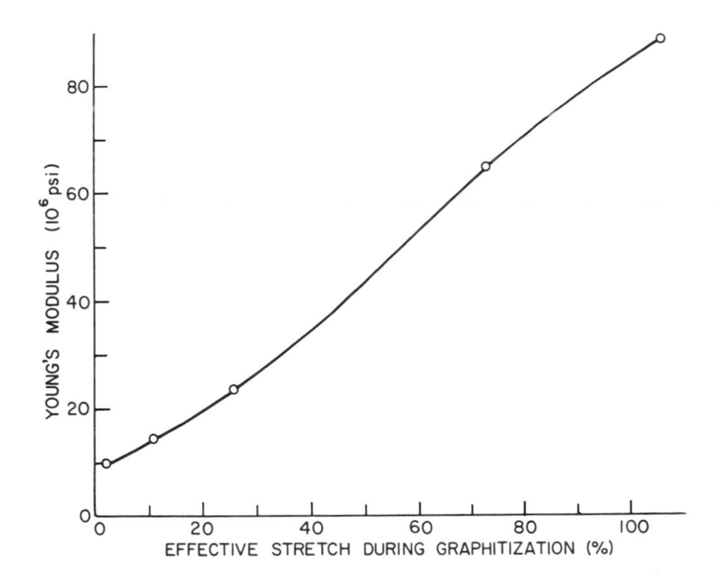

FIG. 12. Effect of stretching during graphitization to 2800°C on graphite-fiber Young's modulus.

was heated for an additional 45 to 50 min to about 2800°C. In the case of the sample that stretched over 100% the yarn tensile stress at the end of the experiment was approximately 80,000 psi. The effective stretch in Fig. 12 is calculated from the reduction in filament cross-sectional area compared with an unstressed but graphitized fiber, taking into account the variation in filament density with modulus (see Fig. 46 in Section V.A); unstressed fibers grew approximately 6% during graphitization.

A similar, but not identical, dependence of Young's modulus on the percentage of stretch was found by Gibson and Langlois [32], who have reported the results of continuous stress-graphitization experiments in which 0.25-sec residence times were used. (In comparing their data with those of Fig. 12 one must take into account the differences in the modulus-density relationship between the two cases; the results are in reasonable agreement when one considers the differences in residence times used and the fact that the starting carbon yarn may have been quite different in the two cases.) Gibson and Langlois showed that Young's modulus is insensitive to residence time in the furnace over the range 0.1 to 0.5 sec. They did find a dependence of Young's modulus on graphitization temperature for a given percentage of stretch (see Fig. 13).

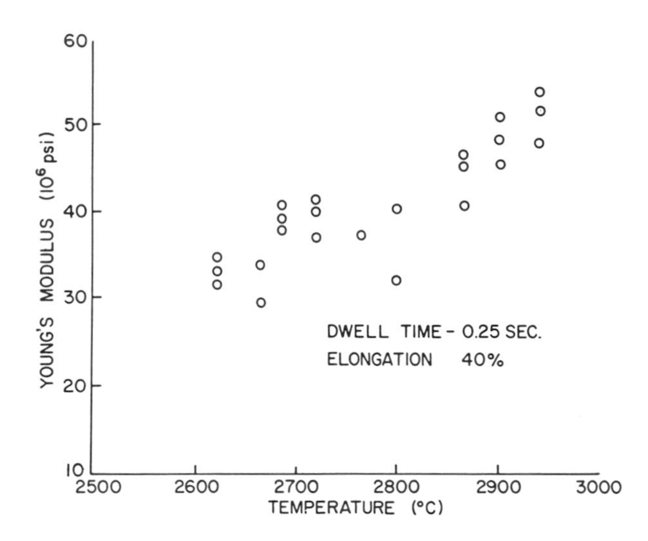

FIG. 13. Effect of graphitization temperature on Young's modulus. From Ref. 32.

The relationship between tensile strength and Young's modulus is given in Section V (Fig. 47). The relationship between Young's modulus and the degree of preferred orientation of graphite layer planes is discussed in Section IV (Figs. 20 and 21).

IV. THE STRUCTURE OF CARBON FIBERS

The structure of carbon fibers has been studied by electron diffraction; X-ray wide- and small-angle diffraction; transmission, scanning, and surface-replica electron microscopy; and light microscopy. These techniques provide information about structural features ranging in size from a few angstroms to thousands of angstroms. This section describes studies of carbon-fiber structure in which each of these techniques is used and reviews the important results obtained by each technique. It ends with the description of a model of the carbon-fiber structure that best fits the observations.

The single most important feature of the high-modulus carbon-fiber structure is the high degree of preferential alignment of graphitelike crystallites parallel to the fiber axis. A nearly perfectly oriented carbon fiber is an extreme form that, therefore, is a particularly simple example of carbon-fiber structure. (An isotropic fiber is another, not very interesting, extreme form.) Electron-diffraction patterns of carbon-fiber internal structure are visually easy to interpret, compared with X-ray diffraction patterns, because of their very small diffraction angles. A good electron-diffraction pattern of a highly oriented fiber exhibits "at a glance" all of the important qualitative structural features on the scale of interatomic distances. We therefore begin with a description of the electron-diffraction pattern of a very highly oriented carbon fiber.

A. Electron-Diffraction Patterns

A selected-area electron-diffraction pattern from a fragment of a very highly oriented rayon-base carbon fiber is shown in Fig. 14 [62]. The fiber axis was normal to the incident electron beam. The following interpretation is that given by Fourdeux, Perret, and Ruland [63], who obtained the pattern, and is based on theoretical work by Ruland and Tompa [64], who calculated the (hk) intensity distributions for nongraphitic carbon fibers with preferred orientation.

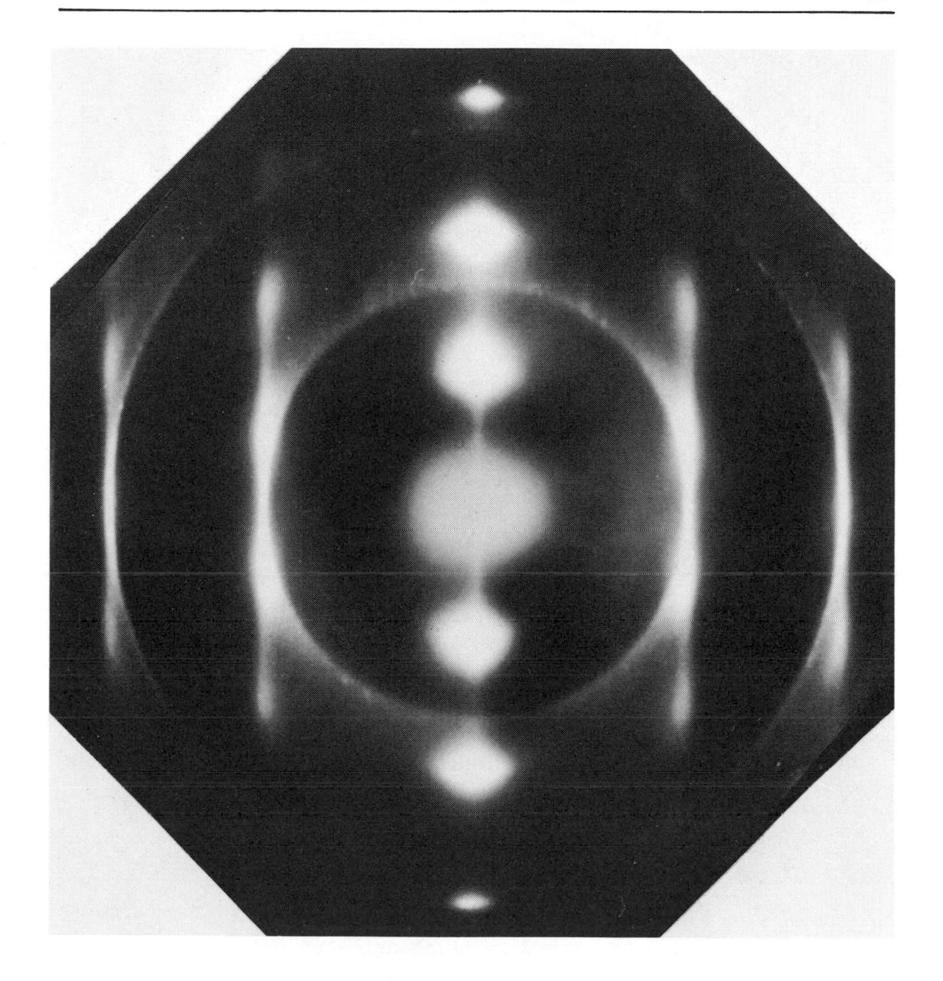

FIG. 14. Transmission-electron-diffraction pattern of a highly oriented rayon-base carbon fiber. Incident beam approximately normal to fiber axis, which is horizontal on the page. From Ref. 62.

When a beam of electrons (or X-rays) is incident (unit vector $\underline{S}_0$) on a sample, the resulting scattering intensity in any direction, $\underline{S}$, is given by an intensity function $I(\underline{s})$, where $\underline{s}$ is the scattering vector in reciprocal space:

$$\underline{s} = \frac{\underline{S} - \underline{S}_0}{\lambda} . \tag{1}$$

By definition, the magnitude of s is given by

$$s = \frac{2 \sin \theta}{\lambda}, \tag{2}$$

where θ is the Bragg angle and λ is the wavelength. The distribution in reciprocal space of scattering intensity is derived from the distribution in real space of matter in the sample through a Fourier transform. An electron-diffraction pattern gives a "picture" of a nearly planar section, passing through the origin, of this intensity distribution because the Ewald sphere of reflection [65] possesses a radius of $1/\lambda$, which, for electrons, is large in comparison with the size of the region in reciprocal space (range of s values) usually studied. In the present case the incident electron beam is normal to the fiber axis, and the fiber structure is cylindrically symmetrical; therefore the distribution of scattering intensity may be expressed as a function of only two spherical coordinates in reciprocal space, s and ϕ, where ϕ is the colatitude angle measured with respect to the fiber axis. Because both the incident beam and the scattered beam are very nearly normal to the page in Fig. 14, s can be thought of as the vector from the center to any point in the pattern, and hence s and ϕ are the polar coordinates of this point.

To learn how $I(s, \phi)$ is derived from the oriented, polycrystalline structure of a carbon fiber, we begin with the intensity distribution for scattering from a single crystal of hexagonal graphite. This distribution is a three-dimensional array of "spots" in reciprocal space, which, when rotated about the c axis, can be displayed as a single planar section through the c axis (Fig. 15). Except for the (00ℓ) spots, all of the $hk\ell$ interferences are now in the form of hoops concentric with the c axis. Each hoop is generated from a spot of intensity distribution surrounding each ($hk\ell$) reciprocal lattice point. The sizes of the spots indicate (qualitatively) the relative intensities of the various ($hk\ell$) interferences. Thus the pattern of Fig. 15 is the ideal electron-diffraction pattern resulting from a multitude of large graphite crystallites all stacked with a common c-axis direction but randomly rotated about that axis, when the incident beam is parallel to the basal planes. It does not yet correspond to Fig. 14, however.

In most cases carbon fibers are nongraphitic. They are composed of "crystallites" of turbostratic graphite, rather than of true three-dimensional graphite crystals. Provided that the layers remain parallel, the fact that

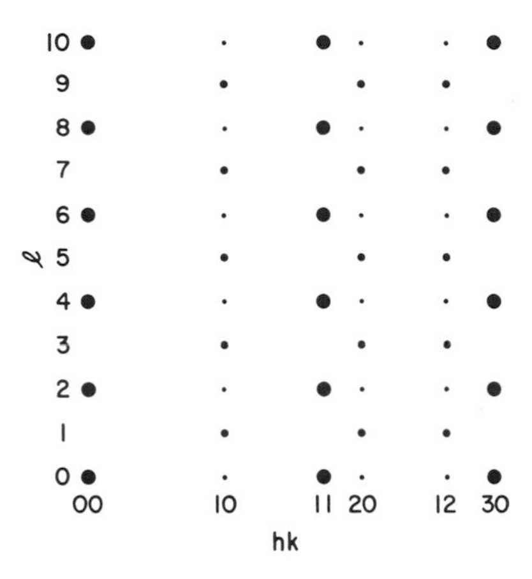

FIG. 15. Planar section through the c axis of the intensity distribution in reciprocal space for a graphite single crystal, represented as a cylindrical average by rotation about the c axis. From Ref. 66.

they are otherwise stacked randomly does not affect the (00ℓ) interferences. However, random stacking significantly alters the (hkℓ) interferences: the hexagonal nets of which each graphite layer is composed diffract independently, like two-dimensional diffraction gratings, and each (hk) row of spots degenerates into a continuous (hk) rod parallel with the c axis. Since the layers are randomly oriented with respect to a rotation about the c axis, the intensity distribution for a turbostratic graphite "crystal" is obtained by rotating the (hk) rods about the c axis. Thus a series of concentric (hk) cylinders, rather than (hkℓ) hoops, is formed. A section through the c axis of the resulting intensity distribution is shown in Fig. 16. If the turbostratic crystals are partially graphitized (as are those comprising the highly stretched fiber of Fig. 14), the (hk) rods are modulated to possess an intensity distribution intermediate between those of Figs. 15 and 16. Finally, a rotation of Fig. 16 will allow the pattern of Fig. 14 to be deduced, as described in the next paragraph.

Highly oriented carbon fibers are composed of "crystallites" of turbostratic graphite whose c axes are all normal to the fiber axis but otherwise lie in all directions in the transverse plane. The intensity distribution in reciprocal space is obtained by rotation of the (00ℓ) spots and the (hk) cylinders of Fig. 16 about the fiber (horizontal) axis. The electron-diffraction

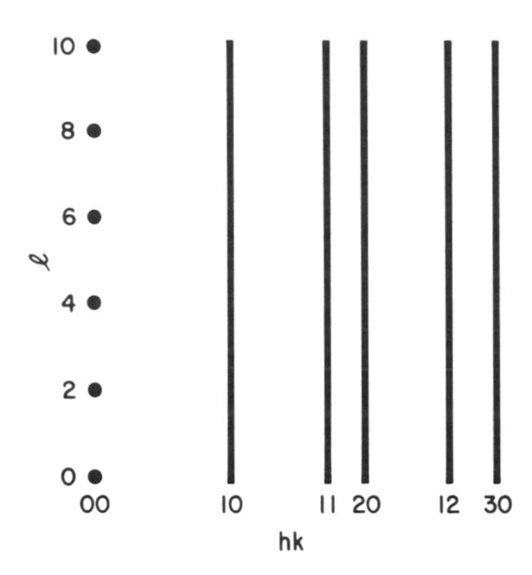

FIG. 16. Planar section similar to Fig. 15, but for a single flake of ideally turbostratic graphite. From Ref. 66.

pattern (Fig. 14) is, approximately, a planar section through this axis of the three-dimensional intensity distribution.

The (00ℓ) spots appear at $\phi = 90°$ and show natural broadening in the s, or radial, direction owing to finite stack height L_c. They also have some broadening in the ϕ direction because of finite layer size L_a, but most of the spreading in that direction has another cause. In the real fiber the graphite planes are not all perfectly parallel with the fiber axes; the consequence of this arrangement can be visualized in Fig. 16 by imagining it to be rocked slightly about an axis normal to the figure, leading to most of the observed ϕ spreading of (00ℓ).

In the planar section of the rotated intensity distribution, the (hk) cylinders produce the (hk) circular bands, each with a pair of (hk) tangential streaks (modulated in this case — see below). The inner radius of the bands is sharp and corresponds to s_{hk}, the inner radius of the (hk) cylinder, inside of which the intensity is zero. The bands are roughly circles because, over a wide range of rotation angles, the intersection of the (hk) cylinders with the plane of the figure is nearly a circle, and one is looking at the walls of the cylinder nearly edge on.

The associated pair of streaks arises from the right and left walls of the (hk) cylinders as seen from the side. Outside the streaks the intensity is zero, but between them it merely falls off quickly with increasing s, corresponding with the interaction of the cylinders outside the wide range mentioned above. The intensity in the streaks is greatest near the (horizontal) axis of rotation and falls off with distance as a result of both cylinder rotation and rocking.

The highly oriented carbon-fiber pattern of Fig. 14 shows yet another significant feature of the streaks. The intensity does not fall off steadily but is modulated. The fiber is partially graphitized, resulting in an actual situation intermediate between Figs. 15 and 16 with modulated (hk) cylinders leading to modulated streaks. This partial graphitization is nearly undetectable by X-ray diffraction.

B. Wide-Angle X-Ray Diffraction Patterns

1. Experimental Techniques

Three techniques are commonly employed to study wide-angle X-ray diffraction by carbon fibers: (a) the flat-film technique, (b) the cylindrical-film-camera technique, and (c) the diffractometer technique.

a. Flat-Film Technique. The flat-film technique is used to obtain a conventional fiber pattern from a vertically oriented bundle of fibers. A collimated, monochromatic X-ray beam (usually CuK_α line obtained by using a nickel filter) impinges perpendicularly on a bundle of fibers, and the scattered radiation is recorded on a flat film placed a distance of 3 or 5 cm behind the fibers. Patterns of the sort shown in Fig. 17 result. The most important features of this pattern are the following: (a) the (002) and (004) diffraction arcs situated near the "equator" and (b) the "ring" due to the (10) diffraction band. Both of these features show effects of preferred orientation of the graphitic layers and are qualitatively similar to those of the electron-diffraction pattern (Fig. 14). Line broadening, due primarily to the small size of the turbostratic crystallites, is evident in Fig. 17. The fiber axis in Fig. 17 is 90° to that in Fig. 14.

The flat-film technique evidently cannot be used to study the entire distribution of scattering intensity that could be studied by X-rays of the wavelength employed. For example, the (002) scattering from all crystallites, whose layer normals lie within approximately 13° (the Bragg angle) of the fiber axis, cannot be observed because the incident beam is

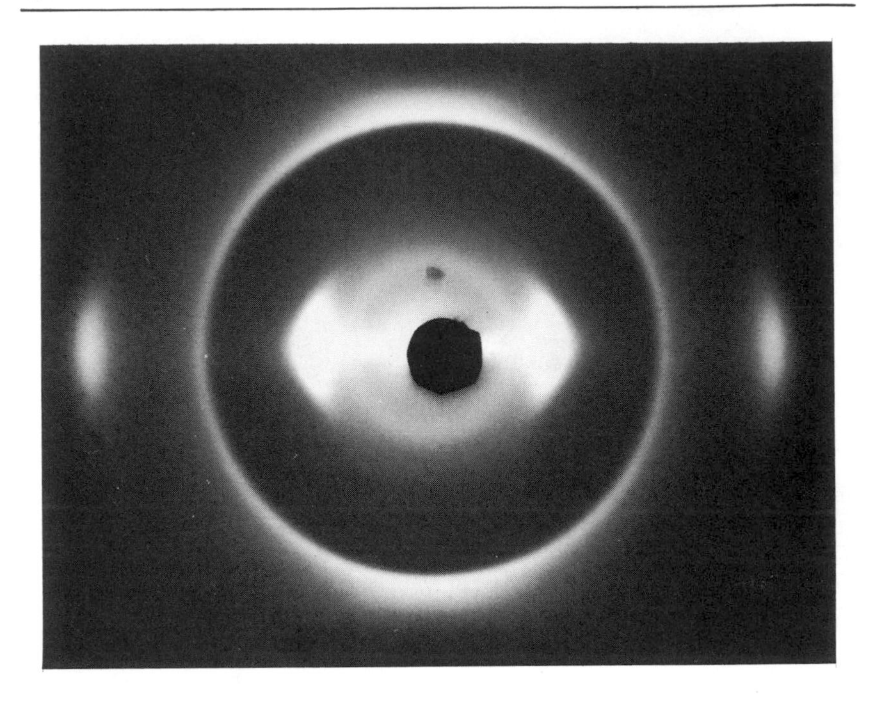

FIG. 17. Wide-angle X-ray diffraction fiber pattern of high-modulus rayon-base carbon-fiber bundle. Fiber axis vertical. Nickel-filtered CuK_{α} radiation. Sample-to-film distance 3 cm. Courtesy of S. L. Strong.

perpendicular to the fiber axis. The angle ϕ between a crystallite layer normal and the fiber axis is 90° for a "perfectly oriented" crystallite, and the scattered intensity appears on the equator. An (002) reflection at the meridian position (directly above or below the center) corresponds to crystallites for which $\phi = 13°$. The relationship between the angle γ measured along the (002) arc on the film and the angle ϕ is [67]

$$\cos \phi = 0.974 \cos \gamma \tag{3}$$

Microdensitometer measurements from the film may be made along the (002) arc (as a function of γ) and across the arc (as a function of Bragg angle θ) to obtain the intensity distributions required to calculate the degree of preferred orientation and the average layer-stack height L_c. Microdensitometer measurements across the (10) or (11) "rings" can be used to obtain estimates of average layer size L_a; however, because of the effects of preferred orientation on the intensity distribution within these rings (Section IV. B. 2), the procedure is by no means straightforward [64].

Before leaving this section, the reader should be apprised of a technique used by Curtis, Milne, and Reynolds [68] to obtain a flat-film diffraction pattern from a single filament of carbon fiber. They used an X-ray microscope fitted with a toroidal mirror to focus an X-ray beam to a spot about 10 μm in diameter. With this technique they were able to measure changes in the preferred orientation of a carbon filament resulting from the application of a tensile load (Section IV. B. 2).

b. Cylindrical-Film-Camera Technique. The cylindrical film camera is used to obtain a conventional powder pattern from carbon fibers.

If the fibers are maintained as a collimated bundle, parallel to the camera axis, the entire equatorial region of both forward- and back-scattered intensity can be studied. Very accurate measurements of line peak position (hence interplanar distances) and line shapes (hence crystallite sizes) can be obtained with a cylindrical camera. However, information contained in the radiation scattered far from the equatorial plane is not recorded. Furthermore, in the practical case of carbon fibers, little information can be obtained from the back-scattered region. The most prominent lines are the (002), (004), (006), (10), and (11) interferences; the (006) line is usually very weak.

If the fibers are ground into a powder and packed into the sample holder with random particle orientations (not a trivial accomplishment), a true powder pattern results. The observed lines now give information concerning spherically averaged structural parameters. So far, however, such spherical averaging has proved to be less interesting than the determination of structural features that depend on preferred orientation in the fiber. An additional drawback to the powdered-fiber technique is that internal strains in the fiber are sometimes altered as a result of grinding.

c. Diffractometer Technique. A diffractometer method for studying the wide-angle X-ray scattering from carbon fibers has been described by Ruland [69]. The diffractometer utilizes Cu radiation and a proportional counter with pulse-height discrimination to record only the scattering from a narrow-wavelength band. The sample, if in the form of continuous yarn, is wound onto a square frame in such a way that all fibers are oriented as nearly parallel as possible. The scattering geometry in relation to the fiber-sample orientation is shown in Fig. 18. The plane of the sample is the y-z plane. The scattering vector $\underline{s} = (\underline{S} - \underline{S}_0)/\lambda$ is parallel to the x axis; thus the z axis always coincides with the normal to the Bragg planes. Due to cylindrical symmetry in the fibers, the scattering intensity $I(\underline{s})$ may again be expressed as a function of only two spherical coordinates: s, the magnitude of the scattering vector equal to $2 \sin \theta/\lambda$, and ϕ, the angle between $\underline{s}$ and the fiber axis. The Bragg angle θ is chosen as in all diffractometers: the counter is rotated about the y axis through 2θ while the sample is rotated through θ. The value of ϕ is independently chosen by rotating the sample in its own plane, that is, about the x axis.

With this device the entire scattering region in reciprocal space attainable with the given wavelength radiation can be investigated because all values of ϕ from 0 to 90° are obtainable for any desired s value. This device has the advantage that one can, for example, set the diffractometer to record the (002) scattering and vary ϕ to obtain directly $I_{002}(\phi)$, which is, approximately, the orientation distribution function for graphite-layer normals with respect to the fiber axis (see next section). Another advantage of this geometry is that the correction for absorption in the sample is independent of ϕ.

2. Methods for Calculating Structural Parameters from X-Ray Data

The most important parameters of carbon-fiber structure that can be determined from wide-angle X-ray scattering are (a) the degree of preferred orientation of graphitic layer planes parallel to the fiber axis; (b) the average height L_c of the stacks of layers; and (c) the average breadth L_a of the graphitic layers. Since the degree of preferred orientation in carbon fibers determines to a large degree their physical properties and since no other form of carbon exhibits such high degrees of preferred orientation of this type (layer normals perpendicular to the axis of symmetry), a major portion of this section is concerned with the determination of preferred orientation. The determination of the other structural features will be discussed later.

a. Degree of Preferred Orientation. The (002) as well as the higher order [e.g., (004)] diffraction arcs are due to Bragg reflections from the parallel stacks of graphite layer planes. The distribution $I_{002}(\phi)$ of (002)

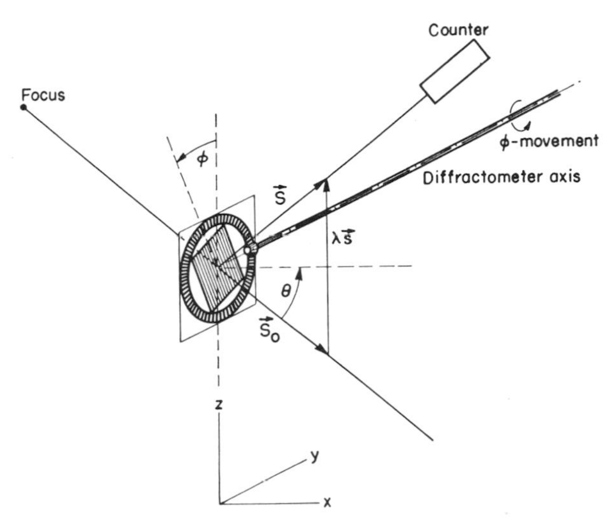

FIG. 18. Scattering geometry for X-ray diffractometer technique utilizing collimated fiber sample. Value of ϕ (the angle between the normal to the Bragg planes and the fiber axis) is chosen independently by rotating the sample holder about the x axis (normal to sample plane). From Ref. 69.

scattering intensity as a function of ϕ, the colatitude angle measured with respect to the fiber-axis direction, is determined by the weighted distribution of graphite crystallite normals with respect to the fiber axis: that is, since ϕ is the direction of the scattering vector $\underline{s}$ and since appreciable (002) scattering intensity occurs only when the direction of $\underline{s}$ nearly coincides with that of the layer normals, $I_{002}(\phi)$ is (approximately) proportional to the number of graphite crystallites (weighted average according to crystallite size) per unit solid angle, the layer normals of which are oriented at an angle ϕ with the fiber axis [67]. This orientation distribution function peaks at 90° and is qualitatively similar to that of extruded graphites. An example of the orientation distribution function for a single PAN-base carbon fiber (with and without applied load) is shown in Fig. 19.

The exact determination of the distribution of crystallite normals from the distribution of X-ray scattering intensity is complicated by broadening effects. These have been discussed in detail by Ruland [69]. Each graphite crystallite gives rise to a distribution of scattering intensity surrounding its (002) lattice point which is broadened in the ℓ direction (radial, or s,

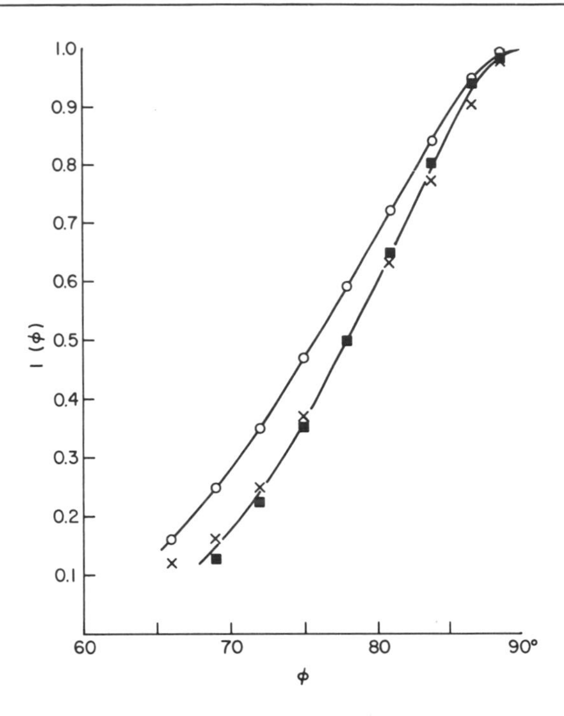

FIG. 19. Orientation distribution function for graphite-layer normals with respect to the fiber axis in a high-modulus PAN-base carbon fiber. Upper curve, no load; lower curve, 5g load applied. From Ref. 68.

broadening) due to finite L_c and also in the hk direction (circular, or ϕ, broadening) due to finite L_a. The L_c broadening can be corrected for if the total integrated intensity (over the full line width in s) is measured for each value of ϕ. This integrated intensity $I_{002}(\phi)$ is not necessarily proportional to the peak intensity, since the radial broadening is a function of ϕ (Section IV.B.3.d). The circular broadening of the (002) spots, provided the broadening is due solely to crystallite-size effects, can be estimated from the broadening in the (hk) bands. This broadening, along with that due to the slit system and to the tilt angle of the fibers on the frame, is "subtracted" from the measured $I_{002}(\phi)$ by well-known deconvolution methods to yield a corrected function that characterizes the orientation distribution of graphite-crystallite normals. (This corrected orientation distribution function is designated $g_{002}(\phi)$ by Ruland [69]; however, the more common notation, $I_{002}(\phi)$, is used here.)

A numerical index of the degree of preferred orientation can be determined from $I_{002}(\phi)$ (after making the above corrections). Following Price and Bokros [70], and also Gasparoux et al. [71], an orientation parameter R_{0z} can be calculated from the equation

$$R_{0z} = \frac{\int I(\phi) \sin^3 \phi \, d\phi}{\int I(\phi) \sin \phi \, d\phi} . \tag{4}$$

This parameter varies from 2/3 for isotropic fibers to 1 for perfectly oriented fibers. This preferred-orientation parameter is useful for the calculation of various physical properties, such as thermal expansion coefficients [67], magnetic susceptibilities [69], and elastic constants [72].

Another preferred-orientation parameter has been proposed by Ruland [69] and applied to a calculation of carbon-fiber Young's moduli (see Section V.B). This parameter, designated q, is obtained by fitting the measured orientation distribution function $I_{002}(\phi)$ to the following expression:

$$I_{002}(\phi) = \frac{(1 - q^2)}{1 + q^2 - 2q \cos \phi} \tag{5}$$

For the extruded-graphite type of preferred orientation, which is exhibited by carbon fibers, q is negative, ranging from -1 for perfect orientation to zero for the isotropic case. Ruland has found that Eq. (5) approximates well the $I_{002}(\phi)$ function for all rayon-graphite fibers except those possessing extremely high degrees of preferred orientation.

For routine relative measurements of the degree of preferred orientation of a carbon-fiber sample, the orientation half-width is commonly employed. The intensity of the (002) diffraction is measured as a function of ϕ (or of γ in the case of the flat-film technique), and the resulting bell-shaped curve is recorded (see Fig. 19). The half-width of this curve, measured at the point of half-maximum intensity, is then given in degrees ϕ (or γ).

b. Crystallite Dimensions L_c and L_a. Two structural parameters, L_c and L_a, are commonly measured by analysis of the line shapes of the principal diffraction lines of a powdered carbon or graphite sample. These parameters represent the apparent average dimensions of the coherent crystalline domains parallel (L_c) and perpendicular (L_a) to the crystallite c axis. The word "apparent" has been emphasized because the diffraction line shapes actually depend on other crystal-structure imperfections in addition to crystallite size. For a general discussion of the problems of determining these structural features in carbon and graphite by X-ray diffraction, the reader is referred to the review by Ruland in Volume 4 of this series [66].

The parameter L_c is determined from the integral breadth $B_{00\ell}(2\theta)$ of the (00ℓ) lines and the X-ray wavelength λ through the use of the Scherrer formula,

$$L_c = \frac{\lambda}{B_{00\ell}(2\theta)\cos\theta} . \tag{6}$$

Broadening of the (00ℓ) lines can also be caused by disorder in the parallel stacking of layers. Although disorder effects can, in principle, be separated from size effects, in practice this separation is difficult [66]. Disorder effects probably predominate in lower temperature carbon fibers, and size effects probably predominate in higher temperature graphite fibers.

The dependence of L_c on crystallite orientation can be studied if the fibers, instead of being prepared as a powdered sample, are maintained as a parallel bundle (flat-film technique or diffractometer technique). The (002) or (004) line width is then measured as a function of ϕ , the angle between the fiber axis and the layer-plane normal.

The parameter L_a is most simply determined from the half-peak width $B_{1/2}(2\theta)$ of the (10) or (11) bands according to the formula given by Warren and Bodenstein [73] for random-layer lattices:

$$L_a = \frac{1.77\lambda}{B_{1/2}(2\theta)\cos\theta} . \tag{7}$$

Since this formula assumes random orientation of the layers, a powdered sample is normally used with the cylindrical-film-camera technique. If collimated fibers are used in conjunction with a diffractometer, a proper spherical average of the (10) or (11) scattering intensity must be obtained [69].

Broadening of the (10) and (11) bands may also result from lattice strains within the layers, the most likely cause (in the case of nongraphitic carbons) being lattice vacancies. However, this cause of broadening is not considered important even in nongraphitic carbons if they have been heat-treated to reasonably high temperatures [74]. High-resolution electron-microscopy evidence shows that stress-graphitized rayon-base carbon fibers have a relatively perfect two-dimensional layer structure [62].

Even if one assumes that lattice strains do not contribute importantly to the broadening of (hk) bands, difficulties still remain relative to the interpretation of the parameter L_a. It has been well established, through studies with phase-contrast electron microscopy on carbon blacks [75] and on carbon fibers [62, 76-78], that continuous graphitic layers can extend over distances very much greater than L_a, since they undergo bending and twisting (Section IV.D.1.a.). Hence L_a is best interpretated as the average linear dimension of the "straight parts" of the graphitic layers.

For carbons heat-treated to low temperatures, where layer diameters are less than 30 Å, the Warren analysis is not very suitable, and a more complex analysis, such as that given by Diamond [79], is required.

Fourdeux, Perret, and Ruland [80] have been able to measure both the apparent layer dimensions parallel to the fiber axis ($L_{a\parallel}$) and perpendicular to the fiber axis ($L_{a\perp}$). Their method utilizes a theoretical treatment of the effect of preferred orientation on the (hk) interferences developed by Ruland and Tompa [64].

3. Experimental Observations

a. General Structural Features and Preferred Orientation. All carbon fibers made so far by the pyrolysis of an organic precursor fiber are nongraphitic and nongraphitizable. (Some very highly stress-graphitized rayon-base carbon fibers and some PAN-base carbon fibers treated with graphitization catalysts have shown partial or complete three-dimensional graphite development.) The graphitic layers, although they may be stacked in a very nearly parallel arrangement, are randomly oriented with respect to rotation about the layer normals — that is, the structure is turbostratic [69, 81-83]. This structural characteristic is determined from the fact that the two-dimensional (10) band is not resolved into the (100) and (101) lines characterizing three-dimensional crystalline graphite, and the same condition applies for the (11) band. This behavior holds true even for carbon fibers heat-treated to very high "graphitizing" temperatures (near 3000°C). Very highly stress-graphitized fibers may show partial graphitization (see Fig. 14).

Layer sizes L_a are between 30 and 125 Å, depending on heat-treatment temperature, preferred orientation, and, to a small degree, the nature of the precursor. Apparent layer-stack heights L_c vary between about 30 and 85 Å, depending on the same factors.

The general structural features described here are quantitatively similar to those of bulk forms of nongraphitizable carbons — for example, cellulose carbon [84], glassy carbon [85], and vitreous carbon [86]. The most distinguishing structural feature of carbon fibers is their high degree of preferred orientation, a characteristic that accounts in great measure for their excellent mechanical properties in the axial direction.

The relationship between the degree of preferred orientation and Young's modulus for both rayon- and PAN-base carbon fibers is shown in Fig. 20. Most of the small discrepancy between the curves for rayon- and PAN-base fibers would be removed if one corrected Young's moduli for porosity, since the PAN-base carbon fibers are denser than the rayon-base carbon fibers. It is evident from the data that routine measurements of orientation half-width can predict Young's modulus with reasonable accuracy only in the low-modulus region.

A careful study of the orientation-modulus relationship has been carried out by Ruland and co-workers [80, 88]. They have measured the preferred

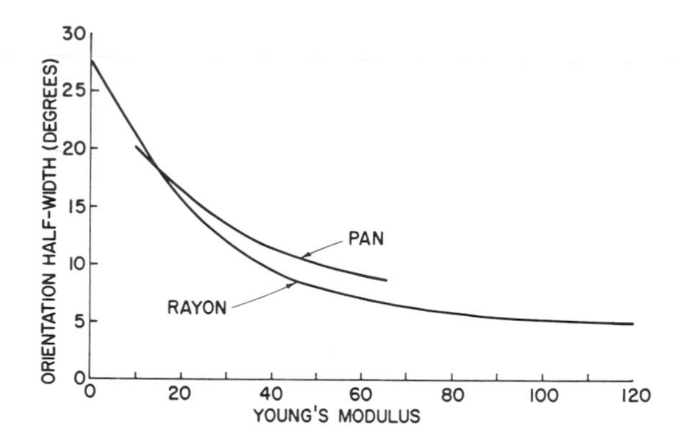

FIG. 20. Orientation half-width as a function of Young's modulus for graphitized rayon-base carbon fibers and PAN-base carbon fibers. From Refs. 2 and 87.

orientation of a variety of experimental and commercial fibers (mostly rayon-base ones, but including two PAN-base fibers and one pitch-base fiber) and plotted values of Young's modulus against the orientation parameter q (Fig. 21). In this figure, since different samples have different degrees of porosity, ranging from 10 to 30%, the Young's modulus values were all corrected to correspond with that of fully dense material. The curve shown in this figure is a theoretical calculation (see Section V. B. 2), which is seen to fit the data very well for all carbon fibers. Except for partially graphitized fibers, whose moduli fall below the curve, the modulus of carbon fibers seems to be a function only of the degree of preferred orientation.

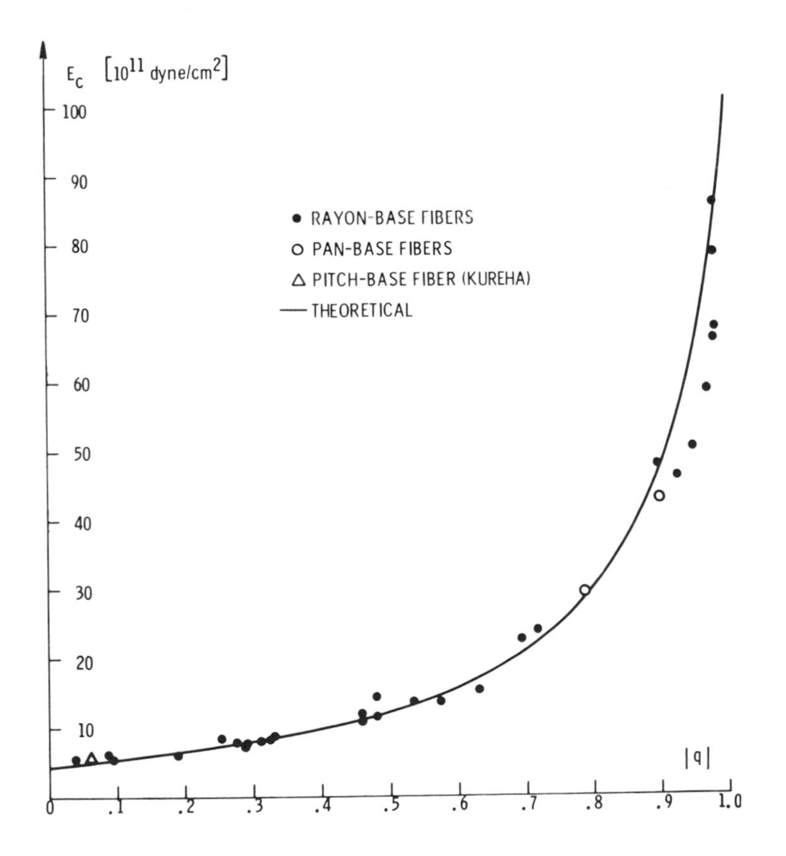

FIG. 21. Young's modulus corrected for porosity, E_c, as a function of preferred orientation q; curve is based on theoretical model. From Ref. 80.

b. Dependence of Preferred Orientation on Heat Treatment. Preferred orientation in carbon fibers can be increased in either of two ways: stretching or high-temperature heat treatment. A combination of both methods is normally required to achieve very high degrees of preferred orientation. The effect of stretching on the modulus of rayon carbon fibers has been covered in Section III. E.

A study of preferred orientation versus heat treatment for carbon fibers derived from two kinds of rayon has been carried out by Ruland [69]. When no stretching was performed during heat treatment, all fibers were of the low-modulus variety (i.e., below 10×10^6 psi). Pyrolysis of the pure rayon was carried out entirely in an inert atmosphere. Figure 22 shows the results of this study. Preferred orientation in the starting rayon is destroyed with the destruction of the cellulose structure (between 240 and 280°C) except for a very slight residual orientation that persists to 1000°C. Above 1000°C, a reorientation occurs, although the magnitude of the preferred-orientation parameter q never achieves the values of 0.8 and above required for high-modulus fibers unless the fibers are stretched during graphitization.

In Ruland's study of rayon-base carbon fibers no direct correlation between preferred orientation in the starting material and that of the final carbon fiber was found. However, an earlier study by the author [81] of rayon tire yarns that had varying degrees of orientation showed that the more highly oriented rayons led, ultimately, to the more highly oriented carbon fiber. Since Ruland found that molecular orientation was almost completely destroyed in the early stages of pyrolysis, he suggested [89] that an oriented microfibrillar structure in the raw material might persist through low-temperature pyrolysis and that this structure would then influence the growth of graphitic layers at temperatures above 1000°C. However, to date, the identity of such microfibrils and their persistence through pyrolysis have not been demonstrated for rayon. An essentially similar suggestion has been made by Watt and W. Johnson [58] in the case of PAN-base carbon fibers; however, although microfibrils are known to exist in the raw material, their influence during the course of pyrolysis on the growth of graphitic layers has not been experimentally demonstrated. In the case of PAN, molecular preferred orientation is not destroyed during pyrolysis as it is in rayon. Since much higher degrees of preferred orientation of graphitic layers are achieved after carbonization at, for example, 1000°C, one naturally suspects that this relatively high preferred orientation is at least partly due to the molecular orientation of the PAN starting material.

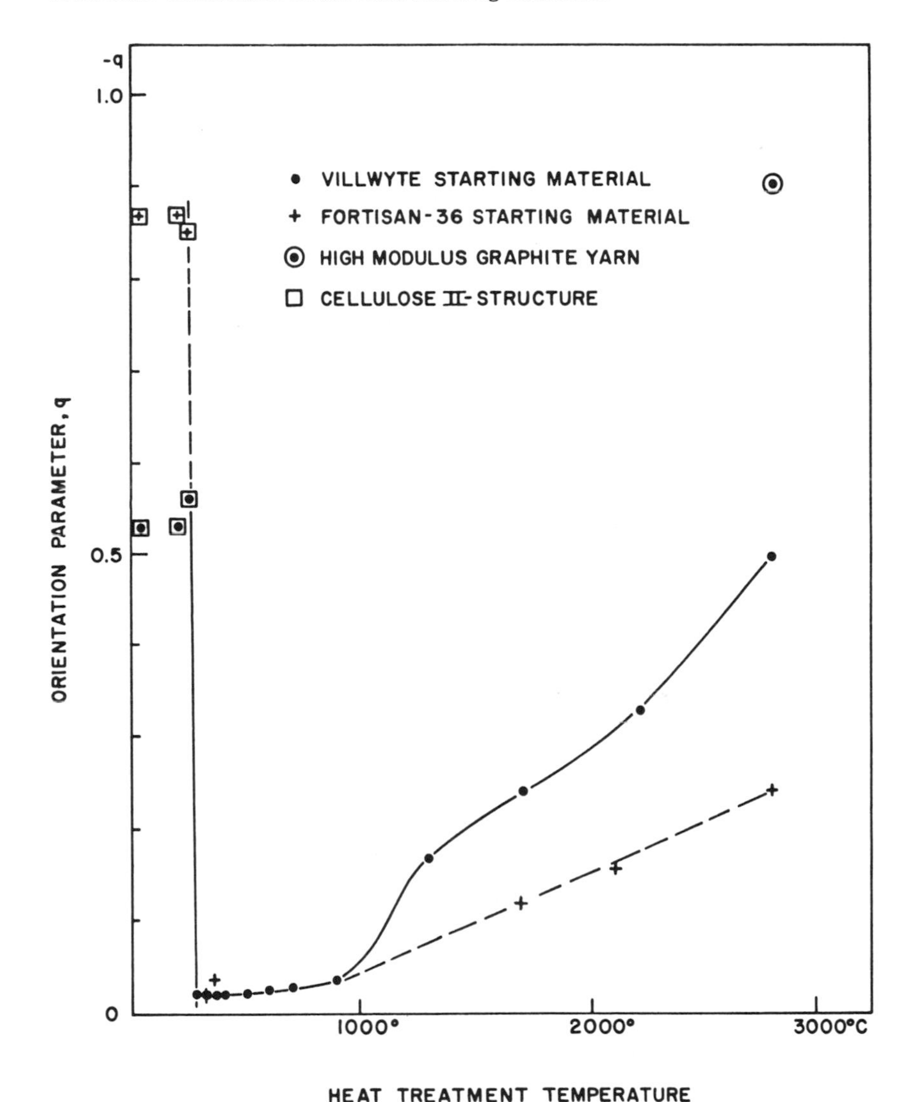

FIG. 22. Preferred orientation in low-modulus rayon-base carbon fibers as a function of heat-treatment temperature. Note single point for high-modulus fiber, which was stretched during graphitization. From Ref. 69.

c. Factors Influencing L_a. The crystallite-breadth parameter L_a depends primarily on heat-treatment temperature. To a smaller extent, it depends also on the degree of preferred orientation and on the precursor material.

The dependence of L_a on heat-treatment temperature is given in Table 1 for PAN-base carbon fibers heat-treated at temperatures between 1000 and 3000°C. One column of data is for RAE carbon fibers studied by Eastabrook and Seed [90], and the other is for the fibers studied by Shindo [82]

TABLE 1

Crystallite Sizes for PAN Carbon Fibers for Various Heat-Treatment Temperatures[a]

Heat-treatment temperature (°C)	Crystallite diameter L_a (Å)	
	Eastabrook and Seed [90]	Shindo [82]
1000	30	39
1500	—	56
2500	90	110
3000	125	200

[a] From Ref. 58.

Although Shindo's values are higher than those of Eastabrook and Seed, the discrepancies are serious only at 3000°C, where experimental conditions are always difficult to control. A monotonic increase in layer size occurs with increasing heat-treatment temperature, from 30 to 40 Å at 1000°C to over 125 Å at 3000°C. D. J. Johnson and Tyson [91] measured only slightly smaller values of L_a, 70 Å for a 2650°C heat-treated PAN fiber. These data represent spherical averages of L_a obtained from powdered samples.

Fourdeux, Perret, and Ruland [80] have measured separately the L_a dimensions parallel and perpendicular to the fiber axis. Their results are shown in Table 2 for two grades of commercial PAN-base carbon fibers and several grades of rayon-base carbon fibers. Estimates of approximate heat-treatment temperatures are given. The table shows that the layers are

TABLE 2

Layer-Plane Dimensions Parallel ($L_{a\parallel}$) and Perpendicular ($L_{a\perp}$) to the Fiber Axis for Several Grades of Carbon and Graphite Fibers[a]

Fiber type	$L_{a\parallel}$ (Å)	$L_{a\perp}$ (Å)	q	Heat-treatment temperature (°C)
PAN-base fibers:				
Type II	35	34	0.787	1250
Type I	119	64	0.902	2700
Rayon-base fibers	59	50	0.230	2800
	81	54	0.820	2800
	91	65	0.980	2900
	130	65	0.982	2900

[a] Data from Refs. 80 and 92.

longer in their axial dimensions than in their transverse dimensions and that this difference increases with increasing degree of preferred orientation.

Since L_a is a measurement of the length of the straight portions of the layers, the possibility must be considered that the layers are actually considerably longer than $L_{a\parallel}$. This possibility, in fact, has been confirmed by high-resolution electron microscopy (Section IV.D.1.a). Table 2 shows that very-high-modulus rayon-base carbon fibers of constant heat-treatment temperature, but varying preferred orientation (due to varying degrees of stretch during graphitization), exhibit a constant $L_{a\perp}$ but an increasing $L_{a\parallel}$ due to stretching. Ruland and co-workers attribute this increase to the straightening of wrinkled ribbons of graphitic layers.

d. Factors Influencing L_c. The crystallite height parameter L_c depends on heat-treatment temperature and orientation of the crystallites with respect to the fiber axis. To a small extent, it also depends on the degree of preferred orientation and on the precursor material.

The crystallite stack height for rayon-base carbon fibers heat-treated above 2800°C is 40 to 50 Å; that of PAN-base carbon fibers heat-treated at 2500 to 2650°C is 50 to 65 Å [80, 82, 91, 93].

The apparent L_c values of carbon fibers of 1200 to 1500°C heat-treatment temperature are less than half those quoted above [45, 80, 94]. However, for such fibers, imperfect layer stacking probably contributes to the broadening of the (002) line (strain broadening); thus the "true" average height of stacks of layers is greater than the calculated L_c. Watt and W. Johnson have suggested that L_c is derived from the size of the microfibrils (or primary fibrils) in the PAN precursor [58]. If this suggestion proves to be correct, the true L_c should show essentially no dependence on heat treatment. Ruland [95] found that the number of layers per stack in rayon-base carbon fibers increases only slightly with heat treatment.

Perret and Ruland [92] have measured the integral breadths of the (002) interferences for both PAN- and rayon-base carbon fibers as functions of the orientation of the crystallites, that is, of the angle ϕ between the layer normals and the fiber axis. Their results are shown in Fig. 23 for PAN type II (carbonized), for Thornel 25 (graphitized), and for PAN type I (graphitized). The ordinate is the breadth of the (002) line expressed in terms of $s = 2 \sin \theta/\lambda$ and is equal to the reciprocal of the apparent L_c according to the Scherrer formula, Eq. (6). The L_c values are greatest for those layer packets that lie most nearly parallel to the fiber axis. This effect is especially pronounced for the Thornel 25 fiber. Perret and Ruland point out that these results are consistent with a structure comprised of long ribbonlike graphitic layers stacked in a parallel arrangement to form microfibrils (Fig. 24). The thickness of the microfibrils, L_c, tends to be greatest where they are most nearly oriented parallel to the fiber axis and tends to be least where the microfibrils branch into off-axis directions.

C. Small-Angle X-Ray Scattering

The small-angle region of X-ray scattering provides information on density fluctuations, which have characteristic dimensions in the range 10 to 1000 Å. One kind of density fluctuation that is among the most amenable to interpretation is that due to sharp changes in density, that is, voids or inclusions. If the voids are all roughly of the same size and shape, analysis of the scattering is relatively easy. If density fluctuations are gradual and irregular, the scattering will be difficult to analyze; however, if such fluctuations are periodic, a Bragg-like maximum will appear and a d spacing can be calculated. Finally, in the very-small angle region multiple scattering will occur and must be accounted for.

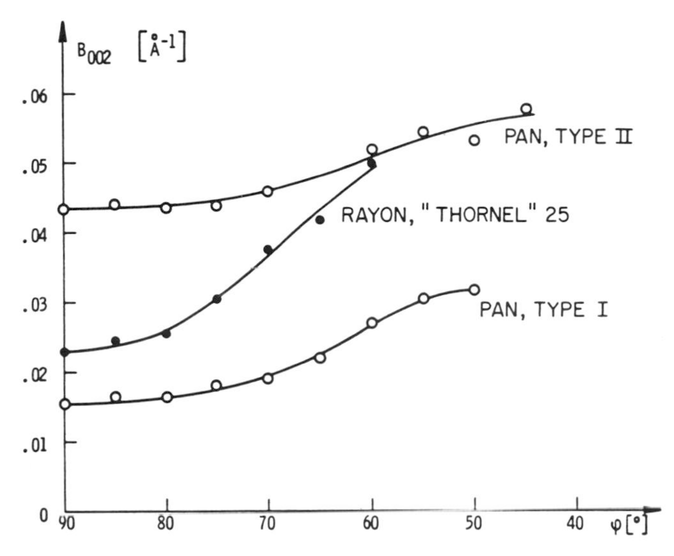

FIG. 23. Breadth of (002) line as a function of angle between layer normal and fiber axis for graphitized rayon-base carbon fibers (Thornel 25) and for types I and II PAN-base carbon fibers. From Ref. 92.

Because of the difficulties of analysis in the small-angle region and because of the specialized equipment required, few people have employed the techniques of small-angle X-ray scattering in studies of carbon. However, recent studies [91, 96-99] on carbon fibers and glassy carbons of medium-to-high heat-treatment temperatures have shown that the pore structure of these materials is ideally suited to analysis by small-angle-scattering techniques. The most important reason for this condition is the long suspected fact [81, 100] that the pores are bounded by graphitic layer planes and hence that a very sharp boundary exists between the pores and the dense material. Furthermore, the pores are only 10 to 15 Å thick and rather uniform in size. Since, in high-modulus fibers, the pores are needle-like and preferentially oriented parallel to the fiber axis, an estimate of their lengths can be made.

1. Small-Angle Scattering from Pores

Photographs taken with a Kiessig pinhole camera of the small-angle scattering from both a low-modulus and a high-modulus graphitized rayon-base carbon-fiber bundle are shown in Fig. 25. Contours of equal intensity

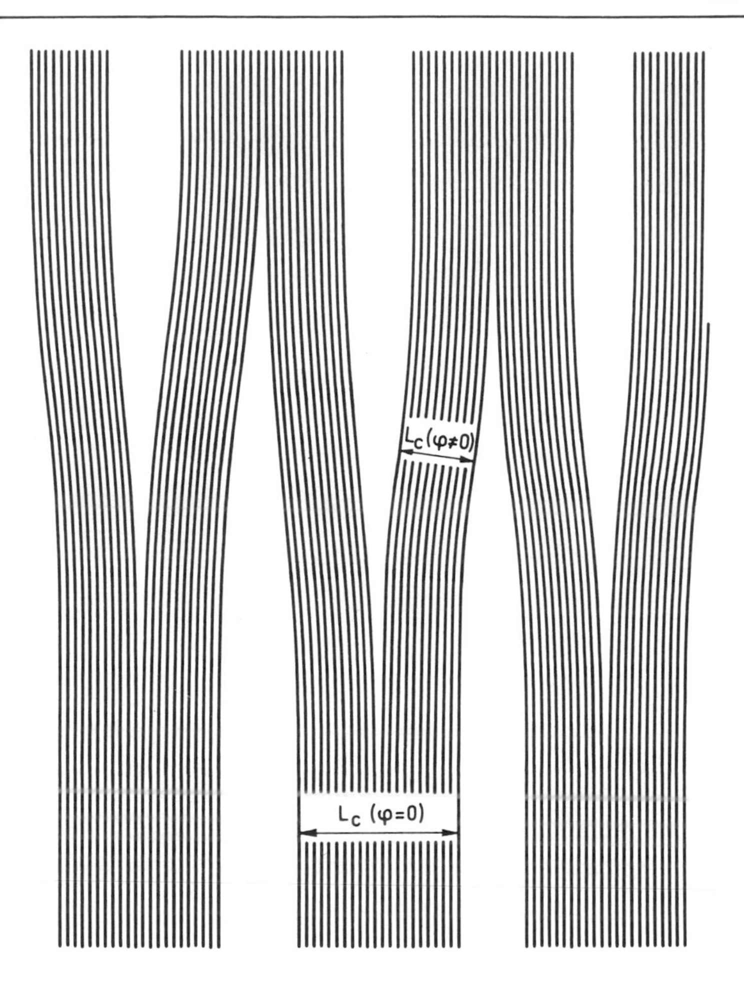

FIG. 24. Schematic representation of the correlation between the thicknesses of layer stacks, L_c, and their orientation with respect to the fiber axis. From Ref. 92.

obtained from two-dimensional microdensitometry of a similar photograph for a type II PAN carbon fiber are shown in Fig. 26. The scattering is evidently strong in or near the equatorial plane out to quite large angles ($\sim 4°$). Deviations of the scattering from the equatorial plane increase as Young's modulus of the fiber decreases, suggesting a relationship between the orientation of pores and the orientation of graphitic layers. Perret and Ruland have demonstrated quantitatively the close correspondence between the degree of preferred orientation of pores and that of graphitic layers for both rayon [97] and PAN [92] carbon fibers.

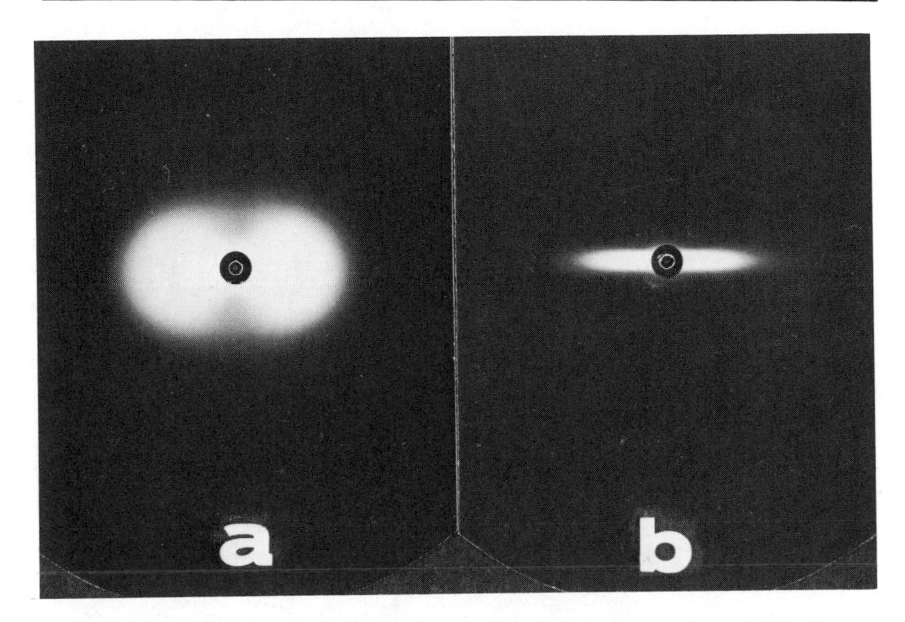

FIG. 25. X-Ray small-angle-scattering photographs of graphitized rayon-base carbon fibers (fiber axis vertical): (a) low modulus (8×10^6 psi); (b) high modulus (71×10^6 psi). Courtesy of R. Perret and W. Ruland.

The dependence of scattering intensity on scattering angle follows Porod's law [101] and therefore shows that a sharp density transition exists between dense material and pores. Analysis of the scattering [91, 97, 98] shows that the average pore diameter is about 10 to 20 Å.

The pores in carbon fibers are thus long and narrow, bounded by graphitic layer planes, and exhibit a preferred orientation intimately connected with that of the graphitic layers. In fibers of high heat-treatment temperature the pores are inaccessible to helium [97]. One question remains: are they platelike or needlelike? The answer seems to be that they are long, flattened needles. If the pores were platelike, as suggested by D. J. Johnson and Tyson [91], the longer of the two dimensions in the transverse plane is unlikely to be greater than the corresponding layer-plane dimension $L_{a\perp}$ (35 to 70 Å). On the other hand, the average length of the pores is probably much greater than $L_{a\|}$, since the layers themselves are much longer than X-ray measurements of $L_{a\|}$ indicate. Furthermore, all electron-microscopy evidence indicates the existence of pores several hundred to several thousand angstroms long (Section IV. D. 1. a).

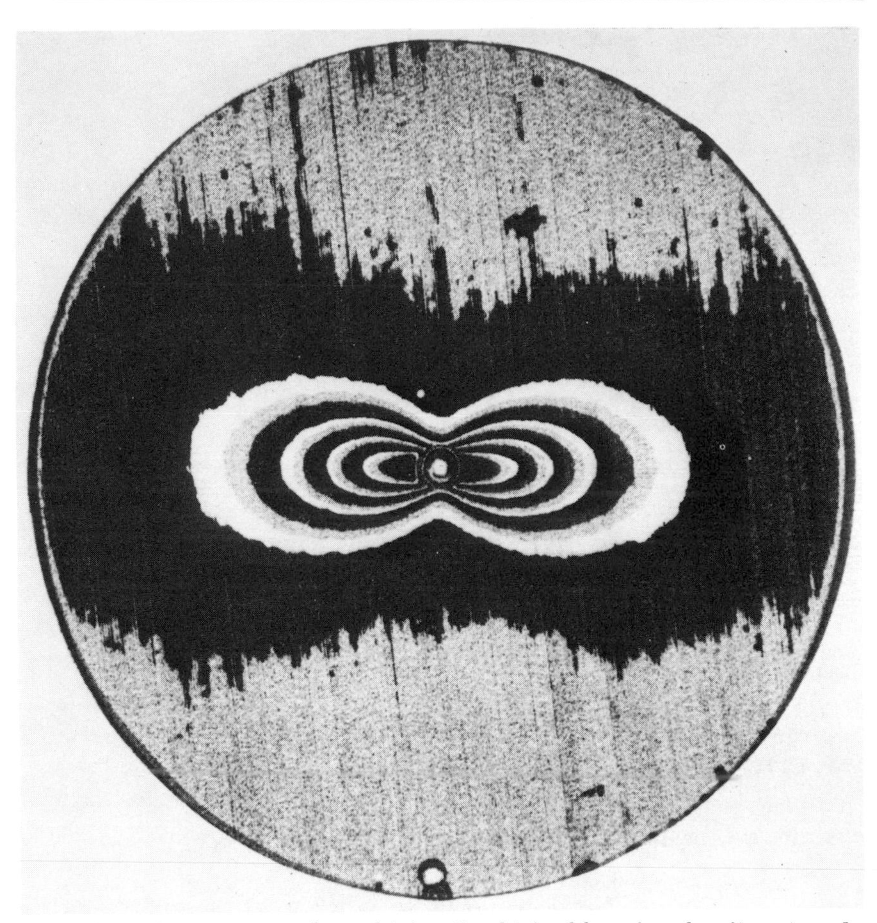

FIG. 26. Contours of equal intensity obtained by microdensitometry of a small-angle-scattering photograph of type II PAN-base carbon fibers. From Ref. 92.

Perret and Ruland have determined (from their small-angle-scattering studies) the lower limits for the lengths of pores in both rayon- and PAN-base carbon fibers [92, 97]. They plotted the product $sB(\phi)$ against s, where $B(\phi)$ is the angular breadth of the scattering intensity measured at constant s and s is the magnitude of the scattering vector, $2 \sin \theta/\lambda$. If the pores are infinitely long, the angular breadth $B(\phi)$ should be a constant independent of s and should depend only on the angular distribution of pore axes. A plot of $sB(\phi)$ versus s should therefore be a straight line through the origin.

However, if the pores are of finite length, sB(ϕ) should level off toward the origin, approaching the value 1/L, where L is the length of the pores. Such a leveling off did not occur in any of the samples studied. Since experimental limitations did not permit measurements to indefinitely small values of s, the authors could obtain only lower limits for the lengths of pores. The results indicate that the pores are at least 100 Å long for the low-modulus fibers and at least 300 or 400 Å long for the high-modulus fibers. The pores are thus at least three times as long as X-ray measurements of $L_{a\|}$ (Table 2).

2. Other Components in Small-Angle Scattering

Three other components in the small-angle scattering from carbon fibers (besides that due to the micropores) have been identified by Perret and Ruland [97]: (a) multiple scattering perpendicular to the fiber axis owing to the size and shape of the filaments, (b) single scattering perpendicular to the fiber axis from density fluctuations in the dense material, and (c) single scattering parallel to the fiber axis because of periodic density fluctuations in rayon in the early stages of pyrolysis.

Multiple scattering perpendicular to the fiber axis is the predominant scattering in the very-small-angle region. Analysis of this scattering permits calculation of the average length of chords intercepted by the periphery of the filament cross section and also of the total filament circumference. Together these quantities characterize the shape and size of the filament. Both of these determinations are found to be in satisfactory agreement with determinations from optical micrographs of filament cross sections [97]. This multiple-scattering component can be effectively eliminated by embedding the fibers in a medium, such as glycerol, that possesses an X-ray-scattering power near that of the fibers [102].

The single scattering perpendicular to the fiber axis contains a component, in both rayon- and PAN-base carbon fibers, that is due to density fluctuations in the material, according to Perret and Ruland [80, 92, 97]. The density fluctuations may be caused by (a) fluctuations in the interlayer spacing or (b) the statistical variation in the size and shape of layers in a given stack (i.e., the terminations of layers within a stack). In many cases the density fluctuations were so high that the second cause was believed to be the principal one. The density fluctuations, which ordinarily decreased with increasing heat-treatment temperature, were enhanced by stretching the rayon-base carbon fibers [80]. Other interesting effects occurred when the fibers were powdered by grinding in a mortar. Fibers of high heat-treatment temperature but low modulus (unstretched) were apparently strained during grinding, increasing the density fluctuations; on the other hand, the stress-graphitized high-modulus fibers showed a reduction in the density fluctuations, indicating that internal strains initially present in such fibers were relieved by grinding.

Single scattering parallel to the fiber axis exhibits maxima in rayon fibers during the early stages of pyrolysis. This component of scattering is apparently due to periodic density fluctuations parallel to the fiber axis with d spacings of about 200 Å. A similar effect has been reported recently for partially oxidized PAN [103]. These fluctuations are attributed to alternating crystalline and amorphous regions in the rayon, which persist or may even be enhanced during the initial stages of pyrolysis to about 300°C [98]. Such an inhomogeneous pyrolysis of rayon could lead to lengthwise periodic imperfections in the final carbon-fiber structure.

D. Microscopy

Although diffraction techniques are invaluable for the determination of the detailed structure of carbon fibers on the scale of 1 to 100 Å and for the quantitative determination of the average size of such structural features, microscopy can reveal certain aspects of the structure that are difficult to obtain by normal diffraction methods. Examples of such structural aspects are (a) the great lengths that can be achieved by both the graphitic layers and the micropores in the direction of the fiber axis, as revealed by electron microscopy, and (b) the preferred orientation of graphitic layers parallel to the fiber surface, as revealed by scanning electron microscopy and polarized-light microscopy. Furthermore, such gross structural features as fibrils several hundred angstroms in diameter, internal flaws of various kinds, and external surface features can be studied only by microscopy.

1. Transmission Electron Microscopy

a. Ultrafine Structure — Microfibrils. All carbon fibers prepared from an oriented precursor fiber exhibit an oriented, fibrous internal texture, provided they have been heat-treated to a sufficiently high temperature. How high a temperature is sufficient depends on the precursor. Rayon-base carbon fibers heat-treated to approximately 1200°C appear nearly amorphous by electron microscopy [104], and the fibrous feature only appears above ~ 2000°C. By contrast, PAN-base carbon fibers heat-treated to 1000°C already exhibit a pronounced texture [83]. In general, the higher the Young's modulus, the more highly oriented the observed texture. This result is illustrated in Fig. 27, which shows the internal structures of several high-modulus fibers from both rayon and PAN. The transverse dimensions of the grains in this texture are typically 50 to 100 Å.

Early observations of an oriented "fibrous" texture in carbon fibers led to the suggestion [81] that the graphitic layers were long and narrow, and that the stacks of layers constituted walls of prismatic-shaped pores. Although "prismatic" is probably not an accurate description of the pore shape, this general picture of the structure has been confirmed and quantified by wide- and small-angle X-ray studies in which, however, only lower limits for the lengths of the pores and of the graphitic layers have been measured (see Sections IV. B. 3. c and IV. C. 1).

Recent studies with high-resolution electron microscopy have shown that the layers are much longer (sometimes thousands of angstroms) than most researchers had suspected. When the phase-contrast technique was employed [105, 106], individual layers within well-defined layer packets were resolved in both rayon- and PAN-base graphitized fibers [76-78, 107, 108]. An example is shown in Fig. 28 for a Thornel 50 fiber. Long, parallel stacks of layers spaced 3.4 Å apart are clearly visible. These layer stacks, which give rise to the dark streaks in the lower resolution micrographs of Fig. 27, have been called primary fibrils [58] or microfibrils [92]. We shall adopt the latter terminology, although we agree with those who warn that no direct evidence exists that these microfibrils are derived from those of the precursor PAN or rayon fibers [91]. A schematic view of these microfibrils as seen in longituditudinal section was constructed by Fourdeux et al. [76] and is shown in Fig. 29.

The microfibrils, such as those shown in Fig. 28, are found to curve more or less frequently and more or less severely, depending on the degree of preferred orientation in the fiber. Abrupt bends of the type that might be expected to occur in twinned crystals of three-dimensional crystalline graphite do not seem to be present; such sharp bends are unexpected in view

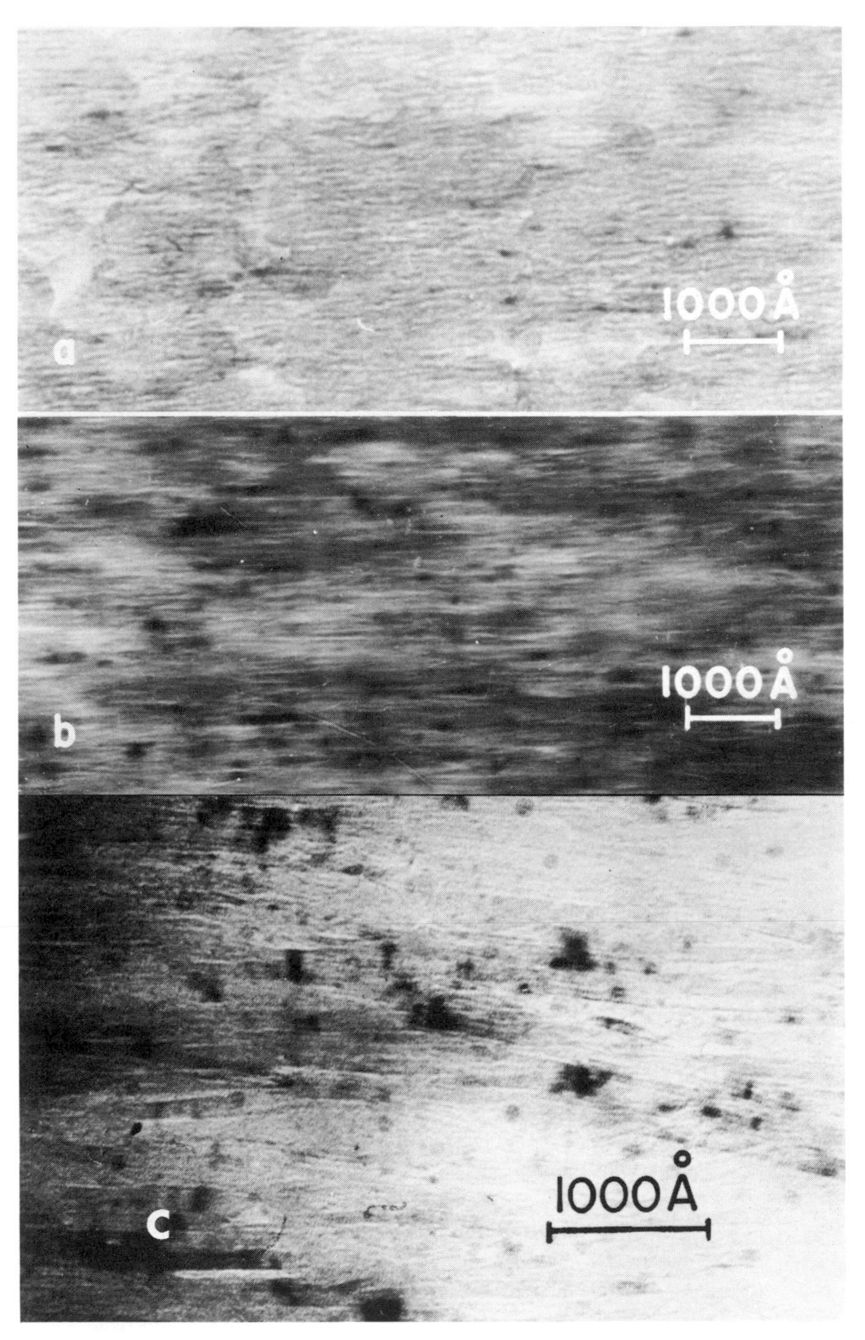

FIG. 27. Transmission electron micrographs of the longitudinal sections of several grades of high-modulus carbon fibers: (a) rayon-base, 60×10^6 psi [104]; (b) rayon-base, 90×10^6 psi [104]; (c) PAN-base, stress-graphitized [99].

of the turbostratic stacking of the layers in the present case. Often one can identify "classical" edge dislocations (terminations of layer planes) like that indicated by the arrow in Fig. 28. Another common feature is the presence of diffuse fringes crossing the microfibril (i.e., parallel to the direction of the scattering vector, (002) in this case). The unusually regular array shown in Fig. 28 is a rotation moiré caused by the superposition of two microfibrils slightly rotated with respect to one another about an axis normal to the page.

Fourdeux et al. [62] have examined the internal structure of very highly oriented rayon-base graphite fibers by high-resolution electron microscopy using dark-field techniques. In Fig. 30 the (002) diffracted beam was used to image those microfibrils that are oriented with their layer planes approximately parallel with the incident electron beam (i.e., we are viewing them edge on, just as in Fig. 28). Frequent rotation moiré patterns are visible. This picture shows that the microfibrils are reasonably uniform in thickness along their length. Close inspection (with the help of the moiré fringes) shows that these thicknesses range from 30 to over 100 Å, or approximately

FIG. 28. High-resolution electron micrograph (phase-contrast technique) of fragment of Thornel 50 filament seen in longitudinal section. Individual graphite layers are resolved. From Ref. 108.

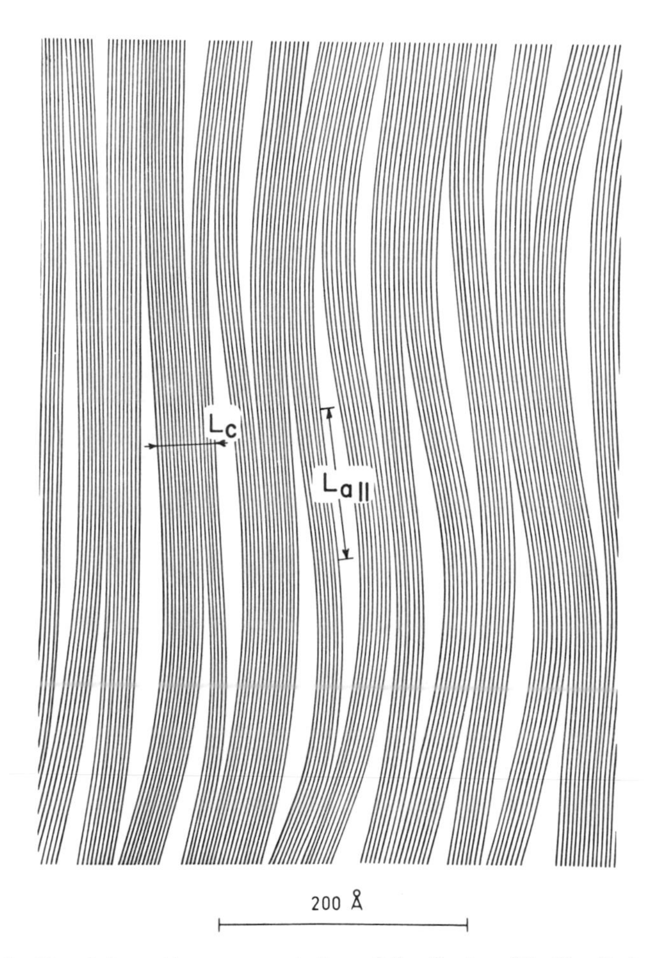

FIG. 29. Schematic representation of the "microfibrillar" structure of high-modulus carbon fibers. Stack size (thickness of microfibrils) is L_c; apparent layer size parallel to the fiber axis is $L_{a||}$. From Ref. 76.

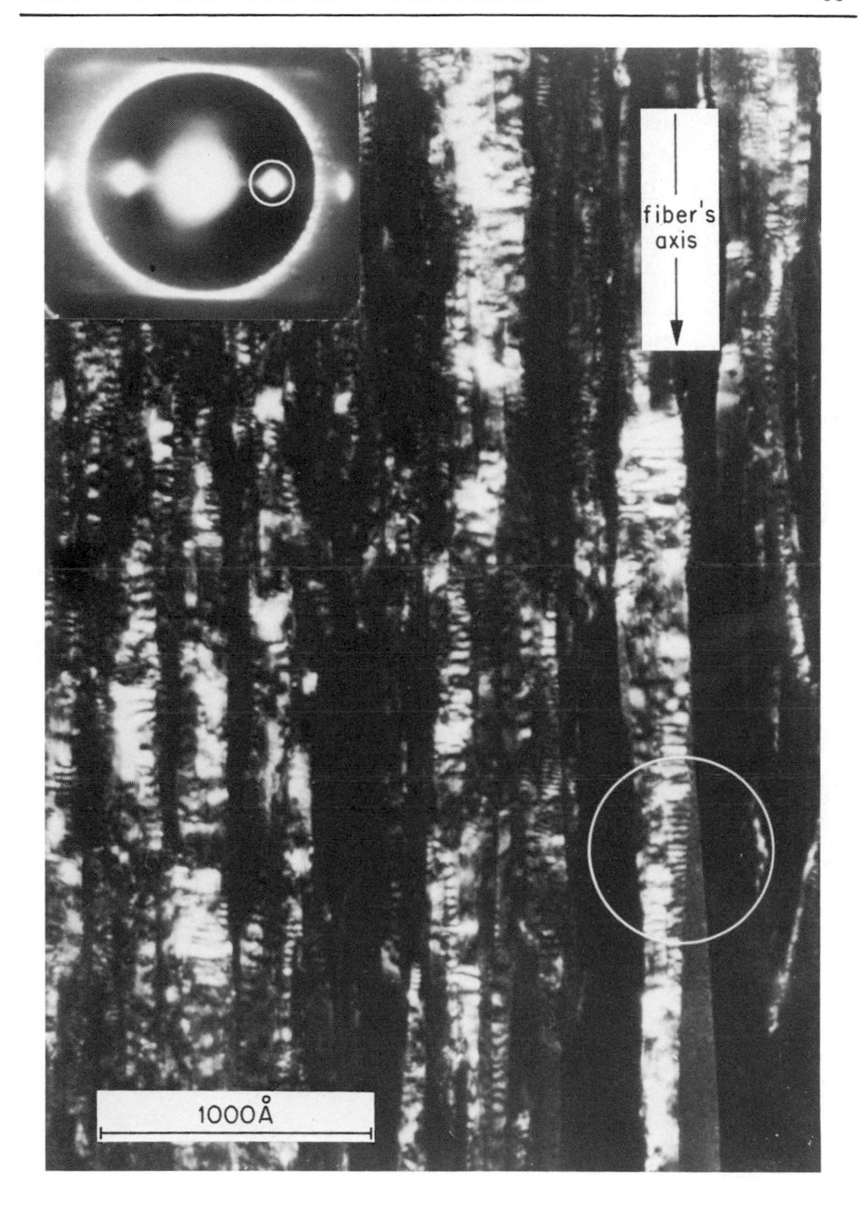

FIG. 30. Dark-field electron micrograph (002) of a highly oriented rayon-base carbon fiber. Circled region shows moiré pattern with spacing corresponding to the tilt angle between the two superimposed microfibrils. From Ref. 62.

the thickness expected from the known L_c values obtained by X-ray measurement. D. J. Johnson and Tyson [91], who previously used this technique to measure the thickness in graphitized PAN-base fibers, found an average value of 65 Å, which is in excellent agreement with their X-ray measurements of L_c for this fiber (60 Å).

By imaging a portion of the (10) diffraction ring, Fourdeux et al. [62] obtained the micrographs of Fig. 31. In this case the microfibrils imaged are those whose layer planes are approximately perpendicular to the electron beam (i.e., parallel with the page). Again, rotation moiré fringes parallel with the scattering vector are evident; their regularity shows that the two-dimensional hexagonal crystal structure of the layers is nearly perfect [62, 80]. These moirés are caused by the rotational disorder of the layer

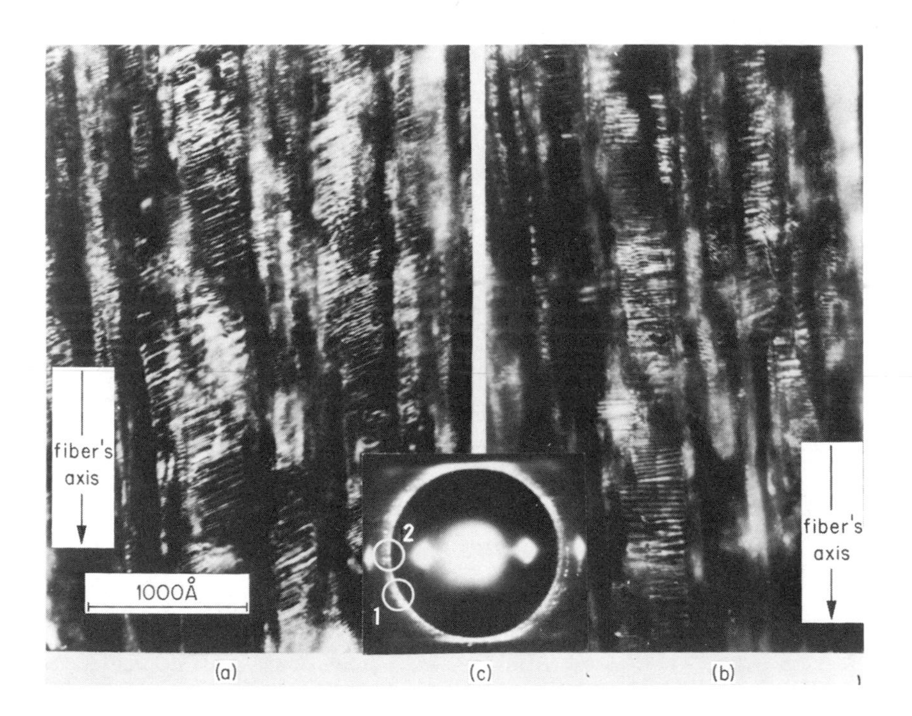

FIG. 31. Two dark-field electron micrographs of same area utilizing circled regions of the (10) band labeled, respectively, 1 (left side) and 2 (right side). Highly oriented rayon-base carbon fiber. From Ref. 62.

stacking in these turbostratic microfibrils; any given moiré pattern is due to the relative rotation between two layers, probably from the same microfibril.

Although considerable evidence is now available to tell us how microfibrils are arranged in longitudinal section, the question remains as to how they are arranged and interconnected sideways, that is, how they appear in transverse section. Fourdeux et al. [80] have considered this question and pointed to the evidence that must be considered. First, a tendency exists for the a planes of microfibrils (composed of lateral edges of ribbonlike layers) to contact the a planes of adjacent microfibrils, rather than the c planes. In other words, there is a tendency for edge-to-edge contact rather than edge-to-face contact. This evidence results from the high probability for the occurrence of moiré fringes in micrographs like those shown in Figs. 28 and 30, and also from the optical anisotropy of cross sections of fibers [109], which will be discussed in Section IV. D. 3. a. Second, the lateral dimensions of the "ribbons" (as seen in Fig. 31) are nonuniform along their lengths. Fourdeux et al. conclude that some ribbons run across two or more microfibrils, tying them together. Additional evidence, to be discussed next, suggests that not just <u>some</u> but <u>nearly all</u> ribbons run across two or more microfibrils and that the layers are, in fact, much wider than they appear in either X-ray measurements of $L_{a\perp}$ (Section IV. B. 3. c) or in micrographs like Fig. 31.

According to a suggestion by Strong [110], Ruland's interpretation of $L_{a\parallel}$ as representing the straight portions of much longer layers parallel to the fiber axis could also apply to $L_{a\perp}$; thus $L_{a\perp}$ represents the (relatively) flat regions of much wider "corrugated" sheets that bend and curve about through the transverse plane. Electron micrographs of transverse sections of rayon-base Thornel 50 fibers are shown in Figs. 32 and 33. The first of these micrographs, which reveals a multiplicity of white dots on a dark background, was interpreted by the present author [104, 111] as revealing a honeycomb of micropores, 60 to 90 Å apart and separated by packets of ribbonlike layers (microfibrils). The second micrograph (Fig. 33), obtained by Harling [107], shows essentially this same structure at high magnification, as revealed by the phase-contrast technique. In the transverse plane the layer packets are rather tortuous (radii of curvature as small as 30 Å are visible) and are frequently branched. Individual layers extend, in this

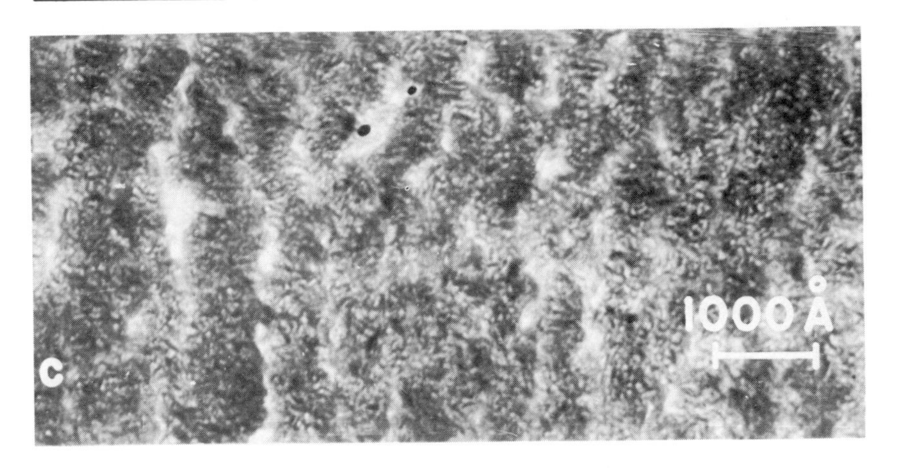

FIG. 32. Bright-field electron micrograph of a transverse microtome section of a rayon-base carbon fiber, Thornel 50, showing micropores (white dots). From Ref. 104.

transverse plane, for distances of several hundred angstroms, although the "straight" regions, corresponding to X-ray values of $L_{a\perp}$ or the widths of the microfibrils as seen in longitudinal section, are usually less than 100 Å. The micropores of Fig. 32 are probably the empty spaces near regions of severe distortion in Fig. 33. Such regions are shown schematically in the "thumb-print" pattern of Fig. 34. Configurations like those shown in Fig. 34 have been observed, on a much larger scale, in optical studies of pitch mesophase formations [112, 113]. The average distance between the pores revealed in Fig. 33 seems to be in reasonable agreement with the 60 to 90 Å found for the low-resolution micrograph of Fig. 32.

Before discussing the possible origins of the microfibrils, we must point out that they are not to be confused with the larger diameter "fibrils" (termed "secondary fibrils" by Watt and W. Johnson [58] discussed in Section IV. D. 1.c. The larger fibrils have diameters greater than approximately 300 Å, whereas the microfibrils are usually 50 to 100 Å in their transverse dimensions.

b. Origin and Significance of Microfibrils. The origin of the microfibrils in carbon fibers is a subject of speculation at the present time. I proposed [81] that the growth and orientation of graphitic layers in rayon-base carbon fibers were directly related to the molecular chain structure of the rayon, but

FIG. 33. High-resolution phase-contrast electron micrograph of a transverse microtome section of a rayon-base carbon fiber, Thornel 50. Individual layers are imaged. From Ref. 107 (in which this micrograph was erroneously attributed to PAN-base carbon fiber).

FIG. 34. Schematic "thumb-print" pattern of microfibrillar structure as viewed in cross section. Empty spaces represent micropore cross sections. "Classical" edge dislocations (edges of layers) are shown.

Ruland [69] questioned this possibility, since he found, in wide-angle X-ray scattering studies, that preferred orientation owing to molecular chain structure was almost completely destroyed during early pyrolysis. Ruland, however, suggested that the oriented growth of layers is influenced by a hypothesized microfibrillar structure of the starting material [89]. Unfortunately microfibrillar structure of the major rayon starting material (Villwyte) has not so far been detected by small-angle X-ray scattering. In the case of PAN-base carbon fibers, however, D. J. Johnson and Tyson [91] have observed 75-Å microfibrils in the precursor fiber, and Tyson [114] has shown that the microfibrillar structure persists throughout pyrolysis.

The significance of the microfibrillar structure is that it determines many of the important physical properties of the fiber. Ruland [98] has shown that the values of Young's modulus of high-modulus carbon fibers can be understood by means of a simple model related to the microfibrillar structure provided the degree of preferred orientation is known (Section V. B). In a similar way, the microfibrillar structure will account for thermal and electrical properties. It also accounts for the high or potentially high tensile strengths of carbon fibers, strengths that are, in fact, limited by the presence of weak boundaries between the larger diameter fibrils or by voids and inclusions (see Sections IV. D. 2. b and IV. D. 3. b).

The microfibrillar structure of graphitized carbon fibers may be responsible for the poor showing of graphitized fibers in compressive tests of composites containing them. According to a suggestion by Smith [115], the compressive strength of the individual filaments may be limited by the shear strengths of those sections of microfibrils whose directions have deviated the most from the direction of the filament axis.

c. Larger Structure Features — Fibrils. The existence in high-modulus carbon fibers of oriented structural units several hundred angstroms in diameter has been reported for both PAN- and rayon-base fibers [83, 93, 104, 116]. These structural units, which have been called fibrils, are composed of the microfibril-and-pore structure described in Section IV. D. 1. a. Their transverse dimensions in PAN-base carbon fibers have been estimated to be 800 to 1000 [83], 250 to 1000 [93], and about 500 Å [117]. Their transverse dimensions in rayon-base carbon fibers have been estimated at 300 to 500 Å [104]. The fibrils have indefinite length and may form a continuous, branched network, as suggested by W. Johnson and Watt [83].

Transmission-electron micrographs showing these fibrils protruding from the ends of longitudinal thin sections of both PAN- and rayon-base carbon fibers are shown in Fig. 35. Perhaps more convincing evidence of this fibrillar structure, however, is shown in replicas of external fiber surfaces or cleaved internal surfaces, which are discussed in the next section.

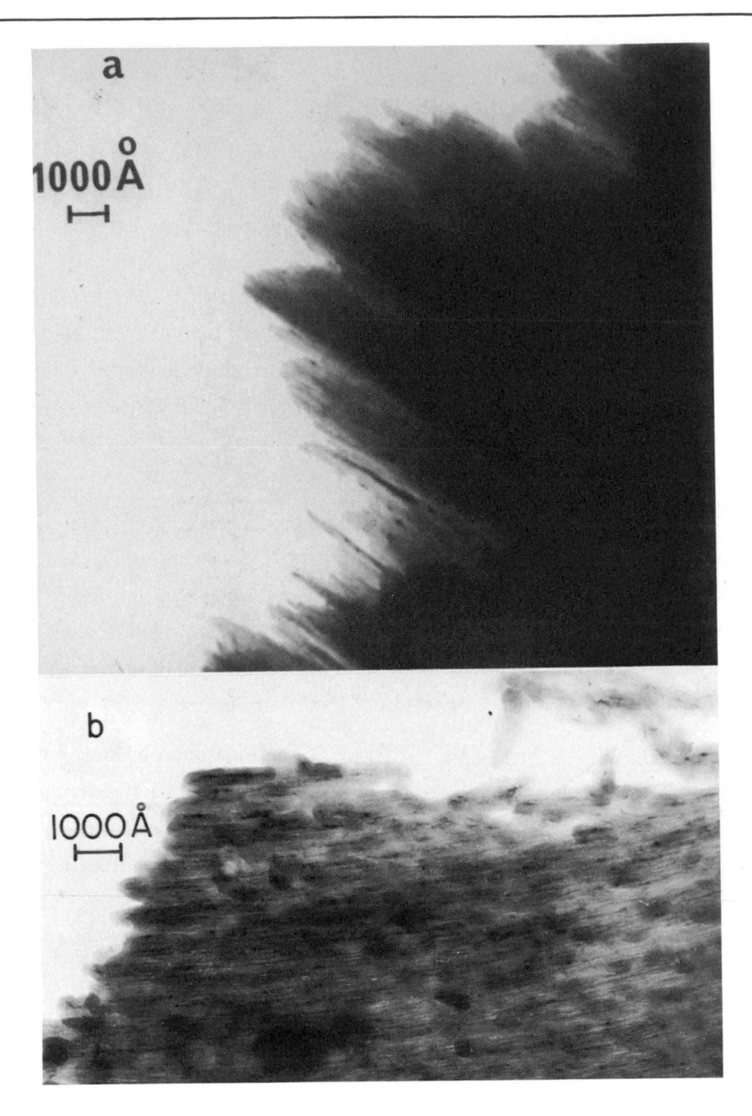

FIG. 35. Electron transmission micrographs showing evidence of fibrils (250 to 1000 Å in diameter) protruding from fragments of high-modulus PAN- and rayon-base carbon fibers: (a) PAN-base fiber, modulus 60 x 10^6 psi [93]; (b) rayon-base fiber, modulus 75 x 10^6 psi [104].

2. Surface-Replica Electron Microscopy

a. Further Evidence of Fibrils. A fibrillar internal structure in high-modulus carbon fibers, in which the fibrils possess transverse dimensions of several hundred angstroms, is evident in surface replicas taken from cleaved longitudinal surfaces. Examples for PAN- and rayon-base carbon fibers are shown in Figs. 36 and 37, respectively. In the case of rayon-base carbon fibers this fibrillar structure seems to "develop" during graphitization, as evident from the ripples in the external fiber surface, which are found only after graphitization. I have found [104] a close apparent relationship between these external surface ripples and the appearance of the transverse-fracture surface; the latter seemed to be composed of broken ends of fibrils, as shown in Fig. 38. The same apparent relationship exists between the external surface ripples and the fibrillar structure in the longitudinal-fracture surface of Fig. 37.

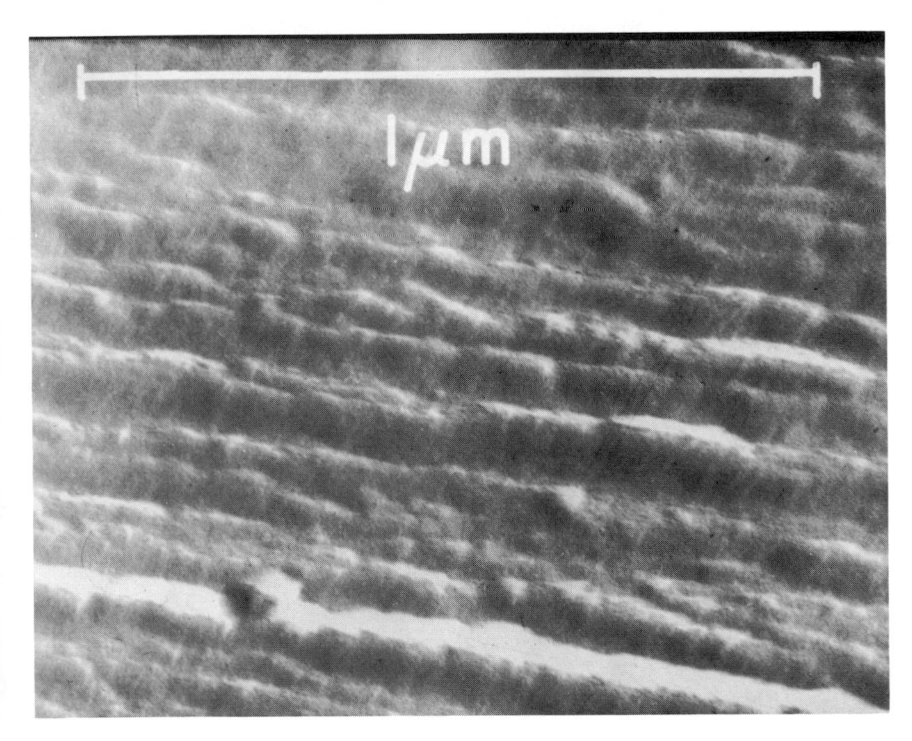

FIG. 36. Surface replica of cleaved longitudinal surface of PAN-base carbon fiber, showing fibrils 250 to 1000 Å across. From Ref. 93.

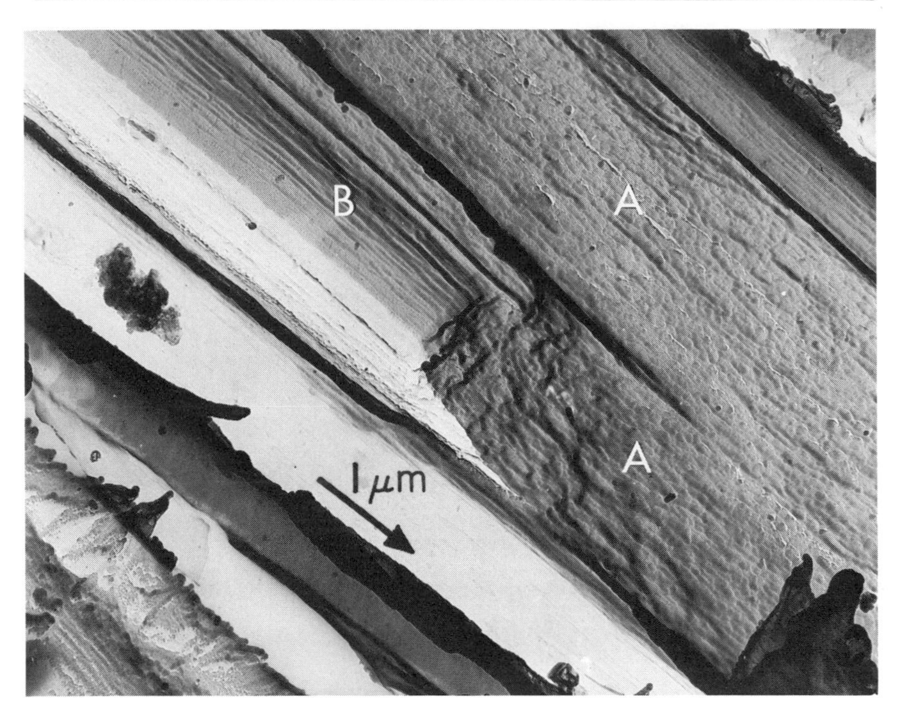

FIG. 37. Surface replica of cleaved longitudinal surface next to external surface of rayon-base carbon fiber, showing fibrils 300 to 500 Å across. Note similarity of fibrillar appearance between fracture surface (A) and external surface (B). From Ref. 118.

Both longitudinal- and transverse-fracture surfaces similar in appearance to those shown in Figs. 36 through 38 have been found in PAN-base carbon fibers by J. W. Johnson and co-workers [116, 119], who used the scanning electron microscope (see Section IV.D.3.a and Fig. 40). They interpret these fracture surfaces as revealing a "fibrillar structure" [116].

b. Origin and Significance of Fibrils. The origin of the fibrillar structure is still a subject of speculation. W. Johnson and Watt [83] and I [104, 111] have suggested that it originates from a fibrillar structure in the raw material that would necessarily possess transverse dimensions somewhat larger than those of the final carbon fiber due to shrinkage, stretching, etc. The very fine 75- to 100-Å microfibrils that are known to exist both in PAN [91] and in some cellulosic fibers [120] cannot account for the larger than 250 Å

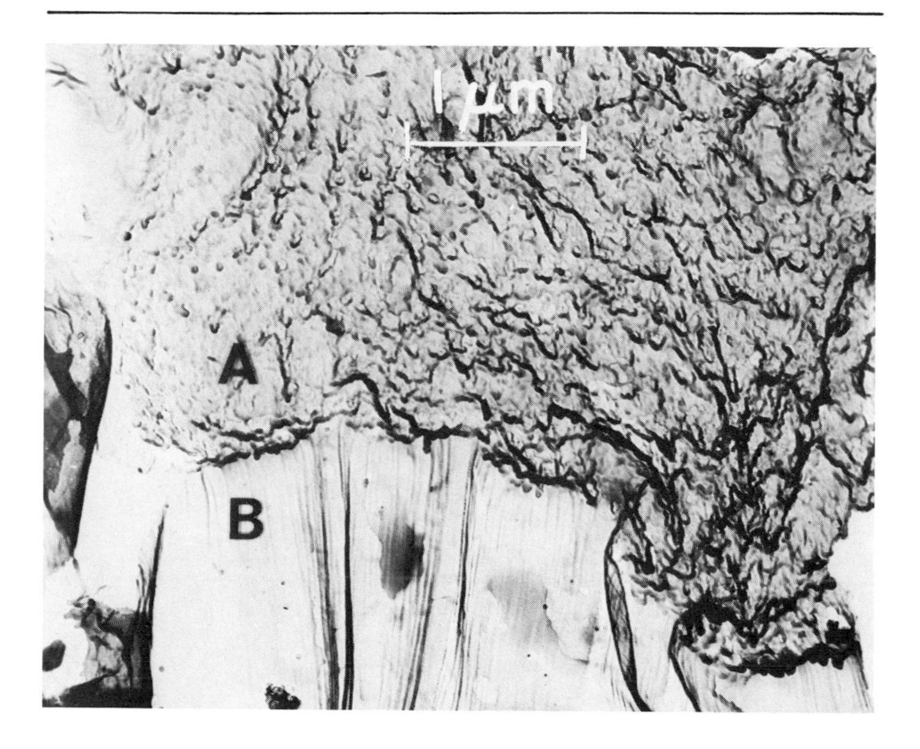

FIG. 38. Surface replica of the transverse-fracture surface of high-modulus rayon-base carbon fiber, showing relationship between texture of fracture surface (A) and that of external filament surface (B). From Ref. 104.

fibrils in high-modulus carbon fibers. In the case of rayon-base carbon fibers, the existence of "bumps" on the transverse-fracture surfaces of low-temperature carbon fibers [104] suggest that the fibrils have some sort of identity even in this relatively amorphous stage, and one could consider this observation to be "evidence" that this fibrillar structure exists even in the precursor fiber. Perhaps more convincing is the structural defect shown in the longitudinal fracture surface of Fig. 39. A pronounced distortion or deviation in the fibrillar structure (at region marked A) is evident. Since it is difficult to imagine that this defect could have occurred during pyrolysis or graphitization, it was probably present in some way in the raw fiber.

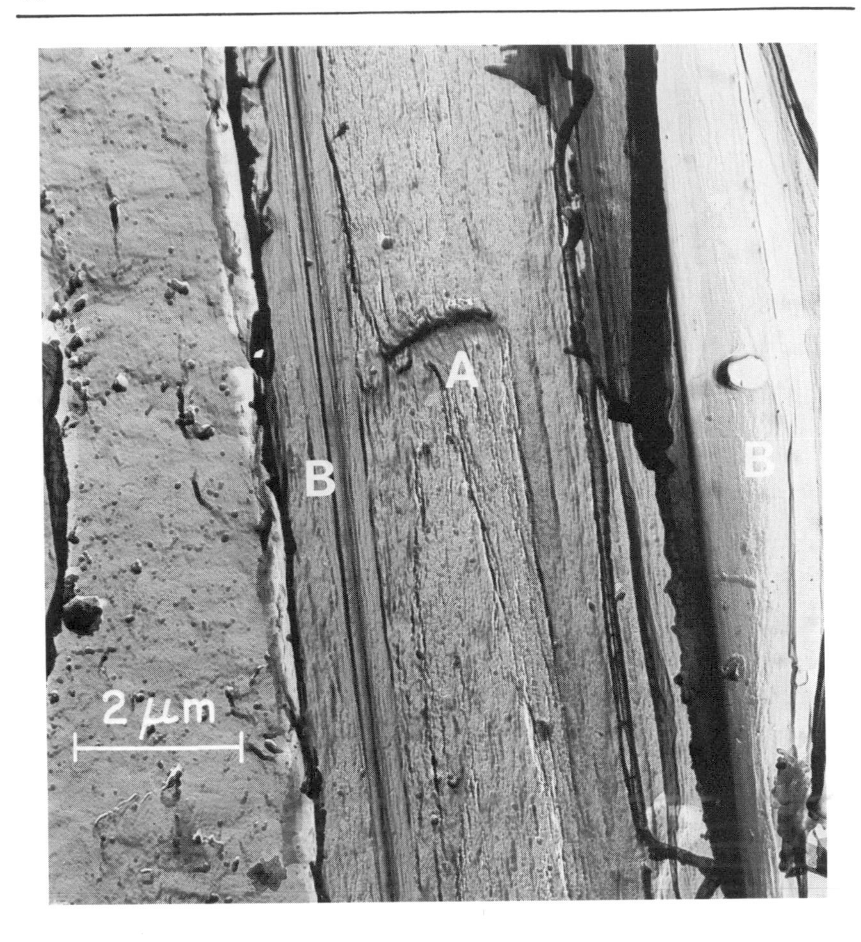

FIG. 39. Surface replica of the cleaved longitudinal surface of rayon-base carbon fiber, showing distortion (A) in fibrillar structure. Region B is external filament surface. From Ref. 8.

Whatever the origin of the fibrils, we consider that they do have important structural significance. Most of the evidence for them comes from the appearance of fracture surfaces. The boundaries between them are evidently weak internal surfaces along which cracks can propagate with comparative ease. W. Johnson and Watt [83] have suggested that they are the strength-determining structural feature in PAN-base carbon fibers; Williams et al. [8, 121] made the same point for rayon-base carbon fibers, citing evidence from the behavior of these fibers in bending. I believe that the fibrillar structure limits the strength of carbon fibers, although gross flaws of the type studied by J. W. Johnson and Thorne [119, 122-125] are also operative and can further reduce fiber strength (see Section IV. D. 3. b).

3. Scanning Electron Microscopy

The scanning electron microscope has been utilized very successfully in the study of fiber morphology, particularly fiber fracture surfaces. In routine use it is capable of achieving nearly as good resolution as that achieved in surface-replica microscopy (100 to 200 Å). Sample preparation is greatly simplified, so that the technique has proved to be highly productive.

a. Lamellar versus Fibrillar Structure; "Tree-Trunk" Preferred Orientation. J. W. Johnson and co-workers [116] have demonstrated that a pronounced lamellar structure, leading to a "tree-trunk" type of preferred orientation as viewed in transverse section [109], can be obtained in graphitized PAN-base fibers.* Their determination of a lamellar structure was based on the appearance of transverse fracture surfaces as revealed by the scanning electron microscope. Fibers prepared from PAN in the conventional manner (oxidation of the fiber to render it infusible, followed by carbonization at about 1000° C and, if desired, graphitization to above 2000° C) exhibited fracture surfaces that appeared fibrillar. However, when the raw PAN fiber was oxidized only for a short time at a high temperature (~ 300° C) and then carbonized and graphitized, a pronounced lamellar structure developed in the core of the fiber. Figure 40 shows such a fiber that has been graphitized to 2260° C. Most of the fracture surface of this fiber shows the normal fibrillar structure, but a central slit-shaped hole is surrounded by a pronounced lamellar structure.

When this type of fiber was graphitized to 2560° C (Fig. 41), the lamellar structure was much more highly developed, and, furthermore, the material with the fibrillar structure had developed a distinct lamellar texture, indicating that it was partially converted to the lamellar type of structure. J. W. Johnson et al. [116] explained this result in the following manner: during the brief oxidation heat treatment an infusible oxidized crust was formed; however, the unoxidized core material was fusible; carbonization of this fiber caused the core material to pass through a fused state, which is known to enhance the development of a highly lamellar, ultimately graphitic, structure

---

*Note added in proof: Recent studies by R. H. Knibbs (J. Microscopy, 94, 273 (1971) and by J. V. Larsen (PhD Thesis, University of Maryland, 1971) have shown that lamellar textures possessing a radial, rather than circumferential, orientation are sometimes seen in PAN-based carbon fibers.

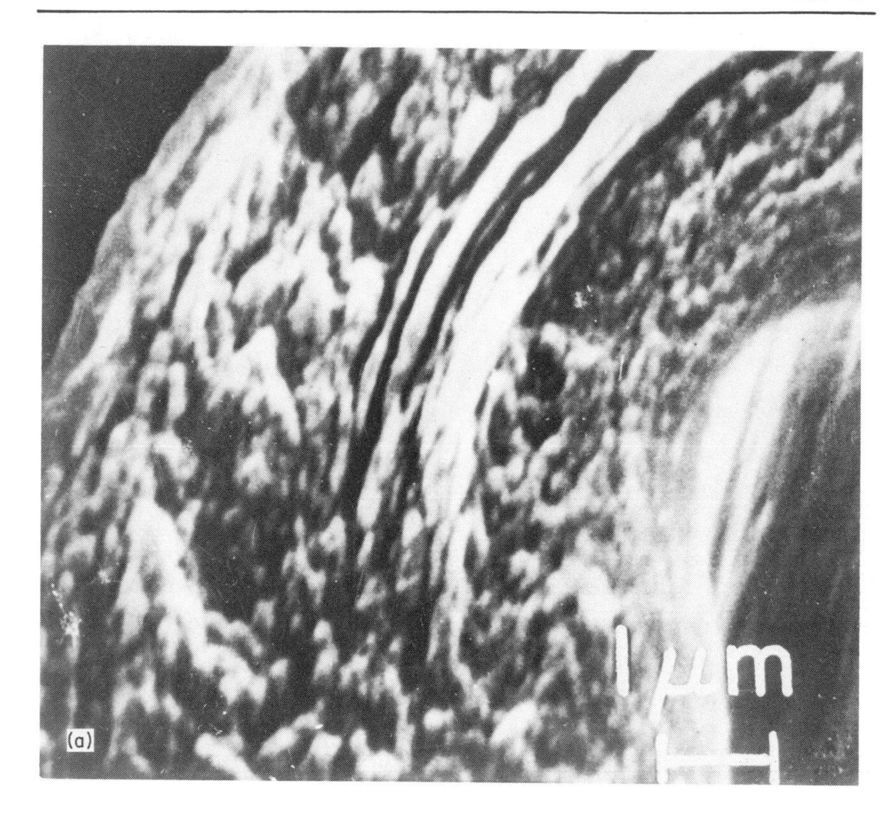

FIG. 40. Scanning electron micrographs of the fracture surface of a PAN-base carbon fiber showing mostly fibrillar structure except for the thin core region. From Ref. 116.

[126]. J. W. Johnson et al. suggested that the growth of the lamellar structure was nucleated on the inside surface of the oxidized crust material, thereby accounting for the annular or "tree-trunk" configuration. Although the outer crust at first formed the normal fibrillar texture, after a sufficiently high-temperature heat treatment, this material, too, converted into a lamellar structure. These authors also found that stress-graphitized PAN-base fibers exhibited structural features similar to these "tree-trunk" fibers. They pointed out that the lamellar structure is evidently a more stable form than the fibrillar one.

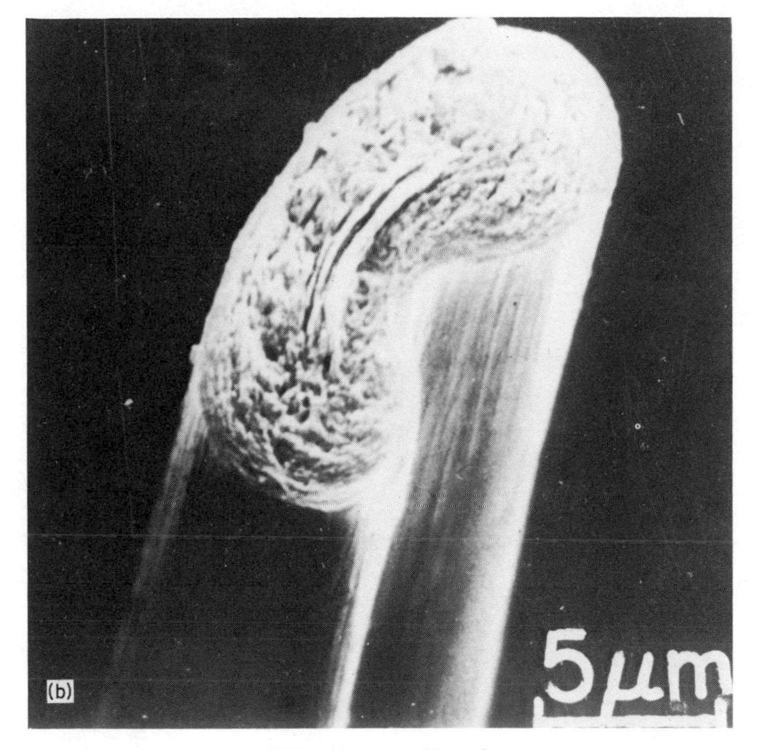

FIG. 40 — continued.

The existence of a "tree-trunk" type of preferred orientation of the graphite layer planes was known previously. It was first detected by Shindo [82] in PAN-base carbon fiber outer layers by electron-diffraction techniques. Butler and Diefendorf [109] used polarized-light microscopy to examine polished fiber cross sections and found that this type of preferred orientation was common in PAN-base carbon fibers. An example is shown in Fig. 42. They found that high-modulus rayon-base carbon fibers exhibited a similar preferred orientation, but to a lesser degree.

The types of fibers studied by Butler and Diefendorf were the "conventional" varieties, which previous workers had shown to possess fibrillar textures. Thus we see that the "tree-trunk" type of preferred orientation may exist even though a predominantly fibrillar texture is present. As this preferred orientation becomes stronger, the fibrillar texture may be expected to convert gradually into the lamellar texture shown in Fig. 41.

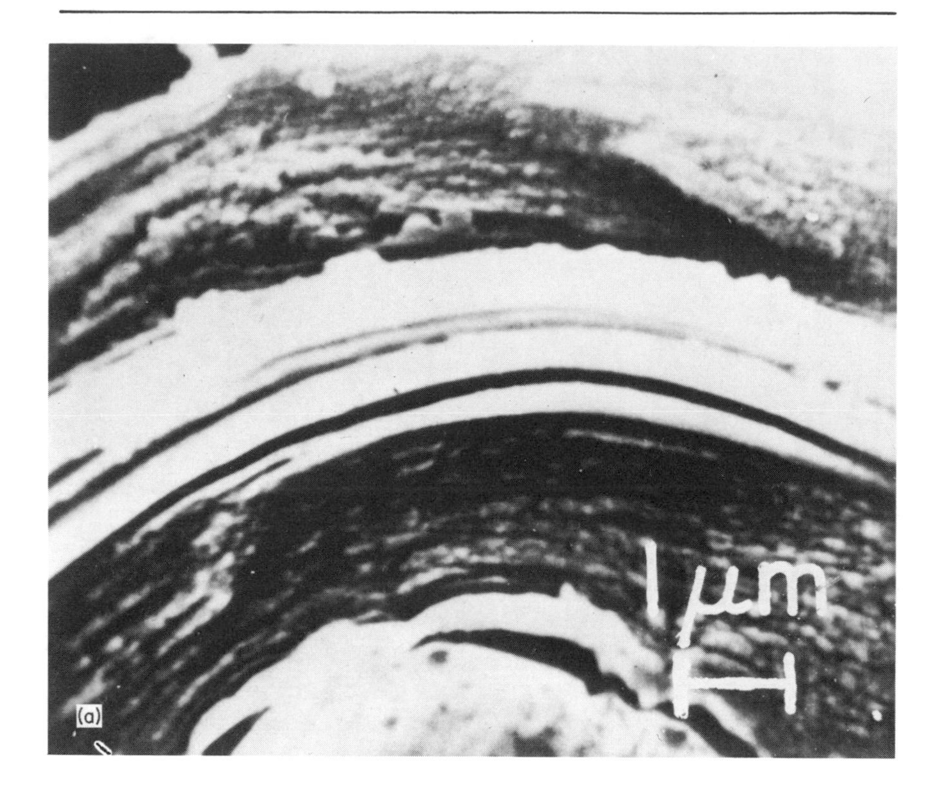

FIG. 41. Scanning electron micrographs of the fracture surface of a PAN-base carbon fiber, showing pronounced lamellar structure in the core region and a partially developed lamellar structure in the remaining surface. From Ref. 116.

The lamellar texture is probably the result of a lesser degree of tortuosity of the graphite layer planes in the transverse direction (cf. Figs. 33 and 34). Thus less tortuosity would be associated with larger transverse-layer dimensions $L_{a\perp}$ (i.e., wider microfibrils) and, in addition, would result in larger regions of preferred orientation and, hence, a lamellar texture.

<u>b. Gross Flaws in Carbon Fibers.</u> J. W. Johnson and Thorne have used the scanning electron microscope to examine the fracture surfaces of individual filaments after strength testing of PAN-base carbon fibers [119, 122]. They were able to distinguish between points of origin of fracture that were

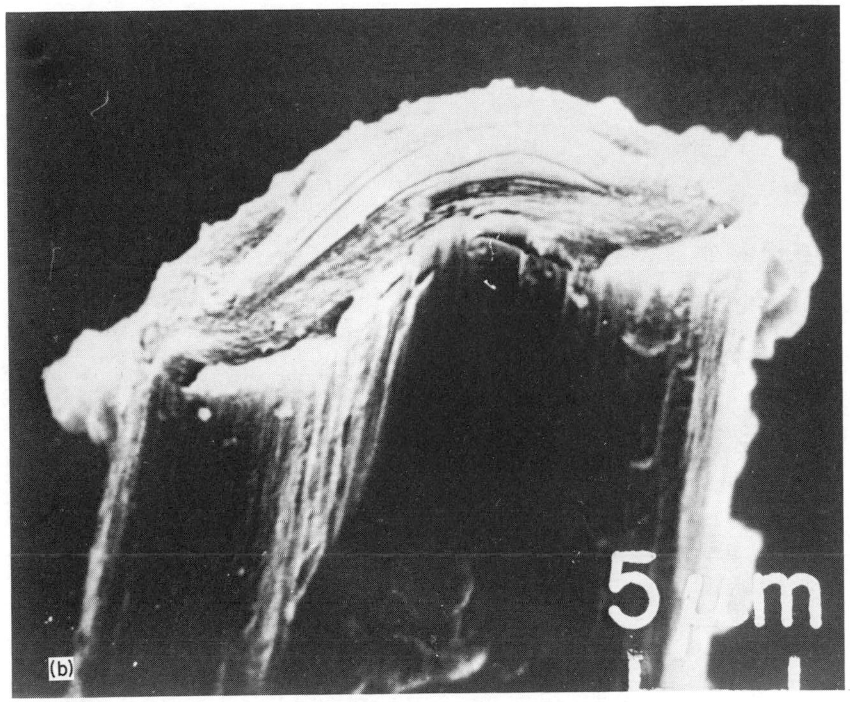

FIG. 41 — continued.

on the fiber surface (Fig. 43) and those that were within the body of the fiber (Fig. 44). In the latter case the flaw was frequently identified as a void. The most severe flaws were found to be gross surface flaws in the "as-processed" carbon fibers; however, after these surface flaws were removed by etching in oxygen, the most severe flaws remaining were internal flaws (voids). A study of the raw material by optical microscopy showed that voids present in the original PAN fibers (Fig. 45) were apparently responsible for the voids in the carbonized fibers [119, 124].

E. Summary Description of the Structure of High-Modulus Carbon Fibers

The basic "molecule" of carbon-fiber structure (analogous to the long-chain molecule in polymers) is the graphitic layer, which is probably thousands of angstroms in the longitudinal dimension and many hundreds of angstroms in the transverse dimension. The boundaries are probably highly irregular. These layers are, in the ideal structure, relatively straight in the longitudinal direction (as shown schematically in Fig. 29) but highly

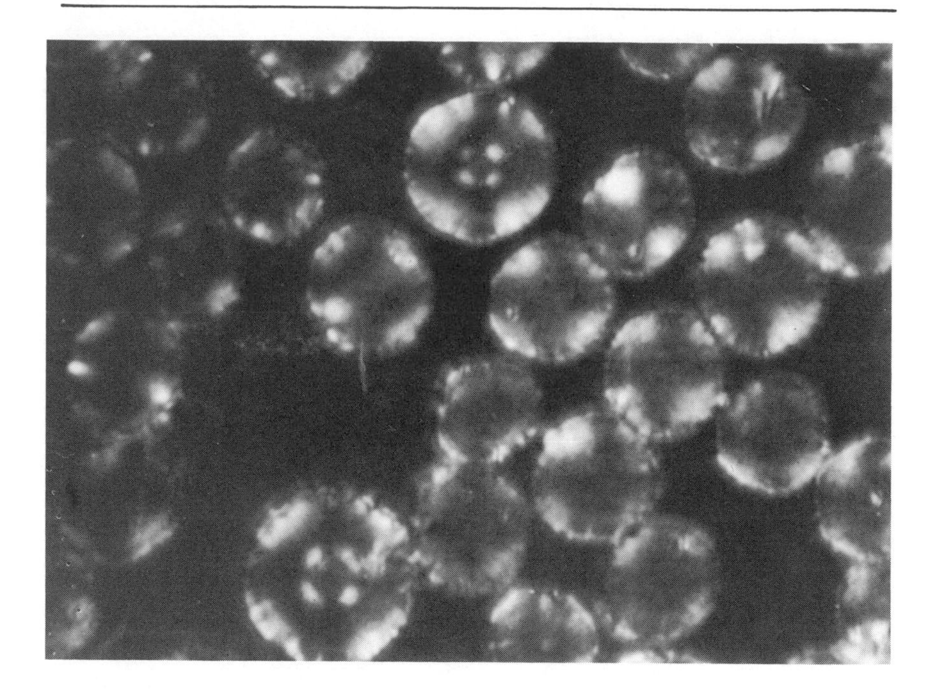

FIG. 42. Polarized-light micrograph of type I PAN carbon-fiber cross sections, showing extinction crosses that indicate the existence of a "tree-trunk" preferred orientation. Crossed polarizer and analyzer are parallel to picture edges. From Ref. 109.

curved and tortuous in the transverse dimension (as shown schematically in Fig. 34), somewhat like an irregularly corrugated metal sheet. The sheets are packed together in turbostratic stacking (3.4+ Å apart) to form a continuous branched network of microfibrils or crystallites. The crystallite dimensions as measured by X-ray diffraction are L_c (microfibril thickness), $L_{a\perp}$ (microfibril width), and $L_{a\parallel}$ (microfibril length). The L_a measurements are actually the dimensions of the "straight" parts of the layers. Microfibril lengths and widths as viewed by high-resolution electron microscopy are considerably greater than the values obtained by X-ray measurements, and, in fact, are almost indefinable, since one microfibril connects smoothly with the next, through curving, twisting, and branching. This connectivity occurs both lengthwise and sidewise (i.e., as viewed both in longitudinal and in transverse section); the only difference between the two views is the

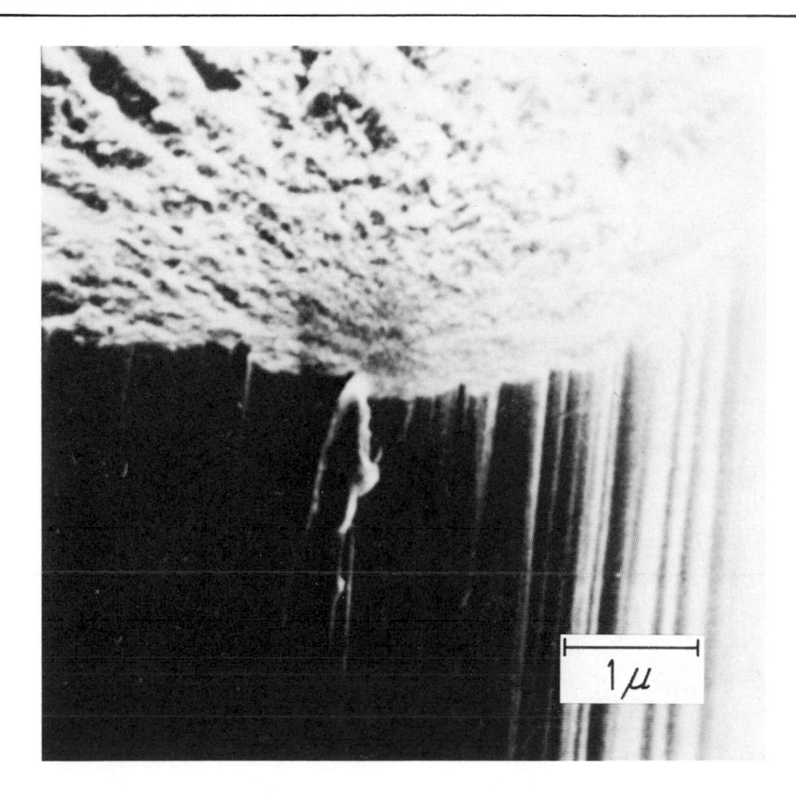

FIG. 43. Scanning electron micrograph of the fracture surface of PAN-base carbon fiber in which fracture initiated at filament surface. From Ref. 119.

frequency of branching and the degree of tortuosity (the latter being greater in the transverse plane). In the ideal case the structure as viewed in the transverse plane is isotropic; however, in some cases there may be a degree of preferred orientation ("tree-trunk") that, when extreme, results in a highly lamellar structure parallel with the external fiber surface.

The bending, twisting, and branching of microfibrils cause mismatching among them. The resulting spaces constitute a system of micropores that are long (as much as thousands of angstroms) and narrow (average transverse dimension between 10 and 20 Å). They possess the same preferred orientation as the microfibrils themselves because they are bounded by the layer planes of the microfibrils. In cross section they probably have a variety of shapes, as suggested schematically by the configurations shown in Fig. 34.

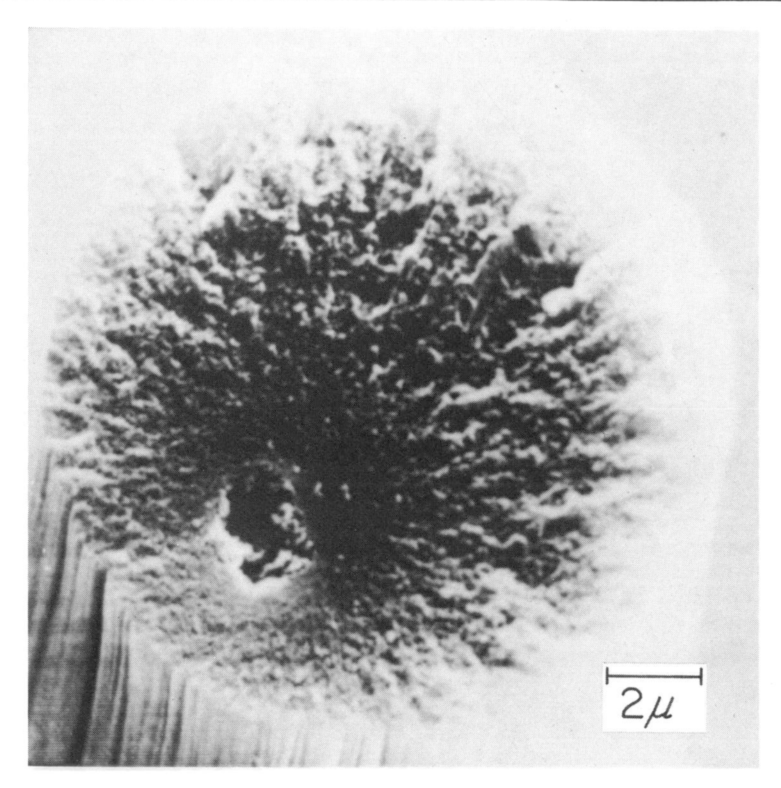

FIG. 44. Scanning electron micrograph of the fracture surface of PAN-base carbon fiber in which fracture initiated from an internal void. From Ref. 119.

A bundle of microfibrils with a diameter of 250 to 1000 Å constitutes a secondary structural unit that has been termed a fibril. These fibrils are made evident primarily by the appearance of fracture surfaces. The lateral boundary between two fibrils may be merely a surface composed mostly of graphitic layer faces (i.e., one through which relatively few connecting graphitic layers pass). Since the microfibrils have a preferred orientation parallel to the fiber axis, the fibrils must contain this same preferred orientation. They may not possess well-defined transverse boundaries, but instead may consist of a continuous, branched network.

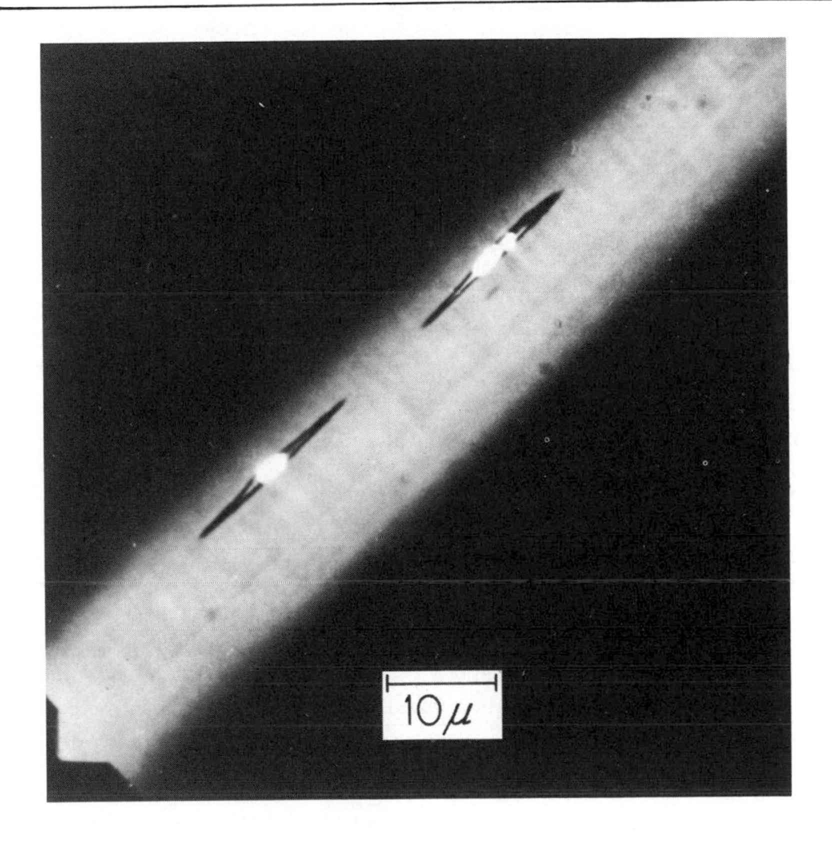

FIG. 45. Optical micrograph showing voids in raw PAN filament believed to be responsible for voids in carbonized fibers. From Ref. 119.

V. PHYSICAL PROPERTIES

A. Density

1. Test Methods

The filament bulk density, defined as the dry weight of the filament (in grams) divided by its geometrical volume (in cubic centimeters), is measured by a fluid-displacement method or by an immersion-weighing method. In either case the fluid must not penetrate accessible filament pores; otherwise the true geometrical filament volume will not be measured. The dry weight (drying is important in the case of fibers with accessible porosity) is determined by ordinary weighing in air and correcting for air buoyancy if the highest accuracy is required.

In the helium-displacement method the volume of the fiber sample is determined by the use of a helium pycnometer. The fiber must be impervious if this method is to yield the true filament bulk density.

Alternatively the fiber volume can be determined by measuring the displacement of mercury in a mercury pycnometer at a pressure of approximately 10,000 psi. Such a pressure is required if the mercury is to fill all interstices between filaments and all pits and valleys on the surfaces of the filaments; however, at this pressure no mercury will penetrate the internal pores, even in carbon fibers of very low heat-treatment temperature, since such pores are less than 30 Å in width. Hence a true geometrical volume is measured.

In the immersion-weighing methods the volume of the fiber sample is obtained by comparing the sample weight in air (corrected for buoyancy) with the sample weight immersed in a liquid of known density. This method requires taking the usual precautions against entrapment of air bubbles in the yarn sample.

These methods usually require sample sizes of between 0.2 and 20 g, depending on the method and the accuracy desired. If only very small samples are available, the fibers can be tested for floating or sinking in a series of liquids until a good density match is obtained. Alternatively one can use a single liquid column that possesses a gradient of densities; the position in the column at which the fiber sample comes to rest indicates its density.

2. Factors Affecting Density

The most important factor affecting the density of graphitized rayon fibers is the amount of stretch to which the fibers were subjected during graphitization. Since the degree of applied stretch determines the fiber's Young's modulus, the fiber density is conveniently plotted against Young's modulus (Fig. 46). These data were obtained from measurements of commercial grades of yarn (Thornel graphite yarns) or experimental yarns prepared by similar processing techniques. Although

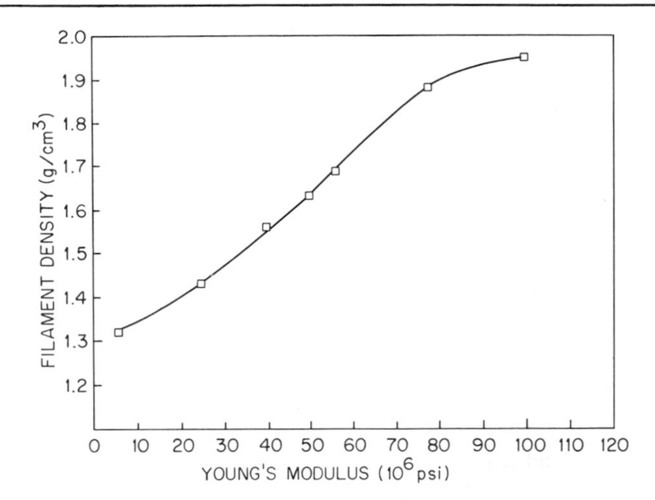

FIG. 46. Density-modulus relationship for stress-graphitized rayon-base carbon fibers. Densities measured by liquid-immersion method. From Refs. 127 and 128.

significant variations in the process would be expected to affect the density, these factors have not been studied in detail.

The wide range of densities represented in Fig. 46 implies a range of porosities from about 40% for low-modulus graphitized fibers to only about 12% for fibers with a Young's modulus of 100×10^6 psi. These porosities, which are all inaccessible (even to helium) [97], are calculated by comparing the filament bulk density with the X-ray density (approximately 2.20 g/cm^3).

Some insight into the processes that lead to the observed density changes with increasing stretch has been provided by the small-angle X-ray scattering studies of Perret and Ruland [129], who found that the average pore diameter is a unique function of the heat-treatment temperature and is not affected by stretching [80]. Since all of the fibers of Fig. 46 were graphitized at the same temperature, they all possess pores of the same average diameter. One may assume, then, that stretching causes the pores to collapse. In turn, the average thickness L_c of microfibrils increases, since the pores are the spaces between microfibrils. Such an increase in L_c was, in fact, observed [129].

B. Young's Modulus

1. Test Methods

Two tensile tests are commonly used to measure the Young's modulus of rayon-base carbon fibers: the single-filament test (used primarily in research and as the basic standard) and the strand test (used primarily for quality control or other routine testing). Both methods utilize an Instron or similar type of constant-rate-of-strain testing machine.

a. Single-Filament Test. The most common method for tensile testing single filaments employs a simple technique for mounting individual filaments on heavy-paper mounting tabs. The tabs, which are precut by a specially designed paper punch, contain a slot or window of length equal to the desired gauge length, usually about 1 in. Preferably, marks or lines on the tab are used to indicate the exact center line of the tab to facilitate accurate alignment of the filament. The filament is first mounted with the use of cellophane tape; it is then permanently cemented in place, usually with a hard-setting wax. The tab is then placed within the grips of the test machines, and the two ends are separated by cutting or burning through both sides of the tab. The elongation-load trace is recorded, and its slope, the apparent compliance of the filament, is measured. (The motion of the crosshead is the apparent filament elongation.) The machine compliance must be subtracted from the measured slope to yield the true elastic compliance of the filament. Machine compliance is best obtained by measuring the apparent compliances of a series of filament samples of varying length and extrapolating the resulting apparent compliances to zero length.

Computation of Young's modulus requires a knowledge of the filament cross-sectional area. Since the determination of an accurate cross-sectional area in each individual filament is often impractical (especially in the case of the irregularly shaped rayon-base carbon fibers), an average value is often obtained from an adjacent sample of filaments to that subjected to tensile testing. One method is to imbed the sample of parallel filaments in epoxy, and, through metallographic techniques, obtain a photomicrograph of the bundle cross section. Areas of individual filaments may then be measured by means of a polar planimeter. A second method, most effective for filaments that are round in cross section, involves the measuring of the filament diameters directly by the use of a microscope fitted with a "split-image" or "image-shearing" eyepiece. In a third method the cross-sectional area can be obtained from a knowledge of the fiber density, yarn weight per unit length, and the number of filaments per yarn bundle.

All of these measurements and techniques require considerable care, and numerous difficulties, which are not discussed here, are encountered. One of the major drawbacks to single-filament testing is the extremely small sample size which is necessarily employed but which may not be representative. A major advantage of single-filament testing is that it tests an intrinsic filament property and permits the determination of property variances rather than averages only.

b. Strand Test. The strand test is used to measure the Young's modulus (and tensile strength) of an entire strand, or yarn, consisting of a bundle of often many hundreds of filaments. The as-received, or "dry," yarn is not normally tested; instead, the yarn is first impregnated with a resin (e.g., epoxy), which is then cured while the yarn is held straight on a frame. (There are two reasons why dry-yarn testing is impractical: (a) since carbon fibers are brittle, direct gripping of the yarn causes serious damage near the grips and (b) frequently a significant but unknown fraction of the filaments may be broken within the gauge section, making test results meaningless.) The impregnated strand of fibers is then mounted in the testing machine by using either rubber-faced grips or by cementing the ends onto heavy metal tabs designed to fit standard grip fixtures on the testing machine. In either case careful alignment is required, since the impregnated strand is very brittle. A gauge length of approximately 10 in. is usually employed. The elongation-load trace is obtained in the usual manner. The machine compliance for this test must be obtained as described in the preceding section for single filaments.

The strand cross section (fiber only, not including the epoxy binder) is obtained from a measurement of dry strand weight per unit length and fiber density (see Section V.A.1). The properties of the resin binder utilized in the strand test may be ignored in calculating both the fiber Young's modulus and the tensile strength because the resin modulus is only approximately 1% of that of the fibers.

The strand test has several advantages over the single-filament test: (a) it involves much larger sample sizes and hence is more representative;

(b) it is faster and cheaper; and (c) it is more relevant to the properties one may expect to obtain when the fiber is incorporated into a composite. The test, of course, requires careful technique, and meaningless results are obtained if significant yarn damage has occurred prior to the formation of the impregnated test specimen.

2. Factors Affecting Young's Modulus

a. Dependence of Young's Modulus on Processing Variables. The dependence of Young's modulus of rayon-base carbon fibers on processing variables has already been discussed in Section III.E. Briefly, the modulus is determined by (a) the amount of stretch imparted to the fiber during the carbonization and/or the graphitization step and (b) the degree of heat treatment of the fiber — that is, the maximum temperature and the duration of the final heat treatment (graphitization step). Different degrees of preferred orientation in the starting material and different degrees of stretch during the initial low-temperature heat treatment would also be expected to affect the final Young's modulus; however, such effects are minor by comparison with the effects of stretching during carbonization and, especially, graphitization.

b. Dependence of Young's Modulus on Fiber-Structure Parameters: Elastic Dewrinkling of Microfibrils. Ruland and co-workers have investigated by X-ray diffraction the structural parameters (preferred orientation, layer size, and layer-stack height) of a variety of stress-carbonized and stress-graphitized rayon-base fibers. They found that the Young's modulus of these fibers correlated very well with the preferred orientation parameter q (see Fig. 21 and related discussion in Section IV.B.3) and did not depend significantly on layer size or stack height [88]. Since layer size and stack height are primarily determined by the degree of heat treatment, Ruland's findings show that fiber modulus is not directly dependent on the degree of heat treatment, but, rather, only indirectly as the heat treatment affects the degree of preferred orientation.

The theoretical curve of E_c versus q, shown in Fig. 21, is based on Ruland's model of elastic dewrinkling of ribbonlike graphitic layers [88]. (E_c is Young's modulus corrected for the microporosity of the fibers; i.e., it is the intrinsic modulus of the matter only.) According to this model, the layers are in the form of long wrinkled ribbons running generally parallel to the fiber axis. Application of a tensile stress to the fiber causes these ribbons to be put into tension. If one considers a tilted section of ribbon, the applied stress, which is parallel to the filament axis, may be resolved into a tensile component acting parallel to the ribbon direction and a shear component acting perpendicular to the ribbon direction. The tensile component elastically extends the ribbon section by an amount determined by the elastic compliance constants S_{11} of the graphite layers ($1/S_{11}$ is the Young's modulus of the graphite crystal parallel to the basal plane). The shear component elastically tilts the ribbon section toward the filament-axis direction by an amount determined by the resistance of the environment to tilting. The elastic compliance constant characterizing the tilting effect is designated k by Ruland. This model leads to the following expression for the reciprocal Young's modulus of the filament:

$$\frac{1}{E_c} = \ell_z S_{11} + m_z k, \tag{8}$$

where

$$\ell_z = \frac{\int \sin^2 \phi \, I(\phi) \, d\phi}{\int \sin \phi \, I(\phi) \, d\phi}, \tag{9}$$

$$m_z = \frac{\int \cos^2 \phi \, I(\phi) \, d\phi}{\int \sin \phi \, I(\phi) \, d\phi}. \tag{10}$$

The parameters ℓ_z and m_z are calculated from a knowledge of the orientation distribution function $I(\phi)$ for graphite layers, ϕ being the angle between the layer normal and the fiber axis (see Section IV. B. 2).

This model is undoubtedly somewhat oversimplified; however, simplicity is one of its virtues. The model has two additional virtues: (a) it is based on the now well-established microfibrillar structure of carbon fibers and therefore makes use of a more realistic notion of the transmission of stresses through the fiber than either the uniform-stress or the uniform-strain model; and (b) the theoretical model fits the experimental data for a wide range of fiber moduli, heat treatments, and precursors (i.e., PAN and pitch, in addition to rayon), all with the use of the same value of the parameter k (for exception, see discussion below). The uniform-stress and uniform-strain models require, on the other hand, the use of shear compliances which are not constants, but which depend on the degree of preferred orientation. The best-fit value of k was found by Ruland to be 0.275×10^{-11} cm^2/dyne.

The compliance constant k clearly has the nature of a shear compliance. If the microfibrils (composed of parallel-stacked ribbons) were able to act independently of their lateral neighbors, k would simply be the shear-compliance constant S_{44} for turbostratic graphite. The value of this constant for irradiated (dislocation-pinned) stress-annealed pyrolytic graphite is 2.5×10^{-11} cm^2/dyne [7]. However, since considerable interaction between neighboring microfibrils is expected (through extensive interlocking), it is not surprising that k is much less than S_{44}. One might, in fact, expect that k should be closely related to the longitudinal-shear-compliance constant S_{44} of the filament, corrected for porosity. The shear stiffness, $C_{44} = 1/S_{44}$, for high-modulus graphite fibers has been found to lie generally between 3×10^6 and 4×10^6 psi, after correcting for porosity (see Section V.C.2). This range of stiffnesses corresponds to compliances between approximately 0.35×10^{-11} and 0.50×10^{-11} cm^2/dyne, in fair agreement with the value of Ruland's k (considering the sensitivity of shear moduli of graphite to structural perfection).

c. Temperature Dependence of Young's Modulus. The temperature dependence of Young's modulus of Thornel 75 has been studied by Cummerow [130], who used a sonic resonance technique. He found that the modulus was very nearly constant from room temperature to 1000°C. (His results assumed that no significant dimensional changes occurred in the filament during heating; recent work by Sarian and Strong [131] shows that dimensional changes may occur in the vicinity of 1000°C.)

C. Longitudinal Shear Modulus

1. Test Methods

The longitudinal shear modulus of rayon-base high-modulus graphite fibers has been measured by the torsion-pendulum method by Williams [132]. A small wire weight was attached horizontally to the lower end of the filament, the upper end of which was fixed to a support. The weight was set into torsional oscillation, and the period was measured. The torsion modulus was computed on the basis of a circular filament whose cross-sectional area was equal to the actual area of the filament. Since the filaments possessed irregular cross sections, considerable scatter in the data was observed, and an unknown systematic error may have been present.

A static test of the torsion modulus was devised by DeCrescente and Richards [133], who mounted the carbon fiber in series with a glass fiber of known torsion constant. By measuring separately the angle of twist of the glass fiber and that of the graphite fiber, they were able to obtain shear stress as a function of shear strain and hence to compute the shear modulus. They, like Williams, assumed the filament cross section to be circular.

The static stress-strain method can also be used to determine filament torsional shear strength by carrying the test to failure.

An indirect measure of the longitudinal shear modulus C_{44} of high-modulus fibers has been obtained by Smith [134], who measured the elastic constants of unidirectional Thornel fiber and epoxy-resin composites by an ultrasonic method based on that used by Markham [135]. Using theories of elastic constants of fiber-resin composites [136-138]. Smith was able to calculate the shear modulus of the fibers themselves, knowing the fiber volume content.

2. Factors Affecting the Longitudinal Shear Modulus. Results of measurements of the longitudinal shear modulus of high-modulus graphite fibers by various methods are shown in Table 3. All data have been corrected for porosity by multiplying by the ratio of X-ray density (2.2 g/cm^3) to actual filament bulk density. Smith's results show that the corrected shear modulus is essentially independent of fiber Young's modulus over the range from 25 x 10^6 to 75 x 10^6 psi. The torsion pendulum and static torsion results on Thornel fibers scatter considerably, but, on the whole, are in agreement with the ultrasonic results. Finally, the two results for type I PAN graphite fiber (Young's modulus 55 x 10^6 to 60 x 10^6 psi) are in reasonably good agreement with the rayon-base Thornel fibers. In summary, the longitudinal shear modulus, corrected for porosity, of high-modulus graphite fibers seems to lie between approximately 3 x 10^6 and 4 x 10^6 psi, independent of modulus and of starting material.

TABLE 3

Longitudinal Shear Modulus of Several Rayon- and PAN-Base High-Modulus Graphitized Fibers

Fiber	Shear modulus [a] (10^6 psi)	Test method	Ref.
Rayon base			
Thornel 25	3.0	Ultrasonic, from composites	134
Thornel 40	3.1	Ultrasonic, from composites	134
Thornel 50	2.9	Ultrasonic, from composites	134
Thornel 50S	2.9	Ultrasonic, from composites	134
Thornel 75	3.0	Ultrasonic, from composites	134
Thornel 50	4.7 ± 1.3	Torsion pendulum	132
Thornel 25	3.2 ± 1.0	Static torsion	133
Thornel 50	1.6 ± 0.4	Static torsion	133
PAN base			
Type I	4.0	Torsion pendulum	139
Type I	4.1	Torsion pendulum	140

[a] Corrected for porosity.

D. Strength

The strength properties of single filaments of rayon-base carbon fibers have been investigated in tension, bending, compression, and torsional shear.

1. Test Methods

The tensile strengths of single filaments are measured by carrying to failure the tensile test for Young's modulus described in Section V.B.1.

Bending tests on Thornel single filaments have been performed by Williams et al. [8], who utilized a loop test originally devised by Sinclair [141]. The filament, supported at its upper end, is formed into a loop, held between two glass slides and lubricated to minimize friction, and stressed by adding small weights to the lower end. The loop dimensions are measured with the aid of a measuring telescope. Theoretical relations are used to compute bending stress and bending strain at the loop apex where the stress is maximal. (Shear stresses are negligibly small.)

Compressive strengths of single filaments were investigated by Koeneman [127], who utilized a technique of embedding the filament in a clear-resin tensile specimen. With the filament oriented in a direction transverse to the resin specimen's tensile axis, application of a tensile load caused the filament, through the resin's Poisson contraction, to become subject to compressive forces that ultimately resulted in its failure.

Torsional shear strength has been measured by carrying to failure the static torsion test for shear modulus described in Section V.C.1.

2. Factors Affecting Strength

a. Discrete Flaws, Fibril Boundaries, Microfibril Boundaries, Kinks, and Dislocations. The strength properties of any material depend, ultimately, on its structural imperfections. Strengths of perfect, flaw-free materials can often be calculated from theory, and such theoretical strengths have occasionally been realized experimentally (e.g., in dislocation-free whiskers). Theories of strength of brittle solids containing flaws are well known (a discussion of these theories as applied to carbon fibers has been given by Williams et al. [8]), but these theories cannot be used to predict strength quantitatively unless a detailed knowledge of the characters, sizes, shapes, and distributions of structural imperfections in the material is available.

The heading of this section lists the important types of flaws that can exist in carbon fibers. Many of these structural features have been described in Section IV. Their possible influences on fiber strength are discussed here.

The influence of discrete flaws (voids, inclusions, and surface imperfections) on PAN-base carbon-fiber strength has been extensively studied by J. W. Johnson and co-workers [119, 122, 125, 142]. They find that discrete flaws primarily control the tensile strengths of PAN-base carbon fibers that have not been heat-treated above 1000 to 1200° C; such fibers behave in a completely brittle manner. Graphitized PAN fibers do not behave in such a brittle manner [142] and are less flaw sensitive [119]. This decrease in flaw sensitivity is consistent with the well-known lack of notch sensitivity of conventional graphites [11]. However, the same flaws apparently do operate in both carbonized and graphitized PAN fibers [143]. In the case of rayon-base fibers, Strong [33] has shown that the poor strengths of stress-graphitized fibers derived from high-tenacity rayons (tire yarns or Fortisan 36) are probably due to their high internal void contents. In the case of graphitized fibers derived from Villwyte rayon (which is relatively low in voids [104]), the fiber strengths show dispersions [87] similar to those exhibited by graphitized PAN fibers [143], suggesting that strengths may be governed by similar types of flaw in both cases.

The influence of fibril boundaries on carbon-fiber strength has been discussed by several investigators [8, 83, 93, 104, 116]. The fact that fibril boundaries do play a part in the fracture process is evident from the appearance of the longitudinal- and transverse-fracture surfaces (Figs. 35 through 40), in which a characteristic fibrillar structure is always evident. Williams et al. [8] have suggested that localized plastic deformation leading to interfibrillar boundary failure precedes ultimate fiber failure. That severe plastic distortions in fibrils can occur has been shown in electron micrographs (Fig. 35b).

Plastic deformation of fibrils must, in turn, be accomplished by plastic deformation of the microfibrils and failure of intermicrofibrillar boundaries. Since microfibrils are frequently separated by needlelike micropores, one must assume that regions of contact between adjacent microfibrils might be disrupted by shear or tensile stresses, which would be concentrated at the sharp edges or tips of the micropores.

Plastic deformation of the microfibrils will most easily occur by layer-plane shear [144]. Bends and kinks in the microfibrils will tend to straighten out under longitudinal tension; they will tend to become more severe under longitudinal compression. One would therefore expect that the tensile strength would be higher than the compressive strength. Microfibrils will deform also under longitudinal shear stresses. However, all of these processes will be restricted by the sideways connectivity between adjacent rotated microfibrils (Section IV.E). Transverse-shear deformations of the microfibrils, which occur more easily, result from application of transverse-tensile, transverse-compressive, or transverse-shear stresses.

The mechanism of shear deformation of the microfibrils is probably a dislocationlike process [144, 145] in the sense that basal slip would tend to take place by many successive processes, each occurring in a small region. However, one questions whether such dislocations are very similar to basal dislocation lines that have been observed in crystalline graphites for the following reasons: (a) they must be very short, since flat regions of the layers are only 50 to 100 Å wide; and (b) they must be very wide (width of core region) because the interlayer-bonding energy, already very low for crystalline graphite, is even lower for turbostratic graphite. Perhaps the defect would be better described as a dislocation patch than a dislocation line.

In summary, fracture in high-modulus graphitized fibers (those not exhibiting purely brittle behavior) occurs along interfibrillar boundaries, but is preceded by localized plastic deformations. Plastic deformation probably involves both (a) slippage between fibrils and (b) basal shear deformation within microfibrils. Both of these processes become more likely as the fibrillar structure becomes more pronounced (by high-temperature treatment and by stretching) because the microfibrils become larger, straighter, better aligned, less "crosslinked," and structurally more perfect. Thus these two plastic deformation processes will be difficult to study separately.

b. Tensile Strength. The tensile strengths of stress-graphitized rayon-base carbon fibers are shown in Fig. 47 as a function of their Young's moduli. Differences in Young's modulus are due to differences in the amount of stretch applied to the fibers during graphitization. Although most of the samples were

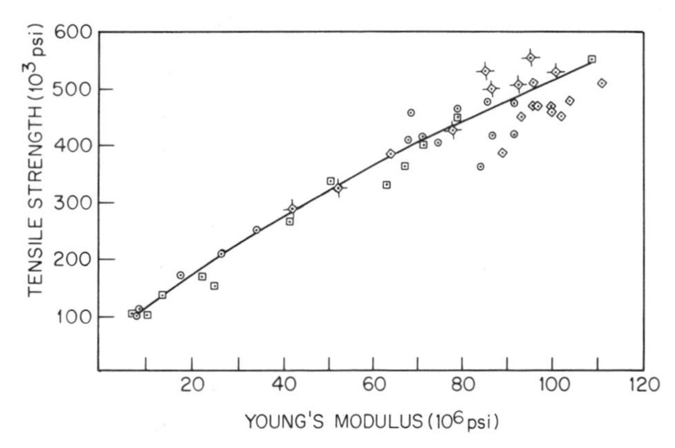

FIG. 47. Relationship between tensile strength and Young's modulus of stress-graphitized rayon-base carbon fibers (2-cm gauge length). From Ref. 87.

graphitized near 2900° C, a considerable number were graphitized at lower and higher temperatures, and no tendency for such samples to deviate from the behavior of the 2900° C samples was observed. The microfibrillar structure, as revealed by electron microscopy, is well developed by 2000° C (the lower limit for graphitization temperatures) and does not visibly change at higher temperatures; on the other hand, no microfibrillar structure is visible in low-modulus rayon-base carbon fibers of heat-treated to less than 1500° C [104].

One can explain the results of Fig. 47 in the context of two theories of strength, the Griffith theory and the Marsh theory [8]. As the microfibrillar structure becomes more highly oriented (due to stretching), the modulus increases; both theories of strength predict that the fiber should then become stronger.

According to Griffith's theory for brittle materials, strength σ should be proportional to the square root of Young's modulus E. In the present case σ clearly increases faster than $E^{0.5}$ (a log-log plot suggests the dependence is closer to $E^{0.7}$). The discrepancy could be attributed to two effects: (a) reduction in the imperfections in the microfibrils due to the plastic dewrinkling caused by stretching [146] and (b) reduction in the sharpness of large internal voids and surface flaws, also caused by stretching. The first of these effects is equivalent to an increase in the surface energy created during fracture (γ in Griffith's theory), and the second is equivalent to a reduction in crack length (c in Griffith's theory).

Another approach to explaining the dependence of σ on E is that favored by Williams et al. [8], who suggested that these fibers probably behave in accordance with Marsh's theory of fracture of certain "brittle" materials in which fracture is supposedly preceded by local inhomogeneous plastic deformation [147, 148]. According to Marsh's theory, σ is directly proportional to E. Such a dependence is only a rough approximation of the observed behavior shown in Fig. 47. However, I suggest here that the discrepancy (i.e., reduction in σ/E with increasing E) may be due to the fact that the fibrillar structure is becoming more pronounced. As the microstructure becomes more highly oriented, the effective "crosslinking" between fibrils decreases, leading to easier slippage between them. Localized plastic flow leading to localized stress concentrations and, finally, fracture can occur at lower average strain levels.

Of two fibers possessing the same degree of preferred orientation and hence the same modulus, the fiber possessing the more pronounced fibrillar structure (resulting from a higher temperature of heat treatment) is the weaker. For example, when a low-modulus rayon-base carbon fiber heat-treated to less than 1500°C is graphitized without stretching, its modulus increases slightly, but its strength is reduced to approximately two-thirds that of the low-temperature fiber. Again, if one compares the strength of a stress-graphitized rayon-base carbon fiber having a modulus of 40×10^6 with that of a type II ($\sim$ 1400°C heat treatment) PAN-base carbon fiber of the same modulus, the latter may be nearly twice as strong as the former; yet if these same two precursors are stress-graphitized at high temperatures to a very high modulus level (e.g., 75×10^6 psi), both exhibit nearly equal breaking strains of approximately 0.5% [87, 149, 150].

c. Nonlinear Stress-Strain Behavior in Bending. The Sinclair loop test (Section V.D.1) has been used by Williams et al. [8] to study the bending behavior of single rayon-base carbon filaments. They found that all high-modulus (stress-graphitized) fibers exhibited linear stress-strain behavior, with a slope equal to the tensile modulus, up to approximately 0.5% strain. Beyond that point an apparent "yield" phenomenon occurred: the apparent stress and apparent strain continued to increase, but with a markedly reduced slope. This stress-strain behavior was, in one sense, reversible; thus, as the load was removed, a similarly shaped curve was followed, but at lower stress levels. The authors showed that a stranded stainless-steel wire subjected to a similar loop test exhibited similar stress-strain behavior; beyond a certain degree of bending, the individual strands began to separate, causing a change in the mode of deformation of the wire as a whole. They suggested, therefore, that the nonlinear character of carbon-fiber bending was due to its fibrillar structure — that is, failure of the bonds between adjacent fibrils occurred at high stresses, permitting them to separate and deform independently. They assumed that this new mode of deformation mostly involved buckling or other deformations of the fibrils on the compression side of the filament. (Almost identical behavior has been found by Jones and J. W. Johnson [142] for high-modulus PAN-base graphite fibers. These investigators found, by scanning electron microscopy, direct evidence of buckling on the compression side of the fibers.)

The yield point in bending was found to occur at a stress of approximately 0.5% of the Young's modulus for most of the fibers studied (which were in the Young's modulus range of 40 x 10^6 to 60 x 10^6 psi). Although this yield phenomenon apparently resulted from a buckling-type deformation on the compression side, it did not lead to catastrophic failure of the entire filament. If the entire filament were subjected to pure compression loading, however, one might expect catastrophic failure to occur at a stress equal to the yield point in bending. Williams et al. pointed out that the tensile failure of these fibers for very short gauge lengths (under 1 mm) occurs at stresses more than 50% higher than these yield stresses. Hence one might conclude that, at least for very short gauge lengths, the tensile strength of these fibers should be at least 1.5 times their compressive strength. Recent experience seems to support this conclusion (see next section).

d. Compressive Strength. Koeneman [127] performed compression tests on single filaments of rayon-base fibers (modulus 40 x 10^6 psi) embedded in epoxy (Section V.D.1). He observed that filament failures occurred through shear along a 45° surface. He did not observe buckling of the filament as a whole, nor did he observe failure of the fiber-matrix interface prior to filament failure. He concluded that compressive failure in composites initiates from the compressive failure of the filaments themselves.

Koeneman estimated that the compressive strengths of these fibers averaged slightly over 360,000 psi, or 0.9% of their Young's modulus. However, his estimate is based on the assumption that the fiber contraction was equal to the transverse Poisson contraction of the epoxy specimen within which the fiber was embedded; since this condition is not likely to be met, the actual fiber compressive strengths were probably much lower than Koeneman's estimates.

Koeneman's conclusion that fiber-matrix debonding is not the cause of compressive failure in a unidirectional graphite-fiber composite has been borne out in recent studies by Beasley [151] on the compressive strengths of unidirectional Thornel fiber and epoxy-resin composites. Beasley found that, beyond a certain minimum fiber-matrix bond strength ($\sim$ 8000-psi shear strength), the compressive strength was insensitive to improvements in the bonding. He also found that sizing agents had no effect. Further, such variables as the composition of the epoxy-resin matrix, the yarn construction and twist, and the fiber Young's modulus were all ineffective in significantly altering compressive strength. The only variable known to affect compressive strength is the processing time-temperature condition used to graphitize the fiber. The greater the time or temperature of graphitization, the lower the compressive strength.

Unidirectional test specimens of 50 vol% Thornel 50S fiber in an epoxy-resin matrix, when prepared under optimum conditions, exhibited compressive strengths of approximately 90,000 psi [151]. If one uses the rule of mixtures to compute the compressive strength of the fiber, the result is 180,000 psi, compared with 330,000 psi for the fiber tensile strength obtained by the strand test. The ratio of fiber tensile to fiber compressive strength in the composite is thus approximately 1.8, a result that is in approximate agreement with the bending-test results given in Section V.D.2.c. Those single-filament-bending experiments suggested that the unexpectedly low compressive strength of high-modulus graphitized fibers is a direct consequence of their fibrillar structure. The ratio of tensile to compressive strength for unidirectional composites of type II (low heat-treatment temperature) PAN-base carbon fibers in epoxy is close to unity. The same condition holds true for low-modulus, low-heat-treatment-temperature rayon-base carbon fibers (grade VYB) [151]. As I have already pointed out, these fibers do not have a pronounced microfibrillar structure, a characteristic that explains their high compressive strengths.

e. Torsional Shear Strength. The only known torsional-shear-strength measurements made on single filaments of rayon-base fibers were made by DeCrescente and Richards [133] by the method described in Section V.C.1. They found that all fibers (including those derived from a PAN precursor) failed at approximately the same level of shear strain: 3%.

E. Electrical Conductivity and Magnetic Susceptibility

The electrical conductivities of stress-graphitized rayon-base carbon fibers have been obtained by measuring the resistance of each filament prior to mechanical testing. In this way electrical conductivity has been found to correlate very closely with the Young's modulus of these fibers, and the relationship is shown in Fig. 48. A straight line passing through the origin adequately describes the relationship of conductivity to modulus for very-high-modulus fibers ($E > 60 \times 10^6$ psi), but a nonlinear relationship obtains for low-to intermediate-modulus fibers.

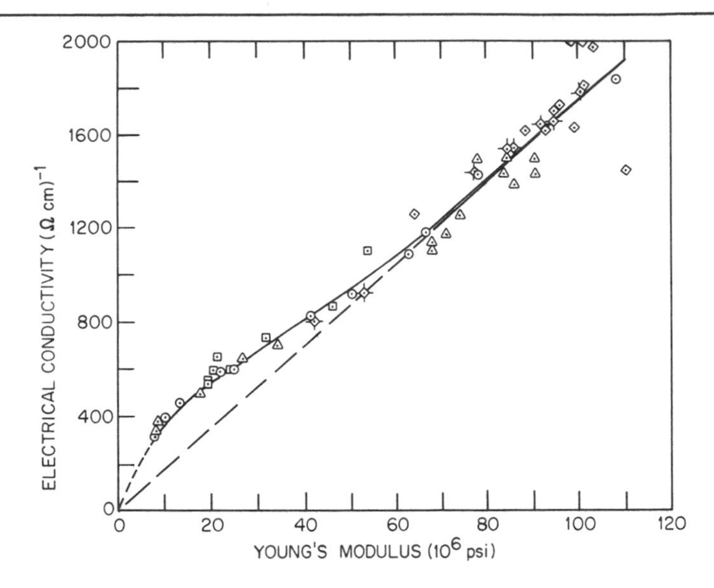

FIG. 48. Relationship between electrical conductivity and Young's modulus of stress-graphitized rayon-base carbon fibers. From Ref. 87.

A distinctly different relationship between electrical conductivity and Young's modulus was found by Sarian and Strong [131] for stress-graphitized fibers annealed at high temperatures subsequent to stress graphitization. For such fibers the electrical conductivities increase with increasing modulus more rapidly than do those of unannealed fibers, as shown in Fig. 49. Evidently the annealing process removes defects that scatter electrons, but do not affect the modulus.

The diamagnetic susceptibility of a series of low-modulus rayon-base fibers heat-treated in batch to various temperatures was measured by Wagoner [152], and the results are shown in Fig. 50. The upper curve is the trace susceptibility $\chi_{\parallel} + 2\chi_{\perp}$, and the lower curve is the anisotropy ratio $\chi_{\perp}/\chi_{\parallel}$, where $\chi_{\parallel}$ is measured with the fiber axis parallel to the magnetic field and $\chi_{\perp}$ is measured with the fiber axis perpendicular to the field. The general increase in trace susceptibility with heat-treatment temperature parallels the growth and perfection of the graphitic layers. The anisotropy of these slightly oriented, low-modulus fibers increases very slowly with heat-treatment temperature. (The increase at 3000° C is attributed to excessive graphitization; very poor fiber mechanical properties were realized.)

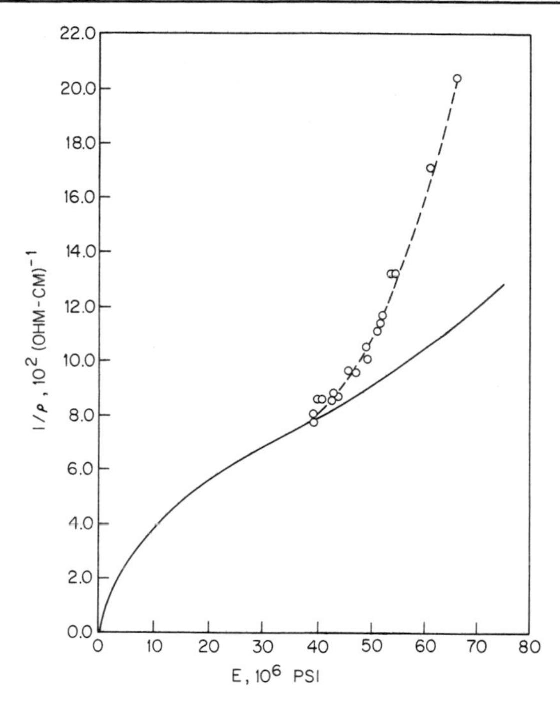

FIG. 49. Relationship between electrical conductivity and Young's modulus of stress-graphitized fibers annealed at elevated temperatures. Solid curve is that of Fig. 48, shown for comparison. From Ref. 131.

The magnetic susceptibilities and anisotropies of a series of Thornel fibers with various Young's moduli have been measured by Scott and Fischbach [153]. Their results are shown in Figs. 51 and 52. The trace susceptibility increases approximately 50% as the Young's modulus is increased from 10×10^6 to 100×10^6 psi, reflecting the straightening of the ribbonlike layers. At the same time the anisotropy ratio increases from about 2 to about 22.

These magnetic susceptibility data show that the magnetic properties are very sensitive to structural perfection and anisotropy in carbon fibers.

F. Thermal Expansion and Thermal Conductivity

The thermal expansion coefficient along the fiber axis was measured by Williams [132], who used yarn samples of low-modulus carbon fibers (VYB) and high-modulus graphite fibers (Thornel). Williams supported the yarn

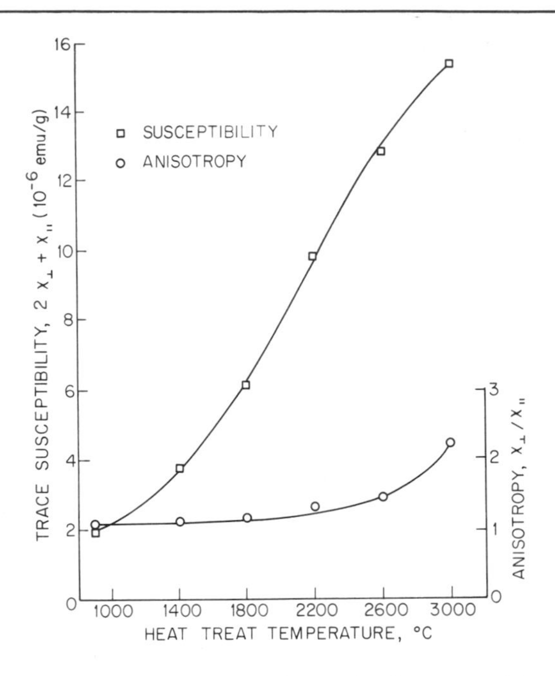

FIG. 50. Diamagnetic susceptibility and anisotropy of low-modulus rayon-base carbon fibers as a function of heat-treatment temperature. From Ref. 134.

sample horizontally between two water-cooled copper posts that were used to electrically heat the yarn to a temperature as high as 1500°C under vacuum. After attaching a small weight to the center of the yarn sample, he was able to observe changes in weight position and hence in yarn-sample length as a function of temperature. Although accurate determinations of the thermal expansion coefficient could not be made, the technique was sensitive enough to show that the thermal expansion coefficient of high-modulus fibers was negative near room temperature, zero near 400°C, and, at higher temperatures, somewhat larger than the thermal expansion of the single crystal along the basal plane (which, of course, is still very low in comparison with that of most materials). Williams' results are shown in Fig. 53.

The intrinsic thermal conductivities of carbon fibers cannot easily be measured. However, since the axial thermal conductivity of high-modulus graphite fibers is much greater than that of resins, it may be estimated

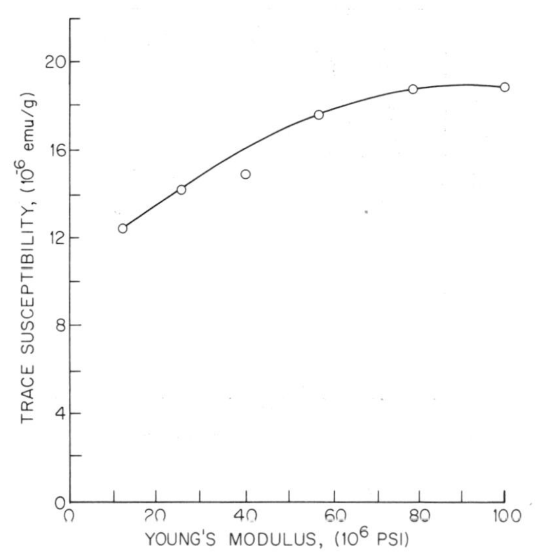

FIG. 51. Diamagnetic susceptibility of stress-graphitized rayon-base carbon fibers as a function of their Young's modulus. From Ref. 153.

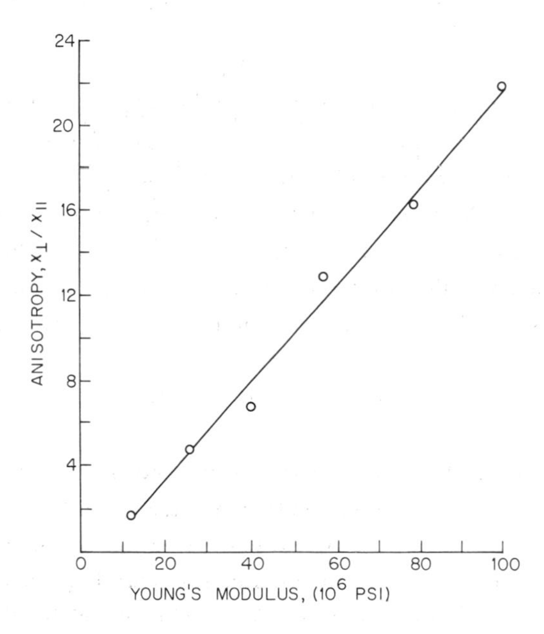

FIG. 52. Susceptibility and anisotropy of stress-graphitized rayon-base carbon fibers as a function of their Young's modulus. From Ref. 153.

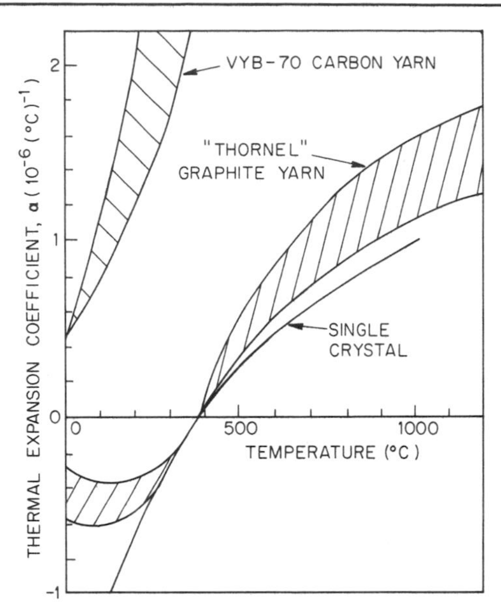

FIG. 53. Temperature dependence of the axial thermal expansion coefficient of low-modulus carbon fiber (VYB) and high-modulus graphite fibers. From Ref. 132.

from the measured thermal conductivity of unidirectional fiber-resin composites. Such a measurement has been made by Blakslee et al. [154] on a Thornel 40 fiber and epoxy-resin composite containing 67 vol% of fiber. The result was 0.13 cal/sec-cm-°C, from which the axial thermal conductivity of the fiber is estimated to be 0.19 cal/sec-cm-°C. This value may be slightly lower than the true value due to the probable existence of frequent fiber breaks in the composite, which interrupt the continuity of the paths for heat flow.

ACKNOWLEDGMENTS

I wish to acknowledge the extensive contributions to rayon-base carbon-fiber technology by many fellow employees of the Union Carbide Corporation, Carbon Products Division, during the past dozen years, particularly those of W. A. Schalamon and A. F. Silvaggi; also the valuable personal contacts with other contributors in the carbon-fiber field, especially W. O. Ruland, W. Watt, and J. W. Johnson.

REFERENCES

1. R. Bacon, A. A. Pallozzi, and S. E. Slosarik, Society of the Plastics Industry 21st Technical and Management Conference, February 1966, Section 8E.

2. W. Watt, L. N. Phillips, and W. Johnson, Engineer, May 27, 1966.

3. A. E. Standage and R. Prescott, Nature, 211, 169 (1966).

4. H. M. Hawthorne, C. Baker, R. H. Bentall, and K. R. Linger, Nature, 227, 946 (1970).

5. S. Otani, Y. Kokubo, and T. Koitabashi, Bull. Chem. Soc., Japan, 43, 3291 (1970).

6. O. L. Blakslee, D. G. Proctor, E. J. Seldin, G. B. Spence, and T. Weng, J. Appl. Phys., 41, 3373 (1970).

7. E. J. Seldin and C. W. Nezbeda, J. Appl. Phys., 41, 3389 (1970).

8. W. S. Williams, D. Steffens, and R. Bacon, J. Appl. Phys., 41, 4893 (1970).

9. R. Bacon, J. Appl. Phys., 31, 283 (1960).

10. D. E. Soule and C. W. Nezbeda, J. Appl. Phys., 39, 5122 (1968).

11. H. H. W. Losty, in Modern Aspects of Graphite Technology (L. C. F. Blackman, ed.), Academic Press, London, 1970, pp. 201-221.

12. S. Yamada, DCIC Report 68-2, Defense Ceramic Information Center, Battelle Memorial Institute, Columbus, Ohio, April 1968.

13. W. H. Smith and D. H. Leeds, Mod. Mater., 7, 139 (1970).

14. A. R. Ubbelohde, D. A. Young, and A. W. Moore, Nature, 198, 1192 (1963).

15. W. F. Kotlensky and H. E. Martens, in Proceedings of the Fifth Conference on Carbon, Vol. 2, Pergamon Press, New York, 1963, p. 625.

16. J. C. Bokros, in Chemistry and Physics of Carbon, Vol. 5 (P. L. Walker, Jr., ed.), Dekker, New York, 1969, p. 1.

17. T. A. Edison, U. S. Pat. 223,898 (1880).

18. W. F. Abbott, U. S. Pat. 3,053,775 (1962).

19. R. Bacon, unpublished work, 1957.

20. Anon., Electrochem. Soc. J., 106, 147C (1959).

21. G. E. Cranch, in Proceedings of the Fifth Conference on Carbon, Vol. 1, Pergamon Press, New York, 1962, p. 589.

22. C. E. Ford and C. V. Mitchell, U. S. Pat. 3, 107,152 (1963).

23. R. B. Millington and R. C. Nordberg, U. S. Pat. 3,294,489 (1966).

24. E. M. Peters, U. S. Pat. 3,235,353 (1966).

25. G. R. Hogg, J. C. Reavis, Jr., and W. E. Russel, U. S. Pat. 3,313,596 (1967).

26. M. T. Cory, U. S. Pat. 3,508,871 (1970).

27. D. L. Schmidt and H. T. Hawkins, in Reinforced Plastics '65, Society of Plastics Engineers Regional Technical Conference, Seattle, 1965, p. 111.

28. R. Bacon, G. E. Cranch, R. O. Moyer, Jr., and W. H. Watts, U. S. Pat. 3,305,315 (1967).

29. R. Bacon and W. A. Schalamon, Eighth Biennial Conference on Carbon, Buffalo, N. Y., June 1967, Paper MI-58.

30. W. J. Spry, British Pat. 1,093,084 (1967); U. S. Pat. 3,454,362 (1969).

31. R. Bacon et al., unpublished technical reports, 1966-1971.

32. D. W. Gibson and G. B. Langlois, ACS Polymer Preprints, 9, 1376 (1968).

33. S. L. Strong, ACS Div. of Organic Coatings and Plastics Chem. Preprints, 31, 426 (1971).

34. G. M. Moutaud and J. L. Duflos, U. S. Pat. 3,322,489 (1967).

35. I. Yoneshiga and H. Teranishi (to Nippon Carbon Co.), Japanese Pat. Specification 2774/70 (1970).

36. P. H. Hermans, Physics and Chemistry of Cellulose Fibers, Elsevier, Amsterdam, 1949.

37. C. V. Mitchell, unpublished work, 1957.

38. R. Bacon, unpublished work, 1960.

39. M. M. Tang and R. Bacon, Carbon, 2, 211 (1964).

40. H. H. W. Losty and H. D. Blakelock, in Second Conference on Industrial Carbon and Graphite, 1965, Society of Chemical Industry, London, 1966, p. 29.

41. R. Bacon and W. A. Schalamon, unpublished technical report, 1968.

42. R. Bacon, W. A. Schalamon, and S. L. Strong, unpublished technical report, 1969.

43. R. O. Moyer, D. R. Ecker, and W. J. Spry, U. S. Pat. 3,333,926 (1967).

44. C. L. Gutzeit, U.S. Pat. 3,479,150 (1969).

45. A. Shindo, Appl. Polymer Symp., 9, 271 (1969).

46. A. Shindo, U. S. Pat. 3,529,934 (1970).

47. S. L. Madorsky, V. E. Hart, and S. Strauss, J. Res. Natl. Bur. Stds., 60, 343 (1958).

48. R. Schwenker and E. Pacsu, Ind. Eng. Chem., 50, 91 (1958).

49. C. L. Gutzeit, U. S. Pat. 3,479,151 (1969).

50. D. R. Moore, S. E. Ross, and G. C. Tesoro, U. S. Pat. 3,527,564 (1970).

51. R. W. Little (ed.), Flameproofing Textile Fabrics, Reinhold, New York, 1947, p. 167.

52. S. L. Madorsky, V. E. Hart, and S. Strauss, J. Res. Natl. Bur. Stds., 56, 343 (1956).

53. W. K. Tang and W. K. Neill, J. Polymer Sci., C6, 65 (1964).

54. W. K. Tang and W. Eickner, U. S. Forest Service Research Paper FPL 84, January 1968.

55. W. A. Reeves, R. M. Derkins, B. Piccolo, and G. L. Drake, Jr., Textile Res. J., 40, 223 (1970).

56. J. V. Duffy, J. Polymer Sci., 15, 715 (1971).

57. S. E. Ross, Textile Res. J., 38, 906 (1968).

58. W. Watt and W. Johnson, Appl. Polymer Symp., 9, 215 (1969).

59. R. Bacon and W. A. Schalamon, unpublished technical report, 1966.

60. W. A. Schalamon and R. Bacon, British Pat. 1,167,007 (1969).

61. R. J. Bobka and R. Bacon, unpublished work, 1965.

62. A. Fourdeux, R. Perret, and W. Ruland, Compt. Rend., 271C, 1495 (1970).

63. A. Fourdeux, R. Perret, and W. Ruland, J. Appl. Cryst., 1, 252 (1968).

64. W. Ruland and H. Tompa, Acta Cryst., A24, 93 (1968).

65. R. W. James, Optical Principles of the Diffraction of X-Rays, Vol. II, The Crystalline State, Bell, London, 1954, p. 8.

66. W. Ruland, in Chemistry and Physics of Carbon, Vol. 4 (P. L. Walker, Jr., ed.), Dekker, New York, 1968, p. 61.

67. G. E. Bacon, J. Appl. Chem., 6, 477 (1956).

68. G. J. Curtis, J. M. Milne, and W. N. Reynolds, Nature, 220, 1024 (1968).

69. W. Ruland, J. Appl. Phys., 38, 3585 (1967).

70. R. J. Price and J. C. Bokros, J. Appl. Phys., 36, 1897 (1965).

71. H. Gasparoux, A. Pacault, and E. Poquet, Carbon, 3, 65 (1965).

72. P. R. Goggin and W. N. Reynolds, Phil. Mag., 16, 317 (1967).

73. B. E. Warren and P. Bodenstein, Acta Cryst., 20, 602 (1966).

74. R. E. Franklin, Acta Cryst., 3, 107 (1950).

75. L. L. Ban and W. M. Hess, in Ninth Conference on Carbon, Boston College, 1969, DCIC, 1969 Paper SS-27, p. 162.

76. A. Fourdeux, C. Hérinckx, R. Perret, and W. Ruland, Compt. Rend., 269C, 1597 (1969).

77. J. A. Hugo, V. A. Phillips, and B. W. Roberts, Nature, 226, 144 (1970).

78. D. J. Johnson, Nature, 226, 750 (1970).

79. R. Diamond, Phil. Trans. Roy. Soc. (London), A252, 193 (1960).

80. A. Fourdeux, R. Perret, and W. Ruland, paper presented at the Conference on Carbon Fibres, Their Composites and Applications, the Plastics Institute, London, February 2-4, 1971.

81. R. Bacon and M. M. Tang, Carbon, 2, 221 (1964).

82. A. Shindo, Studies on Graphite Fiber, Rept. No. 317, Govt. Ind. Res. Inst., Osaka, 1961.

83. W. Johnson and W. Watt, Nature, 215, 384 (1967).

84. H. W. Davidson and H. H. W. Losty, G.E.C. Journal, 30, 22 (1963).

85. S. Yamada, H. Sato, and T. Ishi, Carbon, 2, 253 (1964).

86. J. C. Lewis, B. Redfern, and F. B. Cowlard, Solid State Electronics, 6, 251 (1963).

87. R. Bacon and W. A. Schalamon, Appl. Polymer Symp., 9, 285 (1969).

88. W. Ruland, Appl. Polymer Symp., 9, 293 (1969).

89. W. Ruland, unpublished technical report, 1965.

90. Eastabrook and Seed

91. D. J. Johnson and C. N. Tyson, Brit. J. Appl. Phys. (J. Phys. D), 2, 787 (1969).

92. R. Perret and W. Ruland, J. Appl., Cryst., 3, 525 (1970).

93. D. V. Badami, J. C. Joiner, and G. A. Jones, Nature, 215, 386 (1967).

94. D. H. Saunderson and C. G. Windsor, in Third Conference on Industrial Carbon and Graphite, 1970, Society of Chemical Industry, London, 1971, p. 438.

95. W. Ruland, Summary of Papers, Conference on Fibres for Composites: Strength, Structure, and Stability, University of Sussex, England, 1969, p. 10.

96. R. Perret and W. Ruland, J. Appl. Cryst., 1, 308 (1968).

97. R. Perret and W. Ruland, J. Appl. Cryst., 2, 209 (1969).

98. W. Ruland, J. Polymer Sci., C28, 143 (1969).

99. D. J. Johnson and C. N. Tyson, Brit. J. Appl. Phys. (J. Phys. D), 3, 526 (1970).

100. P. H. Emmett, Chem. Rev., 43, 69 (1947).

101. G. Porod, Kolloid-Z., 124, 83 (1951).

102. S. L. Strong, private communication, 1968.

103. T. Uchida, I. Shinoya, Y. Itoh, and K. Nukada, in Tenth Biennial Conference on Carbon, Lehigh University, Bethlehem, Pa., June 1971, DCIC, 1971, Paper FC-19, p. 31.

104. R. Bacon and A. F. Silvaggi, Carbon, 9, 321 (1971).

105. J. W. Menter, Proc. Roy. Soc. (London), 236, 119 (1956).

106. R. D. Heindenreich, W. M. Hess, and L. L. Ban, J. Appl. Cryst., 1, 1 (1968).

107. D. F. Harling, in Proceedings of the Twenty-Eighth Annual Meeting of the Electron Microscopy Society of America, Houston, Texas, 1970 (C. J. Arceneaux, ed.), Claitor's Publishing Div., Baton Rouge, 1970, p. 468.

108. V. A. Phillips, J. A. Hugo, and B. W. Roberts, in Proceeding of the Twenty-Eighth Annual Meeting of the Electron Microscopy Society of America, Houston, Texas, 1970 (C. J. Arceneaux, ed.), Claitor's Publishing Div., Baton Rouge, 1970, p. 470.

109. B. L. Butler and R. J. Diefendorf, in Carbon Composite Technology, Proceedings of the Tenth Annual Symposium of the New Mexico Section of ASME and University of New Mexico, College of Engineering, Albuquerque, 1970, p. 107.

110. S. L. Strong, private communication, 1970.

111. R. Bacon and A. F. Silvaggi, Abstracts, Eighth Conference on Carbon, Buffalo, 1967; Carbon, 6, 230 (1968).

112. J. Dubois, C. Agache, and J. L. White, Metallography, 3, 337 (1970).

113. J. L. White, J. Dubois, and C. Souillart, European Atomic Energy Community, Euratom Rept. EUR 4094e, 1969.

114. C. N. Tyson, Nature Phys. Sci., 229, 123 (1971).

115. R. E. Smith, private communication, 1970.

116. J. W. Johnson, P. G. Rose, and G. Scott, in Third Conference on Industrial Carbon and Graphite, 1970, Society of Chemical Industry, London, 1971, p. 443.

117. J. W. Johnson, private communication, 1970.

118. W. S. Williams and R. Sprague, unpublished work, 1967.

119. J. W. Johnson, Appl. Polymer Symp., 9, 229 (1969).

120. A. Howsmon and W. A. Sisson, in Cellulose and Cellulose Derivatives, Part I (E. Ott et al., eds.), 2nd ed., Interscience, New York, 1954, pp. 255-264.

121. W. S. Williams and D. Steffens, Abstracts, Eighth Conference on Carbon, Buffalo, 1967; Carbon, 6, 212 (1968).

122. J. W. Johnson and D. J. Thorne, Carbon, 7, 659 (1969).

123. D. J. Thorne, Nature, 225, 1039 (1970).

124. D. J. Thorne, J. Appl. Polymer Sci., 14, 103 (1970).

125. D. J. Thorne, in Third Conference on Industrial Carbon and Graphite, 1970, Society of Chemical Industry, London, 1971, p. 463.

126. J. J. Kipling and P. V. Shooter, Carbon, 4, 1 (1966).

127. J. B. Koeneman, Ph.D. thesis, Case Western Reserve University, Cleveland, Ohio, 1970.

128. L. P. Lowell, private communication, 1970.

129. R. Perret and W. Ruland, in Ninth Conference on Carbon, Boston College, 1969, DCIC, 1969, Paper SS-22, p. 158.

130. R. L. Cummerow, paper presented at the ASTM National Conference on High Performance Fibers — Properties, Applications, and Test Methods, Williamsburg, Va., November 17-18, 1970.

131. S. Sarian and S. L. Strong, Fibre Sci. Technol., 4, 67 (1971).

132. R. Bacon and W. S. Williams, unpublished technical report, 1967.

133. M. A. DeCrescente and R. Richards, private communication, 1970.

134. R. E. Smith, J. Appl. Phys., 43, 2555 (1972).

135. M. F. Markham, Composites, 1, 145 (1970).

136. J. C. Halpin, J. Compos. Mater., 3, 732 (1969).

137. S. W. Tsai, Rept. HPC 68-61, Monsanto Research Corp., St. Louis, Mo., June 1968 (AD 834851).

138. E. Behrens, J. Acoustical Soc. Amer., 45, 1568 (1969).

139. W. T. Brydges, D. V. Badami, J. C. Joiner, and G. A. Jones, Appl. Polymer Symp., 9, 255 (1969).

140. W. N. Reynolds, in Third Conference on Industrial Carbon and Graphite, 1970, Society of Chemical Industry, London, 1971, p. 427.

141. D. Sinclair, J. Appl. Phys., 21, 380 (1950).

142. W. R. Jones and J. W. Johnson, Carbon, 9, 645 (1971).

143. R. Moreton, Fibre Sci. Technol., 1, 273 (1969).

144. S. Allen, G. A. Cooper, and R. M. Mayer, Great Britain National Physical Lab. IMS Rept. 7, November 1969.

145. W. N. Reynolds, paper presented at the Conference on Fibres for Composites: Strength, Structure, and Stability, University of Sussex, England, June 30-July 1, 1969.

146. W. Ruland, unpublished technical report, 1967.

147. D. M. Marsh, Tech. Rept. 161, Tube Investments Research Laboratories, Cambridge, England, 1963.

148. D. M. Marsh, Proc. Roy. Soc. (London), 282, 33 (1964).

149. J. W. Johnson, J. R. Marjoram, P. G. Rose, Nature, 221, 357 (1969).

150. W. Johnson, in Third Conference on Industrial Carbon and Graphite, 1970, Society of Chemical Industry, London, 1971, p. 447.

151. W. C. Beasley, paper presented at the ASTM National Conference on High Performance Fibers — Properties, Applications, and Test Methods, Williamsburg, Va.,, November 17-18, 1970.

152. G. Wagoner, unpublished report, 1965.

153. C. B. Scott and D. B. Fischbach, in Tenth Biennial Conference on Carbon, Lehigh University, Bethlehem, Pa., June 1971, DCIC, 1971, Paper EP-202, p. 321.

154. O. L. Blakslee, A. A. Pallozzi, W. A. Doig, G. B. Spence, and D. P. Hanley, in Advances in Structural Composites, 12th National SAMPE Symposium, Science of Advanced Materials and Process Engineering, Vol. 12, Society of Aerospace Materials and Process Engineering, Anaheim, Calif., 1967, p. AC-6.

CONTROL OF STRUCTURE OF CARBON FOR USE IN BIOENGINEERING

J. C. Bokros, L. D. LaGrange, and F. J. Schoen

Gulf Oil Corporation
Gulf Energy and Environmental Systems Division
San Diego, California

I. INTRODUCTION

A. The Structure of Carbon

The applications of carbon in a variety of situations stem from the fact that the properties of carbonaceous bodies are structure sensitive and that the structure of carbon is variable. Accordingly careful control of the structure of carbons has led to methods by which properties can be fashioned to suit many specialized and differing requirements. To develop an understanding of how such a variety of properties can be obtained from a single elemental solid, and to provide background for the bulk of this chapter, the following paragraphs summarize the essential aspects of the crystal structure, morphology, and surface chemistry of carbon.

In all of the carbonaceous structures to be discussed, most of the carbon atoms are linked together by strong covalent bonds to form planar, hexagonal arrays (Fig. 1a) [1]. One of the valence electrons per atom (a π electron) is free to move within the layers, providing a high, but anisotropic, electrical conductivity. Although the bonding between layers in the ideal graphite structure is of the van der Waals type, indirect measurements show that in real bodies additional linkages (called crosslinks) must exist between layer planes and have a profound influence on the mechanical properties [2]. The elimination of crosslinks is essential to provide lubricity for lubricating graphites, whereas a high concentration of crosslinks is necessary to obtain the highest strength pyrolytic carbons [3].

The individual layer planes in carbon vary in perfection and, in poorly crystalline bodies, are thought to be wrinkled and contain vacant lattice sites. Although the layers in most structures are arranged parallel to one another, forming crystallites (Fig. 1b), the less crystalline structures contain substantial fractions of disorganized carbon (Figs. 1c and d) [4]. The properties of a carbon body like that depicted schematically in Fig. 1d have been found to depend on: (a) the size of the crystallites, (b) the preferred orientation of the crystallites, (c) the density, and (d) the microstructure. The methods used to measure these structural properties have been described by Bokros [5].

A common point of confusion that frequently develops among those not directly concerned with carbon is the distinction between graphite and carbon.

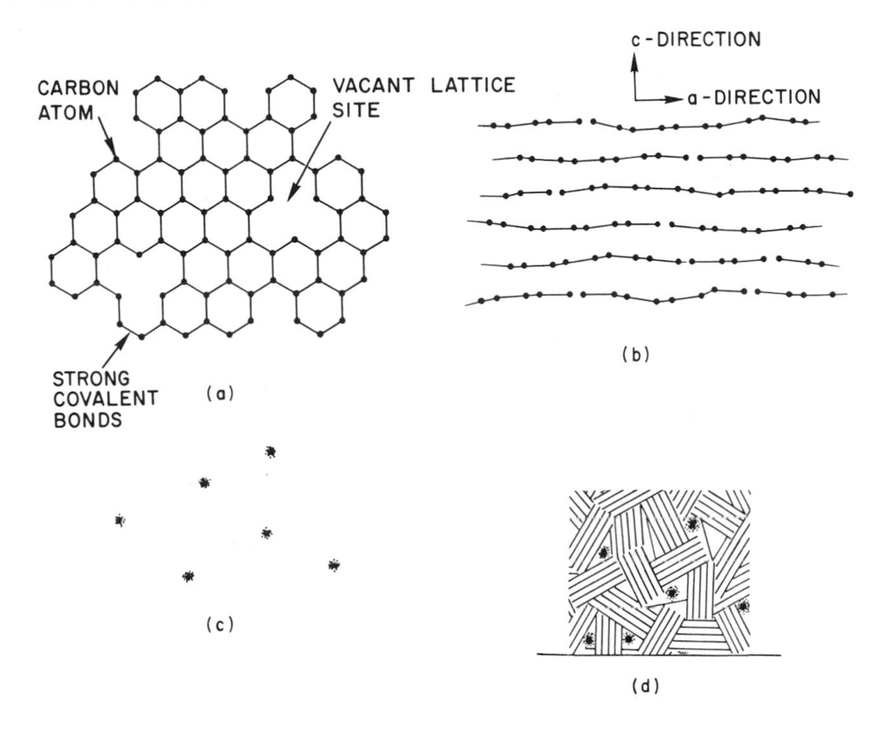

FIG. 1. Schematic diagrams of the atomic arrangements in poorly crystalline carbons: (a) a single layer plane; (b) parallel layers in a crystallite; (c) unassociated carbon; (d) an aggregate of crystallites, single layers, and unassociated carbon.

In graphite the layer planes are large and nearly perfect. Parallel layers in crystallites are ordered with respect to one another, so that the crystal structure is three-dimensional (Fig. 2a) [6]. In pyrolytic carbon, vitreous carbon, and other poorly crystalline materials, the adjacent layers are not ordered with respect to one another, so that the structure possesses only two-dimensional order (Fig. 2b). Such a structure has been called turbostratic [7]. In general, the strongest bulk forms of carbon (except graphite filaments and whiskers) are those that are dense, have small crystallite sizes, and are heavily crosslinked [8].

Because of the strong bonding within the layers and the weaker bonding between layers, the properties of the individual crystallites are highly anisotropic. The extent to which a bulk carbon deposit reflects the anisotropy

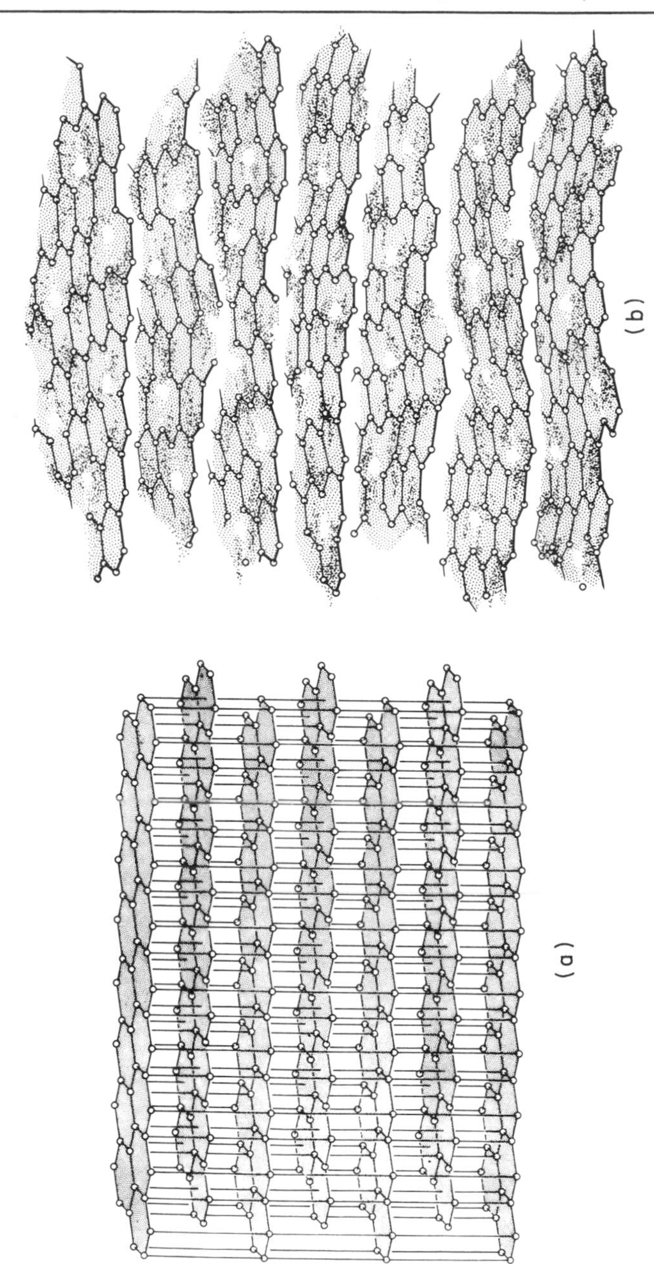

FIG. 2. Schematic diagrams of (a) a three-dimensional graphite lattice and (b) a turbostratic structure.

of the crystallites depends on the degree of preferred orientation of the crystallites. When the degree of preferred orientation is high, the anisotropy of the bulk approaches that of the crystallites. When the crystallites in an aggregate are randomly oriented, the anisotropy of the individual crystallites is perfectly averaged, so that the aggregate is isotropic. For most engineering purposes where complex shapes are required it is desirable to use isotropic materials whose properties do not vary with direction. Examples illustrating the anisotropy of crystallites can be found in Ref. 2 (Fig. 43 for Young's modulus and Fig. 50 for thermal expansion) and Ref. 9 for thermal conductivity.

The microstructure of carbon bodies, like their crystal structure, is extremely variable. Commercial polycrystalline graphites, formed by extruding or pressing mixtures of graphite powders and an organic binder, have microstructures that are laced with pores and microcracks [10]. Their low strength will probably preclude significant application in monolithic parts for use in prosthetic devices requiring high strength or wear resistance.

Pyrolytic and vitreous carbons [11] are impermeable to gases and liquids, and are much stronger than the graphites described here. The strongest isotropic carbons available are those deposited at low temperatures (LTI Pyrolite® carbons, Gulf Oil Corp.). Within this class of materials are carbons with strengths that are three times higher than those reported for vitreous carbons. Because of the superior properties of Pyrolite carbons, a great deal of effort has been devoted to developing these materials for bioengineering applications.

B. The Surface Chemistry of Carbon

The chemical functionality of carbon surfaces has been the subject of considerable study, mainly because of the commercial applications of carbon as an adsorbent and filler for rubber. Many of the pertinent aspects have been summarized in a review by Donnet [12].

Carbon surfaces have active sites, at points of imperfections and at the edges of exposed layer planes, that can participate in chemical reactions. The most common types of chemical functionality formed on carbon surfaces are the oxygen-bearing acid groups that can be obtained by partial oxidation under a variety of conditions. A model of the surface of an oxidized carbon proposed by Boehm [13] (Fig. 3) indicates four primary groups (e.g.,

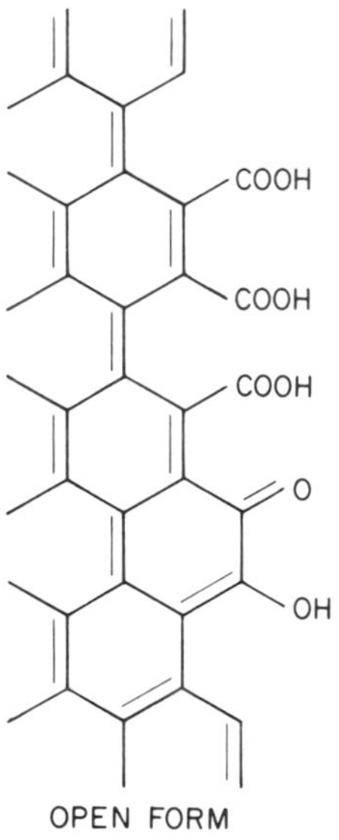

OPEN FORM

LACTONIC FORM

FIG. 3. Models of the surface structure of carbon. After Boehm [13].

carboxyl, carbonyl, hydroxyl, and lactone) whose concentration ratios differ according to the nature of the carbon and the oxidation process [14]. In addition, especially in the pyrolytic carbons deposited below about 1300°C, there are certain to be C-H bonds within the mass and exposed at the surface [12].

The reactivity of carbon surfaces and the wide variations in crystal structure that are possible require that detailed experimentations, if they are to be definitive, be performed only on materials for which the crystal structure, surface topography, and surface condition (as specified by the details of its history) are adequately characterized.

C. Heparin Sorptivity and Thromboresistance of Carbonaceous Materials

A series of experiments initiated in 1961 at the University of Wisconsin led to the development of a variety of new thromboresistant materials — that is, materials that do not induce blood clotting on their surfaces [15-23]. This work revealed that a surface could be rendered thromboresistant by first coating with graphite by dipping in a dispersion of graphite particles in an organic vehicle

(Dag, Acheson Colloids Co., Port Huron, Mich.) and then treating successively in solutions of benzalkonium chloride and heparin. Coatings prepared in this way are called GBH. A comparison of the thromboresistance of GBH-coated plastic and other commonly used biomaterials is shown in Table 1.

TABLE 1

Results of Gott Ring Tests for Various Materials in Canine Venae Cavae (Gott and Whiffen)

MATERIAL	TIME OF IMPLANT	AMOUNT OF THROMBUS IN LUMEN
POLYCARBONATE	2 HR	● ● ● ● ●
POLYPROPYLENE	2 HR	● ● ● ● ●
TEFLON	2 HR	● ● ● ● ●
SILICONE RUBBER	2 HR	● ● ● ● ●
# 304 STAINLESS STEEL	2 HR	● ● ● ● ●
G.B.H. COATED POLYCARBONATE	2 WEEKS	○ ○ ○ ○ ○

[a] The procedure for the Gott test is as follows [17]: Rings of the test material approximately 9 mm long, with a 7-mm internal diameter and 0.5-mm wall thickness, having streamlined leading and trailing edges, are placed in the canine superior or inferior vena cava and fixed in place. The rings are removed when pressure recordings indicate complete occlusion of the ring, or at 2 h for the acute test and 14 days for the chronic test. Results are reported as the visually observed degree of occlusion by thrombus (blood clot) on end-on examination of the excised rings. Each data point (circle) refers to an individual ring implanted in a separate dog. An open circle indicates a completely patent ring and thus a thromboresistant material, whereas a filled circle represents complete occlusion by a thrombus.

The development of GBH coatings stimulated the development of alternate procedures by which heparin could be bonded to a variety of materials [24-36] and prompted serious interest in carbon as a biomaterial.

In 1965 in-vivo exploratory tests of several types of solid carbon were initiated at the Johns Hopkins Hospital. Encouraging results led to a more intensive study of the compatibility of carbon and blood [37-41]. The objective of the resulting program was to determine the relationships between the crystal structure, surface chemistry, and surface topography of carbon surfaces and their compatibility with blood (Fig. 4). Because of the observations of Gott and Whiffen for GBH surfaces, it was originally anticipated that those surfaces that sorbed and retained heparin best would also be the most compatible with blood. The results obtained did not substantiate this prediction.

Twelve different carbonaceous materials were chosen to be representative of the various classes of materials available commercially. Some of the materials were completely permeable, with extremely rough surfaces (Fig. 5). Others were rough but impermeable (Fig. 6a; these carbons have a superficial porosity that extends about 10 μm deep; see Fig. 7 of Ref. 38) or

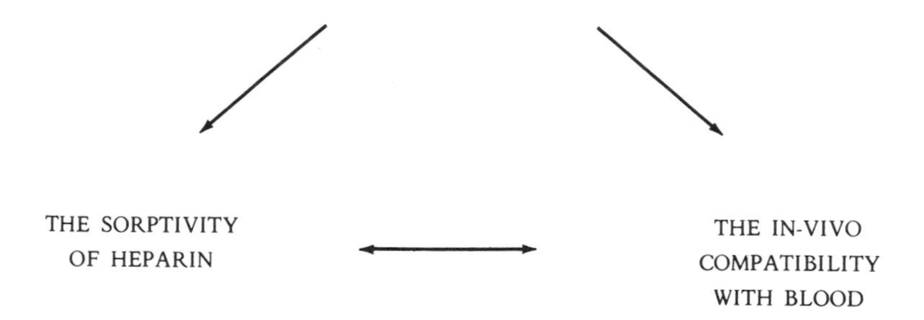
DETERMINE THE EFFECT OF THE
CRYSTALLINE STRUCTURE, SURFACE CHEMISTRY,
AND SURFACE TOPOGRAPHY OF CARBON ON
THE SORPTIVITY
OF HEPARIN
THE IN-VIVO
COMPATIBILITY
WITH BLOOD

FIG. 4. Schematic diagram of research program objectives.

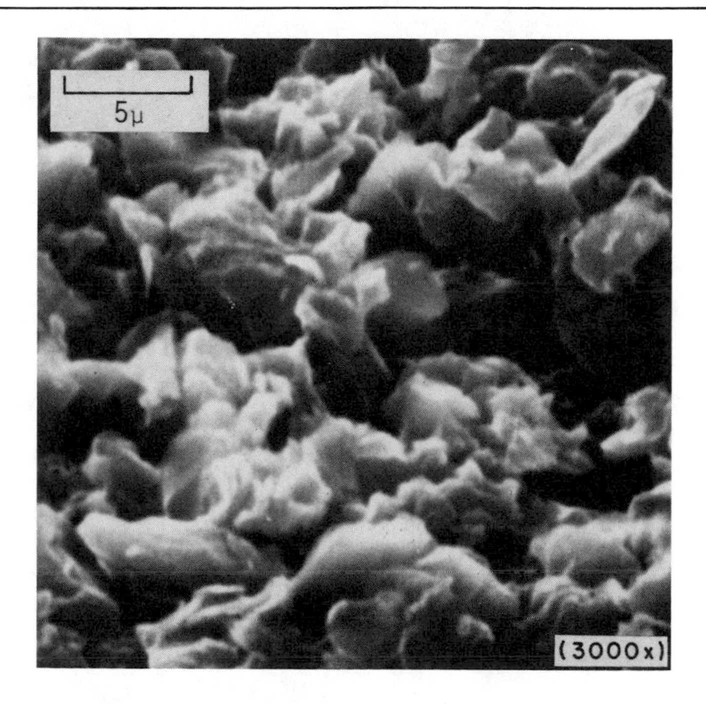

FIG. 5. Scanning electron micrograph of the surface of an ultrasonically cleaned and outgassed surface of Poco type AXZ graphite (this material is permeable).

impermeable carbons polished to remove the surface roughness (Fig. 6b). In some experiments the surfaces were oxidized under controlled conditions to determine the influence of oxygen-bearing surface functionalities, such as those depicted in Fig. 3, on both heparin sorptivity and the in-vivo thromboresistance. The details of the structures of the carbons, the heparin sorptivity, and the corresponding in-vivo thromboresistance have been published [38-40]. Accordingly only the essential aspects of the findings are summarized here.

The results obtained from the study of clean, smooth, impermeable carbon surfaces are depicted schematically in Fig. 7. It was found that clean, smooth LTI Pyrolite carbons have excellent thromboresistance without pretreatment with benzalkonium chloride or heparin. Heparin is adsorbed as a monolayer on these surfaces, and the amount adsorbed is not enhanced by a pretreatment with benzalkonium chloride. Furthermore, the

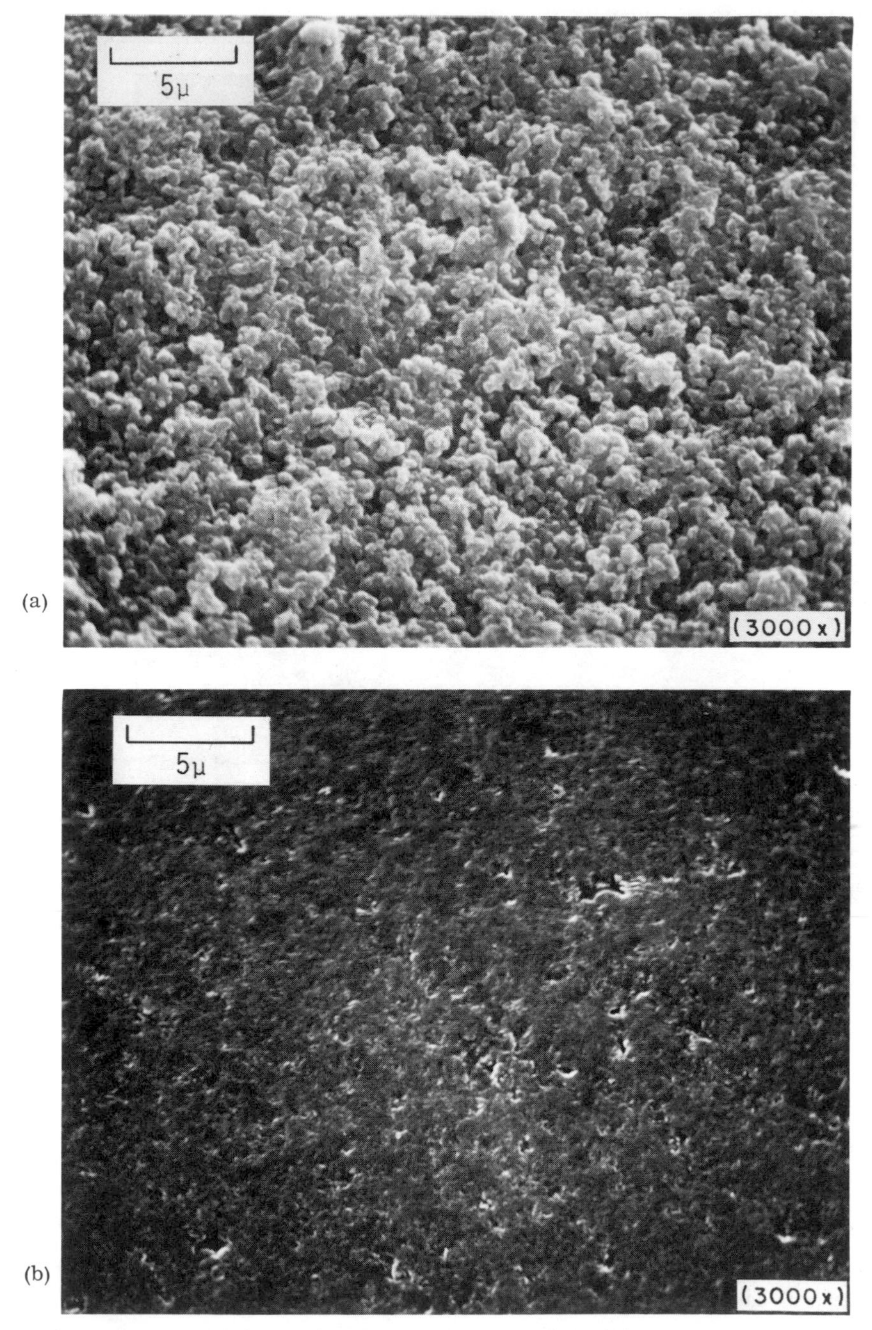

FIG. 6. Scanning electron micrographs of (a) the surface of an as-deposited impermeable LTI Pyrolite carbon and (b) the polished surface of an impermeable LTI Pyrolite carbon.

heparin adsorption did not improve the in-vivo thromboresistance. In plasma (Fig. 9 in Ref. 38) and in vivo (Table 4 in Ref. 41), the heparin was elutriated to surface concentrations of less than 1 $\mu g/cm^2$ in a very short time.

The study of well-characterized surfaces treated with oxygen (hydrophilic sites) indicated that the presence of oxygen functionality on the surface markedly reduced the thromboresistance (Fig. 8; see Table VI in Ref. 38 for details). The presence of chemisorbed oxygen did not detectably influence the adsorption of either heparin or the heparin-benzalkonium complex.

Many of the above results at first seemed inconsistent with data reported for GBH surfaces; the apparent discrepancy was clarified when the effects of surface topography on heparin sorptivity and on thromboresistance were determined. The findings are represented schematically in Fig. 9. It was found that increasing the surface roughness of impermeable carbons increased the heparin adsorption by an amount accounted for by the increase in surface area. The heparin was adsorbed as a monolayer, and the adsorption was unaffected by pretreatments with benzalkonium chloride. In all cases the thromboresistance of rough surfaces was less than that of corresponding smooth surfaces. This is in accord with other reports [42-47]. Using the results depicted in Figs. 7, 8, and 9, it was possible to conclude that mechanisms by which heparin molecules are proposed to be linked to carbon or graphite surfaces through a series of chemical bonds involving benzalkonium chloride must not be correct.

The results obtained from permeable materials are similar to those reported for Dag surfaces. The heparin sorption in porous carbon surfaces was markedly enhanced by a pretreatment with benzalkonium chloride, and, like Dag surfaces, the permeable materials were only significantly thromboresistant if they were treated first with benzalkonium chloride and then with heparin. Pretreatments with heparin alone were ineffective in providing any significant degree of thromboresistance. Increasing either the volume of accessible porosity (Fig. 14 in Ref. 39) or the concentration of the benzalkonium chloride that was used in the pretreatment (Table IV of Ref. 39) enhanced

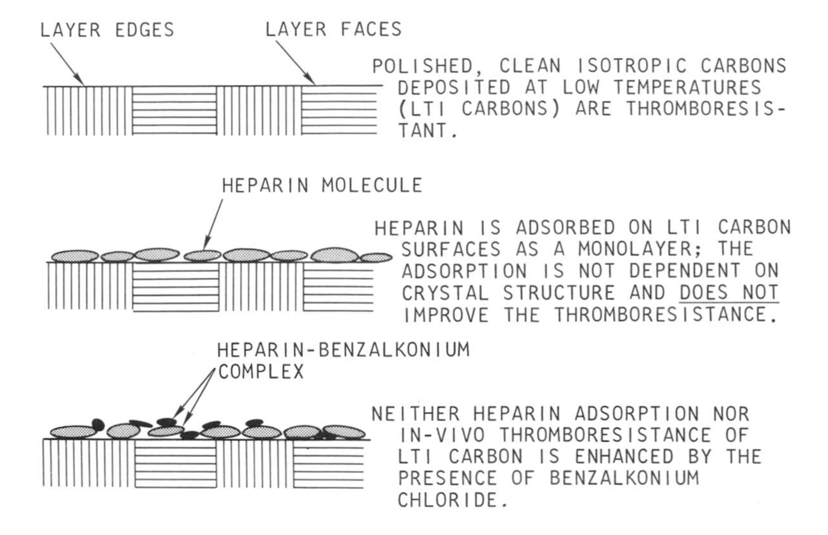

FIG. 7. Results for clean, smooth LTI Pyrolite carbon surfaces.

both the heparin sorptivity and the in-vivo thromboresistance of the porous carbon surfaces.

On the basis of the results summarized here, it was concluded that Dag surfaces must have accessible porosity and that the retentivity of the heparin-benzalkonium complex is mechanical [39]. The mechanism that was postulated is as follows:

1. During the pretreatment the benzalkonium chloride solution enters the accessible porosity.

2. When the specimen is subsequently placed in the heparin-bearing solution, heparin also enters the accessible porosity.

3. The heparin complexes with the residual benzalkonium ion in the pores to form the insoluble, gelatinous quaternary ammonium salt of heparin, where it is trapped and held mechanically.

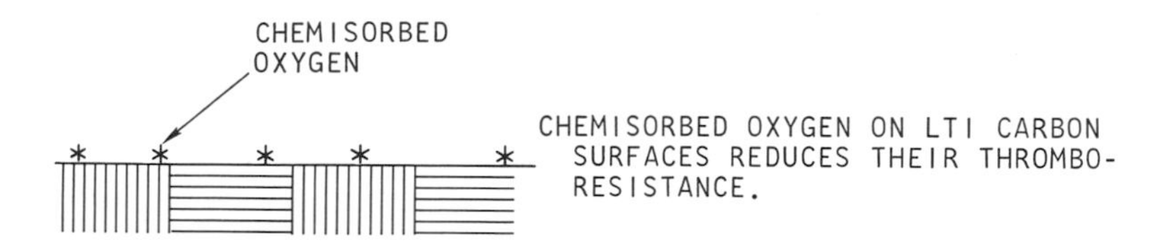

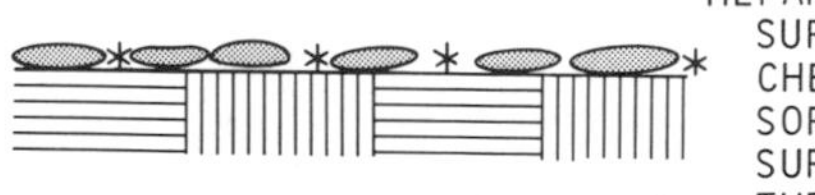

HEPARIN ADSORPTION ON LTI CARBON SURFACES IS NOT INFLUENCED BY CHEMISORBED OXYGEN, AND ADSORBED HEPARIN ON OXIDIZED SURFACES DOES NOT IMPROVE THEIR THROMBORESISTANCE.

FIG. 8. Results for smooth LTI Pyrolite carbon surfaces contaminated with oxygen.

Some of the variations in results reported for GBH coatings [29, 48-54] may be due to variations in the morphology of the porosity in the final, cured Dag coating. It is thought that in the later stages of drying, when the solvent is almost gone, the small, anisotropic graphite flakes compete with one another and the local surface forces to achieve the lowest energy configuration. The extent to which the latter is accomplished depends on the size and shape of the flakes and the rate of drying. Hence variations in the morphology of the porosity in Dag coatings might be expected unless rigorous controls are placed on the application, draining, and drying steps [39].

Milligan, Davis, and Edmark [54] have pointed out that sorption and permeability of Dag-35 coatings are important to the thromboresistance of GBH surfaces prepared from this coating. Their explanation of the role of the porosity in conferring the thromboresistance is, however, not the same as the mechanism proposed above. It seems likely that the thromboresistance of GBH coatings results both because of the inherent compatibility of the graphite flakes in the Dag paint and the added active suppression of thrombus

SURFACE ROUGHNESS DECREASES THE THROMBORESISTANCE OF LTI CARBON SURFACES; ADSORBED HEPARIN (OR HEPARIN-BENZALKONIUM COMPLEX) DOES NOT INCREASE THEIR THROMBO-RESISTANCE.

HEPARIN SORPTION ON CARBON SURFACES WITH DEEP ACCESSIBLE POROSITY IS MARKEDLY IMPROVED BY A PRETREAT-MENT WITH BENZALKONIUM CHLORIDE. SURFACES LIKE THIS ARE ONLY THROMBORESISTANT WHEN THE HEPARIN-BENZALKONIUM COMPLEX IS SORBED IN THE POROSITY.

FIG. 9. Results for rough and porous carbon surfaces.

formation by the sorbed heparin which is elutriated slowly in vivo. Milligan, Davis, and Edmark [54] find that Dag-35 yields the most thromboresistant GBH coatings and also displays greater sorption and higher permeabilities than any other graphite-resin material they studied. Uy and Kammermeyer [34] report that porosity also plays an important role in the sorption of heparin on their TGBH surfaces.

The data depicted schematically in Figs. 7, 8, and 9 indicate two possible approaches in the development of biocarbons for use in intravascular prostheses: (a) the development of carbon surfaces with accessible porosity optimized for the retention of the heparin-benzalkonium complex; and (b) optimization of the crystal structure and surface chemistry of smooth carbon surfaces for in-vivo thromboresistance without heparin. Since the mechanical properties of porous carbon bodies are significantly inferior to those of carbon with few or no large pores, the second approach was considered appropriate. Furthermore, it is possible that the thromboresistance conferred by the benzalkonium-heparin complex trapped in the porosity of a carbon surface might not be lasting. Thus the current program is emphasizing studies of impermeable carbon surfaces, and heparinization treatments have been restricted to those in which heparin molecules are covalently bonded to the surface.

Special emphasis has been placed on pyrolytic carbons deposited from gaseous environments in such a way that their microstructures do not contain growth features (e.g., Pyrolite carbons). These carbons are stronger and tougher than any other bulk form of carbon. They can be applied as coatings to a variety of substrates, including metals, ceramics, and graphite; and using special fabrication procedures, Pyrolite carbon coatings can be placed in a permanent compressive state of stress that enhances the toughness of the finished device. The isotropic varieties can be applied to complex shapes without the danger of delamination and cracking that restricts the use of anisotropic varieties of pyrolytic "graphite" used in aerospace applications for heat shields.

In the remainder of this chapter the structure and pertinent properties of Pyrolite carbons will be discussed, and examples illustrating the rapidly expanding list of biomedical applications will be presented.

II. DEPOSITION, STRUCTURE, AND PROPERTIES OF PYROLITE CARBONS

A. Deposition and Structure

Pyrolite carbons are deposited in a fluidized bed from a hydrocarbon-containing gaseous environment [55]. The variables influencing the structure of carbon deposited in such a process are shown diagrammatically in Fig. 10. By controlling the deposition conditions (i.e., the deposition temperature, the composition of the fluidizing gas, the bed geometry, and the residence time of the gas molecules in the bed) it is possible to control the anisotropy, density, crystallite size, and microstructure of the carbon deposited. When conditions are arranged to allow the formation of growth features in the microstructure, such as those visible in the photomicrograph of Fig. 11a, the deposits are relatively weak; that is, the fracture stress in bending is only about 20,000 psi [3]. If, however, such conditions are

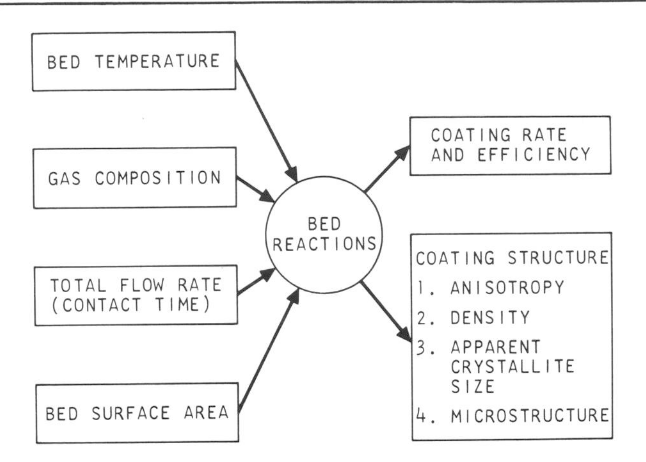

FIG. 10. Diagram relating the four principal coating variables, the pyrolysis reactions in the fluid bed, the deposition kinetics, and the structure of the carbon deposited.

maintained that visible growth features are not formed (Fig. 11b), deposits with fracture stresses approaching 100,000 psi can be obtained [3, 8]. The isotropic carbons deposited at low temperatures without growth features (LTI Pyrolite carbons) are of particular interest because their isotropy allows a wide variety of shapes to be coated.

Using the fluidized-bed process, it is possible to introduce various other elements into the fluidizing gas and codeposit these with carbon. For example, the scanning electron micrographs in Fig. 12 show the polished surfaces of two silicon-alloyed LTI Pyrolite carbon deposits. The carbon shown in Fig. 12a contains 7 wt% silicon; that shown in Fig. 12b contains 16 wt% silicon. The silicon is present as a dispersion of silicon carbide particles, which appear white in the micrographs. The silicon carbide, which has a hardness of 9 on the Mohs scale (diamond is 10), is added to enhance the hardness for applications requiring resistance to abrasion.

B. Mechanical Properties

Kaae [8] has recently measured the mechanical properties of LTI Pyrolite carbons and determined their fracture stress, elastic moduli, and fracture toughness as a function of density. Some of his data are presented in Figs. 13a and 13b. The fracture stress varies from 50,000 psi for deposits

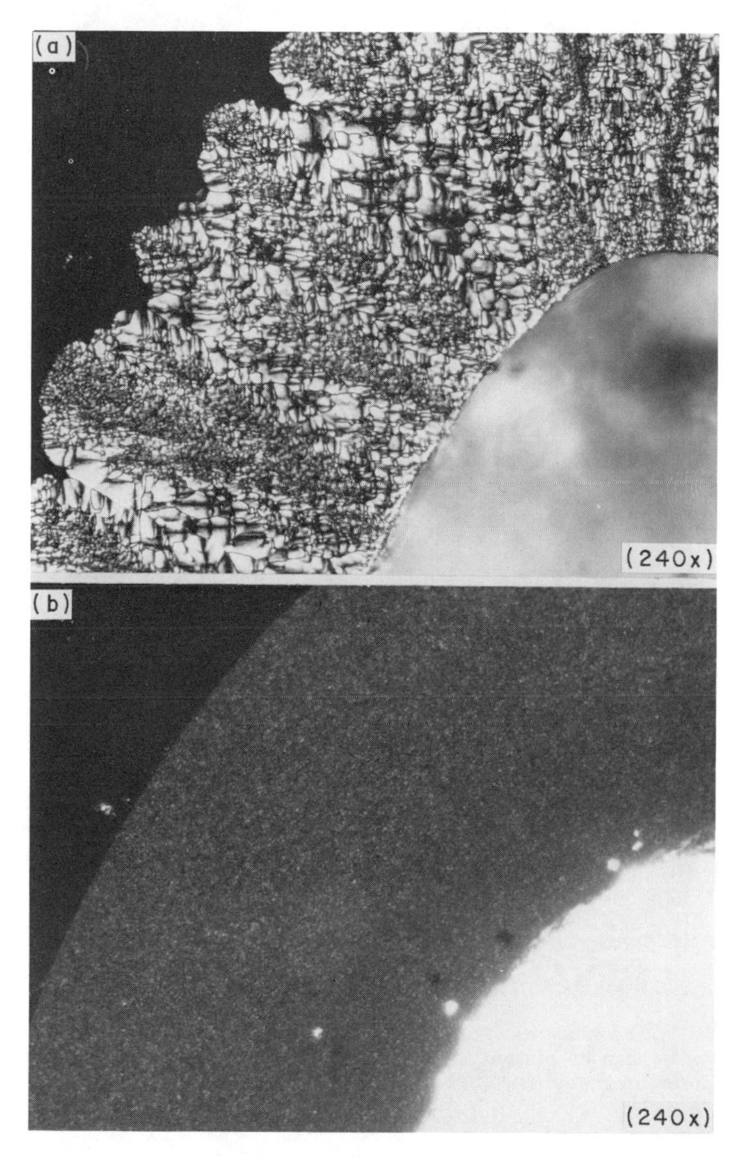

FIG. 11. Microstructures of carbons deposited in a fluidized bed: (a) a granular carbon with distinct growth features; (b) an isotropic carbon without growth features (polarized light). Magnification 240x.

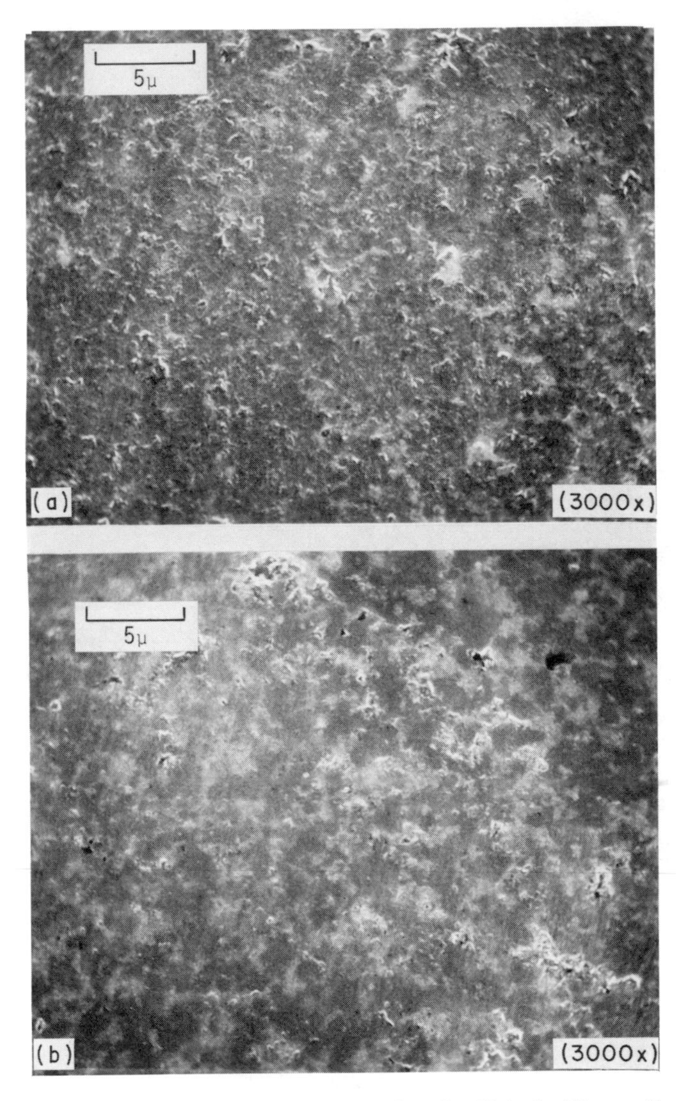

FIG. 12. Scanning electron micrographs of polished silicon-alloyed LTI Pyrolite carbons containing (a) 7 wt% silicon and (b) 16% wt% silicon. The white areas are silicon carbide.

with densities near 1.4 g/cm^3 to about 75,000 psi for deposits with densities near 1.9 g/cm^3. The corresponding moduli increase from 2.5 x 10^6 to 4 x 10^6 psi, which, incidentally, is in the range reported for bone [56]. Kaae

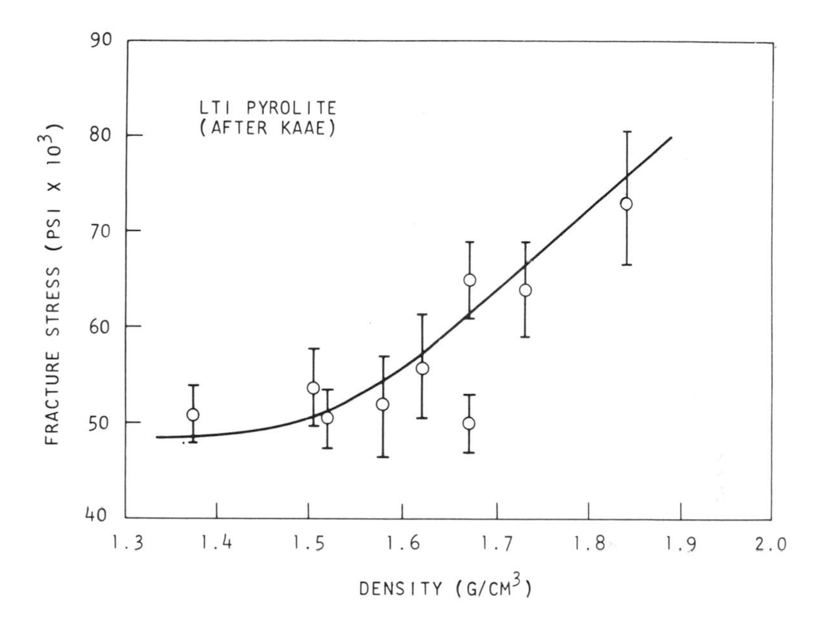

FIG. 13(a). Fracture stress versus density for unalloyed LTI Pyrolite carbons [8].

reports values of 400 to 800 in.-lb/in.^{3} for the strain energy to fracture. These values are compared in Table 2 with those reported for other bulk forms of carbon. The energy to fracture is particularly important in such devices as artificial heart valves and bone and joint prostheses that are subjected to considerable mechanical abuse.

Two additional properties having substantial importance in prosthetic devices are fatigue behavior and wear resistance. Both of these are presently under investigation in our laboratory. It appears that the endurance limit (stress at which no degradation occurs under cyclic loading) of LTI Pyrolite carbons is equal, or extremely close, to the single-cycle fracture stress [60]. This compares with an endurance limit of about half the single-cycle fracture stress for a high-density isotropic graphite [61]. The wear resistance of LTI Pyrolite carbons appears to increase with density in a pattern similar to that of the elastic modulus. It is worth noting that under similar conditions the wear resistance of silicon-alloyed LTI Pyrolite carbon is an order of magnitude greater than that of glassy or vitreous carbon [see Fig. 21(a) of Ref. 60].

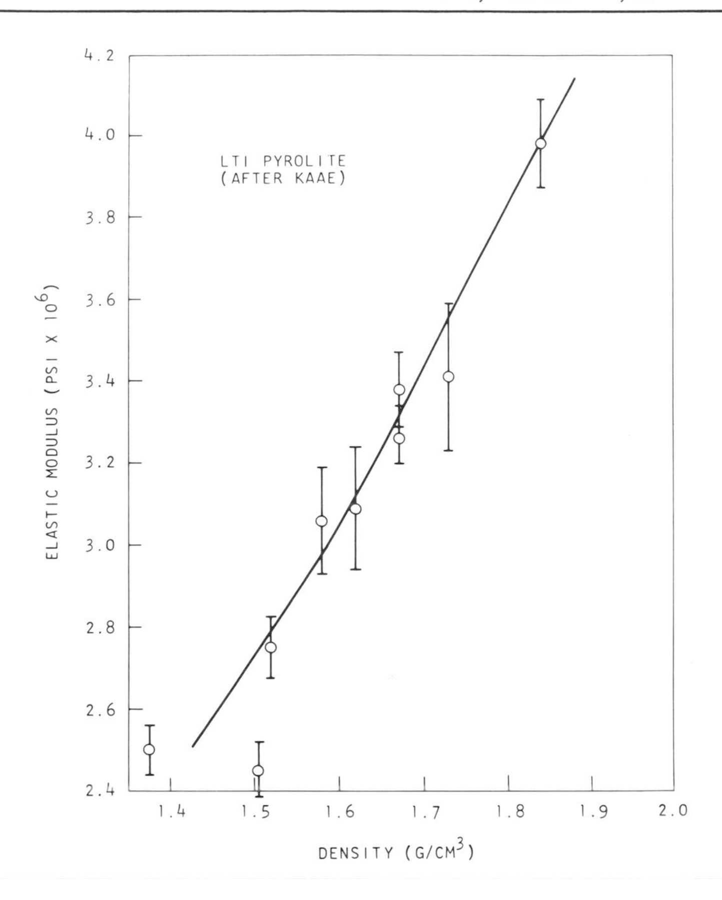

FIG. 13(b). Elastic moduli versus density for unalloyed LTI Pyrolite carbons (after J. L. Kaae, Ref. 7).

The mechanical properties of silicon-alloyed coatings have been determined, and fracture strengths in excess of 100,000 psi have been reported [62].

C. In-Vivo Thromboresistance

A great deal of in-vivo testing has been carried out at the Johns Hopkins Hospital by Gott and co-workers using short tubular test specimens (rings) implanted in the canine inferior vena cava. The test is very severe (see Table 1); details of the procedure have been published [17]. The bulk of

TABLE 2

Strain Energy to Fracture for Various Types of Carbon

Carbon type	Fracture stress σ (10^3 psi)	Elastic modulus E (10^6 psi)	Strain energy to fracture, $\sigma^2/2E$ (in.-lb/in.^{3})
LTI Pyrolite carbon	75	4	700
Vitreous carbon [a]	25	3.5	90
Pyrolytic graphite [b]	20	3.5	60
Strongest polycrystalline artificial graphite [c]	11	1.7	40

[a] Data from Ref. 57. [b] Data from Ref. 58. [c] Data from Ref. 59.

the data from in-vivo tests of carbon surfaces can be found in Refs. 38 and 40. The dependence of the thromboresistance on surface roughness, on the presence of oxygen-bearing functional groups on the surface (hydrophilic sites), and on pretreatment with heparin is revealed in the data summarized in Table 3. Comparison of the data in the vertical columns shows that the thromboresistance for any particular surface is independent of pretreatments with heparin. This is because these particular surfaces did not have a significant amount of surface porosity in which the benzalkonium-heparin complex could be trapped.

Comparison of the corresponding data for the rough, as-deposited surfaces (see Fig. 5) with those for polished surfaces indicates that smooth surfaces are less thrombogenic than rough ones. This is because regions of stasis associated with the roughness provide spots for cell adherence and the nucleation of thrombi.

There is a striking reduction in the thromboresistance of surfaces that were heated in oxygen to provide hydrophilic sites. The treatment employed provides a maximum amount of chemisorbed oxygen ($\sim 1\%$ of the surface sites) [63]; this is much more than would be formed, for example, by steam autoclaving.

The most thromboresistant surfaces were those that were polished and then cleaned by vacuum (10^{-6} torr) outgassing at 900°C. The sterility of the

TABLE 3

Results of Gott Ring Tests Illustrating the Dependence of the Thromboresistance of LTI Pyrolite Carbon on the Surface Smoothness, Oxygen Functionality on the Surface, and Pretreatments with Heparin [38]

TREATMENT BEFORE IMPLANTING	DURATION OF TEST	AS-DEPOSITED		POLISHED	
		VACUUM OUTGASSED	HEATED IN O_2 AT 450°C	VACUUM OUTGASSED	HEATED IN O_2 AT 450°C
BENZALKONIUM CHLORIDE–HEPARIN	ACUTE (2 HR)	● ◐ ◐ ◐ ○ ○ ○	● ● ● ● ◐ ○	◐ ○ ○ ○ ○	● ◐ ○ ◐ ○
	CHRONIC (14 DAY)	○(a) ◐(b) ●(c) ●(c) ◐		● ◐ ○ ○ ○	
HEPARIN	ACUTE (2 HR)			◐ ◐ ○ ○ ○	◐ ◐ ◐ ○ ○
	CHRONIC (14 DAY)			● ◐ ○ ○ ○	◐ ◐ ◐ ◐ ◐
NONE	ACUTE (2 HR)	● ◐ ◐ ○ ○	● ● ● ● ○	◐ ◐ ○ ○ ○	● ◐ ◐ ● ○
	CHRONIC (14 DAY)			○ ◐ ○ ○ ○	

(a) DIED AFTER 7 DAYS
(b) DIED AFTER 12 DAYS
(c) DIED AFTER 2 DAYS

outgassed surfaces was maintained until implantation; then the rings were implanted using the pretreatments listed in Table 3. Of the 15 rings implanted for 2 h, all remained open; 5 showed evidence of minor thrombus formation. Of the 15 implanted for 2 weeks, 12 remained open, and of these 12, only 2 showed minor thrombi.

It is worth noting that wide variations in performance with surface topography and cleanliness are obtained (Table 3), and it should be emphasized that, as for any biomaterial, proper characterization is necessary to avoid spurious and misleading results.

Table 4 compares the in-vivo thromboresistance of the silicon-alloyed carbons with that of unalloyed LTI Pyrolite carbon. Although the data suggest that silicon carbide may itself be thromboresistant, in-vivo tests have shown that it is quite thrombogenic [40].

Efforts to improve the thromboresistance of LTI Pyrolite carbons by covalently bonding heparin to the carbon surface were not successful. The methods used are outlined in Fig. 14. Details of the procedures can be found in Refs. 40 and 41. Although methods by which heparin could be firmly bonded to carbon were developed (evidenced by the retention of from 1 to 3 $\mu g/cm^2$ heparin after washing for 1800 h in a 1-mg/ml water solution of

TABLE 4

Results of Gott Ring Tests Comparing the Thromboresistance of Polished LTI and Polished Silicon-Alloyed (12 wt% Silicon) LTI Pyrolite Carbon[a,b]

TREATMENT BEFORE IMPLANTATION	AMOUNT OF THROMBUS IN LUMEN			
	UNALLOYED PYROLITE		ALLOYED PYROLITE	
	2 HOUR TEST	14 DAY TEST	2 HOUR TEST	14 DAY TEST
BENZALKONIUM CHLORIDE-HEPARIN	◒○○○○	●●○○○	◒◒◒○○	◒◒○○○

[a] All materials were outgassed under vacuum at either 900°C (LTI Pyrolite) or 130°C (silicon-alloyed LTI Pyrolite) and maintained sterile to time of implant.

[b] Black areas in circles depict thrombus in ring at end of test.

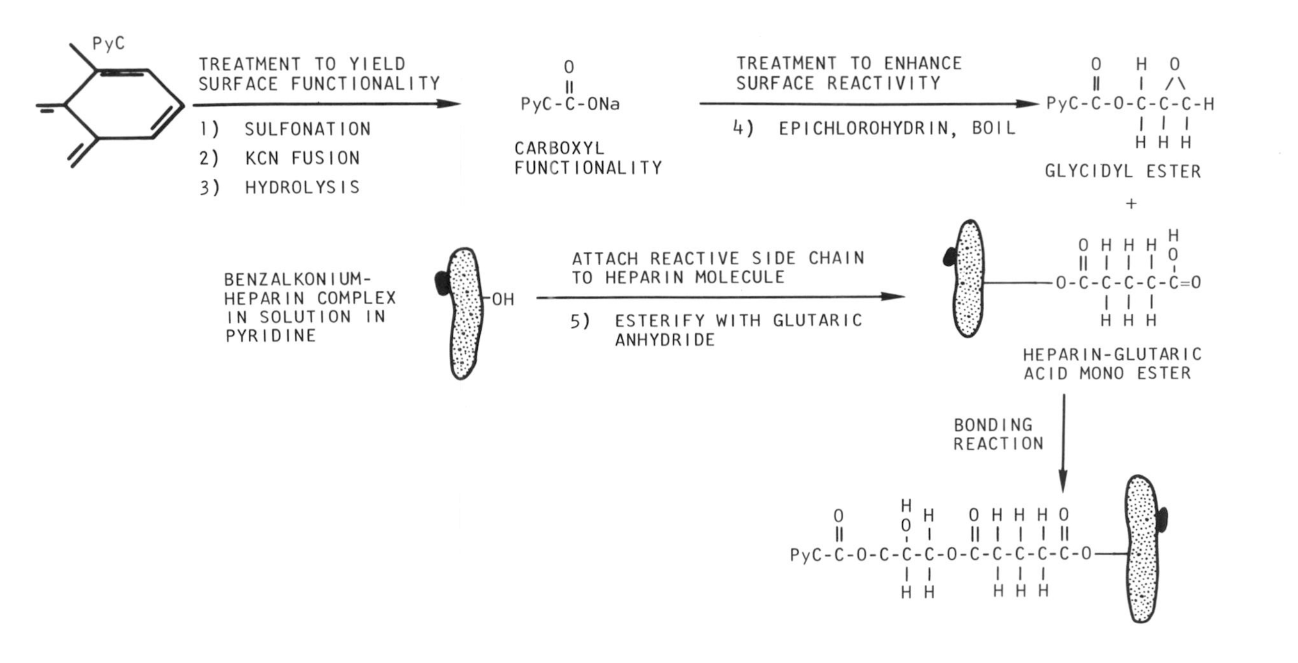

FIG. 14. Outline of procedure for covalent bonding of heparin to pyrolytic carbon surfaces.

sodium laurylsulfate at 40°C; see Figs. 10 and 11 of Ref. 41), no improvement in thromboresistance was obtained (Table 5). The data, in fact, suggest a slight decrease in thromboresistance. The decrease may be due to the oxygen-bearing surface sites that did not react with heparin molecules during the bonding step. It appears that, at least for carbon, covalently bonding heparin to the surface does not necessarily provide antithrombogenic activity. The elutriation of heparin from the heparinized surfaces (Table 5) is in line with similar reports for other heparinized surfaces [31, 64].

D. In-Vitro Blood Compatibility

Although thrombosis on a foreign surface in blood is the most obvious evidence for material incompatibility within the vascular system, there are, in addition, other more subtle but equally adverse effects that can occur. Just as foreign surfaces can affect the proteins that initiate clotting, they can also adversely alter other molecular constituents. Halbert, Anken, and Ushakoff [65, 66] have investigated the effects of various candidate implant materials on plasma proteins. Two types of LTI Pyrolite carbon surfaces were among the materials studied. The specimens consisted of carbon-coated ZrO_2 particles that were tumble-polished, cleaned, and then sterilized either by vacuum outgassing at 900°C or by steam autoclaving. The surface topography of the particles is shown in the scanning electron micrographs in Fig. 15. While the particles appear smooth at magnifications of 48X and 800X, roughness is evident at 8000X. The results from the outgassed and autoclaved samples were identical. There was no effect of exposure to the LTI Pyrolite carbon surfaces on either the plasma enzyme activity (Table 6) or the concentration of the 17 plasma proteins monitored (Table 7). The slight rise in the concentration of lactic dehydrogenase was not judged to be significant. More recent work of Halbert et al. has provided identical results for silicon-alloyed LTI Pyrolite carbon.

To determine whether exposures to either the outgassed or autoclaved LTI Pyrolite carbon surfaces interfered with the integrity of the entire system of plasma proteins involved in the coagulation process, Halbert et al. [65, 66] added an appropriate quantity of calcium chloride to the samples of plasma that were exposed to the test surface and recorded the time required to form a firm clot. The results for the unalloyed LTI Pyrolite carbon are listed in Table 8. Similar tests with silicon-alloyed LTI Pyrolite carbon showed no effect of the exposure on the clotting time.

TABLE 5

IN-VIVO THROMBORESISTANCE AND HEPARIN RETENTIVITY ON ROUGH AS-DEPOSITED LTI CARBON SURFACES

Implant Time (hr)	Standard Benzalkonium-Heparin Treatment[a]: Heparin Surface Conc. ($\mu g/cm^2$) Before	After	Thrombus in Ring	Heparin Covalently Bonded in One Step: Heparin Surface Conc. ($\mu g/cm^2$) Before	After	Thrombus in Ring	Heparin Covalently Bonded in Two Steps: Heparin Surface Conc. ($\mu g/cm^2$) Before	After	Thrombus in Ring
1	2.2	1.0	●[b]	--	--	--	--	--	--
2	--	--	--	0.8	0.5	●[b]	1.6	0.9	◒
2	--	--	--	0.6	<0.1	○	1.4	2.9	○
2	--	--	--	0.6	<0.1	○	--	--	--
4	2.9	2.6	◒	1.4	<0.1	●[b]	--	--	--
24	--	--	--	0.8	<0.1	●	1.4	3.6	○
24	--	--	--	--	--	--	13	1.5	○
170	2.1	0.5	○	0.8	<0.1	●	16	7.5	●[b]
170	--	--	--	0.9	<0.1	○	8.5	4.6	◓
260	--	--	--	1.6	0.4	●[b]	--	--	--
340	2.2	0.6	○	1.3	0.4	○	7.7	4.3	●
340	--	--	--	--	--	--	16	1.9	●
720	3.2	3.6	○	1.7	0.4	◔	8.0	3.5	○
720	--	--	--	--	--	--	15	6.0	○

[a] Outgassed at 900°C in vacuum prior to treatment

[b] Dog died

TABLE 6

Effect of Steam Autoclaved or Vacuum Outgassed LTI Carbon on Plasma Enzyme Activity (After S. P. Halbert)

ENZYME	EFFECT ON ENZYME ACTIVITY AFTER:		
	0 HOURS	24 HOURS	48 HOURS
GLUTAMIC-PYRUVIC TRANSAMINASE	0	0	0
GLUTAMIC-OXALACETIC TRANSAMINASE	0	0	0
ISOCITRIC DEHYDROGENASE	0	0	0
LACTIC DEHYDROGENASE	0	SLIGHT INCREASE	SLIGHT INCREASE
CHOLINESTERASE	0	0	0
AMYLASE	0	0	0
ACID-PHOSPHATASE	0	0	0
ALKALINE PHOSPHATASE	0	0	0

In parallel studies it was found that nonthrombogenic surfaces heparinized via the quaternary ammonium salt tridodecylammonium chloride (TDMAC) produced significant protein denaturation. On the other hand, silicon rubber, which is thrombogenic, turned out to be compatible with plasma proteins. Only the LTI Pyrolite carbon, polycarbonate, silicon rubber, and some acrylamide surfaces were relatively inert toward plasma proteins. References 65 and 66 should be consulted for further details of the results from this careful study.

The data collected by Halbert and co-workers [65, 66] emphasize the requirement that candidate materials must provide a combination of properties to be useful as an intravascular implant material. Not only must the material be thromboresistant, but it must also not alter the formed elements of blood or molecular constituents. At the same time the material must not itself be adversely affected. The review article by Akutsu and Kantrowitz [67] summarizes many of the reported observations of the in-vivo degradation of a variety of polymers. The inert character of LTI Pyrolite carbon, together with its compatibility with blood, is most encouraging.

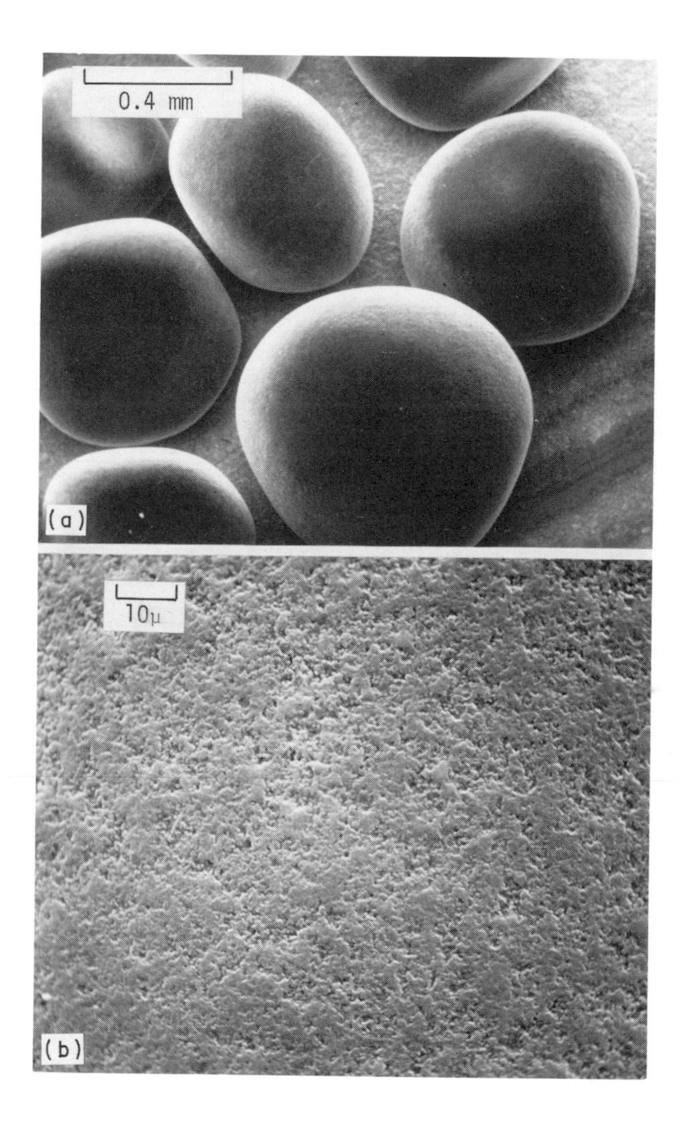
0.4 mm
(a)
10μ
(b)

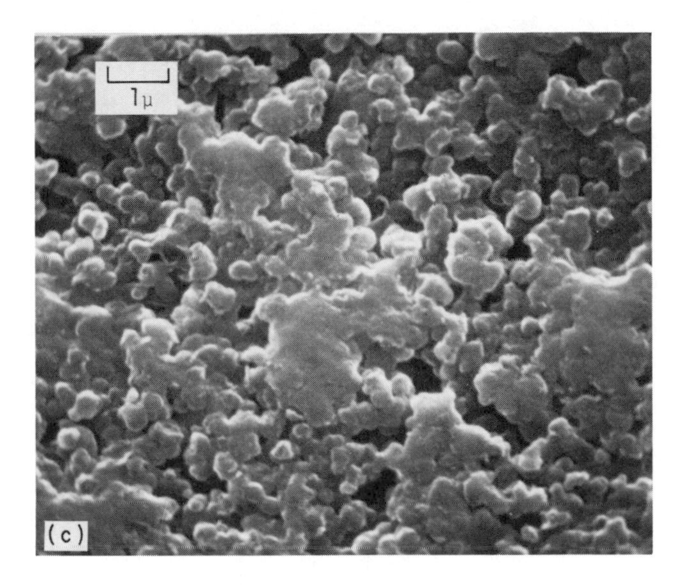

FIG. 15. Scanning electron micrographs of the surfaces of carbon-coated particles tested in plasma by Halbert, Anken, and Ushakoff [65, 66].

Using another approach, Mason and co-workers [68-70] measured the in-vitro compatibility of materials with blood. The test procedure involved taking fresh blood from a donor and dividing it into several portions, each of which was exposed to a test surface. The control surface was siliconized glass, which is known to cause little damage to blood in comparison with most foreign surfaces. After the blood had been in contact with the test surface for a specified time, sodium citrate was added, the anticoagulated blood was centrifuged to remove cellular material, and several tests were run on the platelet-poor plasma. The tests were determinations of the partial thromboplastin time [71, 72], a measure of the activation of clotting; the Stypven time [73, 74], a measure of the release of platelet factor III or other thromboplastic materials; and assays for adenosine triphosphate (ATP) and hemoglobin [75]. Data are reported as ratios of the test values to those obtained when the blood is exposed only to siliconized glass. Values greater

TABLE 7

Effect of Steam Autoclaved or Vacuum Outgassed LTI Carbon on 17 Plasma Proteins (After S. P. Halbert)

PROTEIN	CHANGE IN CONCENTRATION AFTER:		
	0 HOURS	24 HOURS	48 HOURS
TRANSFERRIN	0	0	0
CERULOPLASMIN	0	0	0
IMMUNOGLOBULIN "G"	0	0	0
IMMUNOGLOBULIN "A"	0	0	0
IMMUNOGLOBULIN "M"	0	0	0
ALBUMIN	0	0	0
FIBRINOGEN	0	0	0
PRE-ALBUMIN	0	0	0
β LIPOPROTEIN	0	0	0
β_2 GLYCOPROTEIN	0	0	0
α_1 LIPOPROTEIN	0	0	0
α_2 MACROGLOBULIN	0	0	0
α_1 ANTI-TRYPSIN	0	0	0
HAPTOGLOBIN	0	0	0
HEMOPEXIN	0	0	0
α_1 ACID GLYCOPROTEIN	0	0	0
C3 (β_{1c}/β_{1a} GLOBULIN)	0	0	0

than unity denote good thromboresistance for the first two determinations; for the second two determinations, values less than unity denote less hemolysis than that caused by siliconized glass.

A number of LTI Pyrolite carbon specimens with a variety of surface conditions were exposed to blood in an all-siliconized-glass system for 10 min with gentle mixing. Because the test system contained siliconized glass, the results for the carbon specimens could never be better than results obtained for an all-siliconized-glass system. All of the data, of which those in Table 9 are typical, indicate that both the pure and silicon-alloyed varieties of LTI Pyrolite carbon are at least as compatible with blood as siliconized glass. The values of hemoglobin and adenosine triphosphate release indicate that no significant hemolysis is caused by the specimens. Mason's data place these carbons among the most compatible materials tested [69, 70].

TABLE 8

Calcium Replacement Clotting Times after Exposure of Plasma to LTI Pyrolite Carbon Surfaces[a]

Sample of LTI Pyrolite carbon		Clotting time (min) after		
		0 h	24 h	48 h
Steam-autoclaved:				
Trial 1	Control	11	32	117
	Carbon	11	37	24-28
Trial 2	Control	14	23	116
	Carbon	14	22	64
Vacuum-outgassed at 900° C	Control	14	32	105
	Carbon	14	25	61

[a] Data from Refs. 65 and 66. All tests performed at 37° C.

TABLE 9

Results of Blood-Compatibility Tests with Pure and Silicon-Alloyed LTI Pyrolite Carbons[a]

Coating type	Stypven time[b]	Partial thromboplastin time[b]	Hemoglobin release[b]	ATP release[b]
Silicon-alloyed LTI Pyrolite carbon	1.00	0.99	1.00	0.53
Pure LTI Pyrolite carbon	0.98	1.04	1.10	1.20

[a] Data from Refs. 69 and 70. Each result is the average of six tests. Duration of test: 10 min. Surface preparation: grinding and polishing through 0.25-μm diamond. Sterilization: ultrasonic cleaning in benzene, boiling in benzene for 10 min, boiling in benzene for 10 min, boiling in ethyl alcohol for 10 min. Treatment before test: 30 min in 0.154 formal NaCl.

[b] Expressed as ratios of the value obtained in the tests to those obtained with blood exposed only to silicone-coated glass.

E. Compatibility with Tissue and Blood Cells

As a further measure of the biocompatibility of carbon, Johnsson-Hegyeli [76-78] microscopically observed the interactions between human blood and tissue cells and carbon surfaces. Compatibility between tissue cells and carbon surfaces was determined by cultivating Wish human amnion cells as a monolayer culture directly on the surfaces. The results for a variety of surface conditions are listed in Table 10. With the exception of the surfaces treated with benzalkonium chloride and heparin, all surfaces supported the growth of normal cells. The surfaces treated with benzalkonium chloride-heparin caused destruction of cells. This is most likely due to the presence of benzalkonium chloride.

In studies of surfaces identical with those listed in Table 10, Johnsson-Hegyeli contacted the specimens with citrated (ACD-treated) Red Cross blood at 37°C for various lengths of time, rinsed them in physiological saline, and then observed and photographed the specimens while they were still wet. The results were summarized in Table 11. The absence of aggregates of cells on the surfaces indicates that there was no cell damage [79]. The results are consistent with the data reported by Mason and co-workers [69, 70] and suggest that clean LTI Pyrolite carbon surfaces sterilized by a variety of methods are compatible with both tissue and blood cells.

F. Surface Energy of Carbon Surfaces

Lyman and co-workers [80-83] have pointed out that the surface free energy of a material should be an important factor in determining the interactions between the material and the molecular and formed elements in blood. This accounts for the fact that the reduction in the surface free energy of carbon surfaces by forming deposits with a preponderance of low-energy c faces exposed at the surface improves the thromboresistance [39]. Bischoff [84], however, has suggested that the work of adhesion might be a more pertinent surface property. He feels that, although gross trends may be obtained from correlations with macroscopic surface variables, the approach suffers from the disadvantage that the adsorption of proteins and other materials so significantly alters surface parameters that thorough studies of other factors, such as the microscopic molecular orientations, are more important. In support of the latter view, the attachment of oxygen to 1 out of every 100 sites on a carbon surface had little effect on the contact angle formed with water, but had a profound effect on the thromboresistance of the carbon surface (Table 4 in Ref. 37). Thus the macro and micro approaches both appear to be important and, in fact, often complement one another in developing an understanding of foreign surface-blood interfaces.

Baier [85] has examined the surface properties of polished LTI Pyrolite carbon surfaces that had been ultrasonically cleaned several times in boiling benzene and alcohol, and soaked in saline before in-vivo exposure in the Gott ring test. The exposed surfaces were compared with controls. A minute thrombus was formed on each of the implanted rings (Table 12).

The control carbon surfaces yielded a value of about 50 dynes/cm for the critical surface tension [86], which is considerably higher than thought to be required for good blood compatibility (Fig. 16). The critical surface tension for the surfaces that had been exposed to blood, however, was much lower (28 dynes/cm), well within the apparently biocompatible range found by Baier [85]. After implantation, the contact-angle data for the carbon surface were found to be in all respects similar to those obtained for naturally thromboresistant blood-vessel intima. Using multiple-internal-reflection attenuated infrared spectroscopy, Baier was able to show that the reduction in the critical surface tension was due to a very pure adsorbed protein layer.

Recalling the data of Halbert and co-workers [65-66] (Tables 6 through 8) for the compatibility of plasma protein with similar surfaces, it may be inferred that the carbon surfaces owe their thromboresistance to the fact that they can adsorb plasma proteins without altering them, and that the intermediate protein layer mimics the natural vascular lining. The layer was found to be desorbable, suggesting the establishment of a dynamic equilibrium between the carbon surface and the proteins in the bloodstream. Other illustrations of the role that proteins must play at biological interfaces have been reported [87, 88], and there is considerable recent evidence of the formation of conditioning films of adsorbed protein on foreign surfaces [89-92].

G. Summary of Blood Compatibility of LTI Pyrolite Carbon

All of the results taken together indicate that carbon is a useful material for applications in bioengineering. Properly prepared LTI Pyrolite carbon surfaces have a high level of thromboresistance. Of major significance is

TABLE 10

Summary of Johnsson-Hegyeli Data for Tissue-Culture Studies on Carbon Surfaces[a]

Candidate material No.	Type of carbon	Sterilization	Treatment before test	Results of tissue-culture test	
				Growth rate	Cell-colony appearance[b]
GGA-8	Silicon-alloyed LTI	Shipped in 1:750 benzalkonium chloride	Soaked 1 h in 10 mg/ml heparin; rinsed in normal saline; steam-autoclaved	Complete destruction of cells	
GGA-7	Silicon-alloyed LTI	None	Steam-autoclaved	About 50% of controls	Round, discrete; tendency to curl around edges
GGA-6	Silicon-alloyed LTI	Steam-autoclaved, 250°C for 20 min at 230-psi stream	Resterilized by steam autoclaving	About 50% of controls	Round, discrete
GGA-3	LTI	Shipped in 1:750 benzalkonium chloride	Soaked 1 h in 10 mg/ml heparin; rinsed in normal saline; steam-autoclaved	Complete destruction of cells	
GGA-1	LTI	None	Steam-autoclaved	About 50% of controls	Round, discrete

GGA-2	LTI	Steam-autoclaved, 250°F for 20 min at 230-psi steam	Resterilized by steam autoclaving	About 50% of controls	Round, discrete
GG-4	LTI	Heated at 130°C for 12 h under vacuum	Steam-autoclaved	Less than 50% of controls	Round, balled up, discrete
GGA-5	LTI	Heated to 900°C for 1 h under vacuum	Steam-autoclaved	Slight reduction	Irregularly shaped; confluent
	Control[c]		Steam-autoclaved	Control	Irregularly shaped; confluent

[a] Data from Ref. 76. All carbon surfaces were prepared by polishing one side through 1-μm diamond.

[b] Cellular morphology: normal appearance in all cases.

[c] Glass surface (Bellco tissue-culture glass).

the fact that the thromboresistance of LTI Pyrolite carbon surfaces is inherent in the physical and chemical properties of the material and does not rely on fragile monomolecular layers bonded to the surface of a thrombogenic or otherwise chemically active substrate.

The fact that LTI Pyrolite carbons do not alter plasma proteins or plasma enzyme activity is the likely reason for the good compatibility of carbon with blood cells; the compatibility is a probable consequence of the ability of LTI Pyrolite carbon to maintain an intermediate dynamically adsorbed layer of proteins at the interface. Problems of biodegradation within the body, such as those often encountered when polymers are implanted for long periods, will almost certainly not arise with chemically inert carbon bodies.

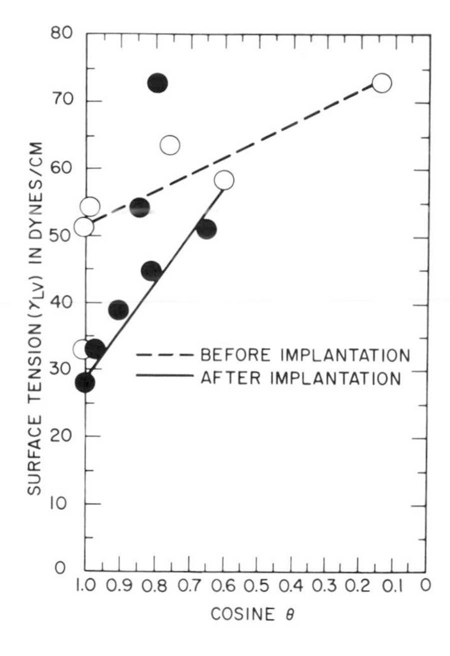

FIG. 16. Surface tension versus contact angle θ for various liquids on polished LTI Pyrolite carbon controls and companion surfaces exposed to canine blood for 2 h in the vena cava. Broken curve: before implantation; solid curve: after implantation. From Ref. 85.

The rigid nature of carbon bodies, which is an asset, for example, in certain artificial-heart-valve designs, is a detriment when flexibility is required. This presents a formidable challenge to bioengineers, who may have to compromise their designs so that real devices that are biocompatible and lasting within the body can be realized.

TABLE 11

Summary of Johnsson-Hegyeli Data for Cell Adherence[a]

Candidate material No.[b]	Cell adherence after exposure to ACD Red Cross blood for 1 h at 37°C
GGA-1	Occasional platelets
GGA-2	Occasional platelets
GGA-3	Occasional erythrocytes and platelets
GGA-4	30 platelets, 5 erythrocytes, and occasional leukocytes per 500x field
GGA-5	10 platelets, occasional erythrocytes, and occasional leukocytes per 500x field
GGA-6	20 platelets, occasional erythrocytes per 500x field
GGA-7	10 platelets, occasional erythrocytes, and some leukocytes per 500x field
GGA-8	40 primary platelets per 500x field

[a] From Ref. 76. [b] See Table 10.

TABLE 12

Results from Polished, LTI Carbon-Coated Rings Tested for 2 h in Canine Vena Cava at Johns Hopkins Hospital[a]

THROMBUS IN LUMEN

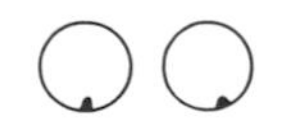

[a] Surfaces of Rings Were Studied by Baier After In-vivo Test.

III. APPLICATIONS OF PYROLITE CARBONS

A. Applications in Prosthetic Heart Valves

The design, testing, and construction of improved prosthetic heart valves present a formidable challenge to bioengineers. A patient undergoing surgery for aortic valve replacement in 1968 had about a 70% chance of surviving for at least 2 years. The corresponding figure for mitral valve replacement was 60%, and that for multiple-valve operation was only 50%. Improved operative procedures and intensive postoperative care have reduced the operative mortality to less than 10%. Detailed descriptions of valve structure and testing, materials for use in the construction of valves, and clinical experience with many of the valves that are currently used have been compiled in the Proceedings of the Second National Conference on Prosthetic Heart Valves [93].

Materials problems that interfere with the successful performance of valves can be grouped into three categories:

1. Thromboembolic complications associated with both the design and materials used in the construction of valves.

2. Biochemical degradation of valve components.

3. Mechanical failure by wear or fatigue of rubbing or flexing components of valves.

Degradation as in categories 2 and 3 is often interrelated since biochemical degradation accelerates fatigue, and rubbing, which removes the products of biochemical degradation, can accelerate the rate of the biochemical reaction by continually exposing new surface to the corroding media.

The use of large surface areas of exposed metal in valves is often quoted as leading to thromboembolic complications. For example, the Journal of the American Medical Association reported that with some older models of mitral prostheses in which the metal of the valve seat and cage was exposed, the incident of embolic complications was in excess of 60% [94]. A cloth covering on the metal can sharply reduce these complications, but other problems associated with fabric wear or uncontrollable tissue proliferation that restricts flow [95] can arise. The degradation of the silicone-rubber balls used in ball valves provides a good example of deterioration that is due to biochemical incompatibility and leads to mechanical failure [96, 97].

1. The DeBakey-Surgitool Aortic Valve

The experimental ball-valve prosthesis shown in Fig. 17 is an example of an approach to the solution of the problems described here. All components used in the construction of the experimental valve, except the sewing ring, are coated with thromboresistant, silicon-alloyed LTI Pyrolite carbon. Since none of the surfaces are heparinized, such complications as those described by Saltzman [98] and Halbert et al. [65] can be avoided. Furthermore, the good thromboresistance demonstrated by the LTI Pyrolite carbon suggests that the use of anticoagulants will not be required.

FIG. 17. An experimental all-carbon ball-valve prosthesis (the DeBakey-Surgitool aortic valve). The hollow ball, metal cage, and seat are all coated with from 0.4 to 0.5 mm of silicon-alloyed LTI Pyrolite carbon.

The cage of the valve consists of a ductile-metal substrate onto which about 0.4 mm of silicon-alloyed LTI carbon has been deposited (Fig. 18). The latest balls (a cross section is shown in Fig. 19) are constructed by hollowing out a solid graphite ball so that, after coating, the density of the ball matches that of blood. A threaded plug is cemented in the hole through which the tool was inserted for the hollowing-out procedure. The ball is then baked to convert the cement to carbon in the screw threads, thus fixing the plug. (The earliest versions were made by brazing two hollowed-out halves of graphite together and then depositing the carbon on the surface. The new method simplifies manufacture and quality-control procedures.)

In the early version the silicon-alloyed LTI Pyrolite was deposited on the hollow balls in two layers; currently the coating is deposited in a single layer [99]. The coefficient of thermal expansion of the substrate graphite is chosen so that, during cooling after coating, the outer carbon layer is forced into a compressive state of stress, which enhances the toughness of

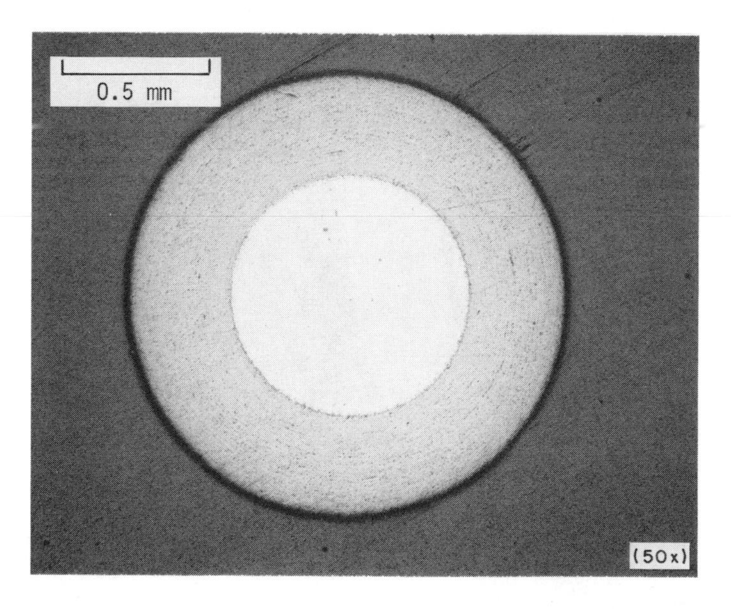

FIG. 18. Cross section showing the silicon-alloyed LTI Pyrolite carbon on a ductile-metal cage.

the finished ball [99]. The balls have been tested to 15,000 psi hydrostatically (the limit of the equipment) without failure. When loaded in compression between parallel steel platens, fracture (the first crack) occurs at loads in the range 300 to 1000 lb (Fig. 20), that is, more than an order of magnitude

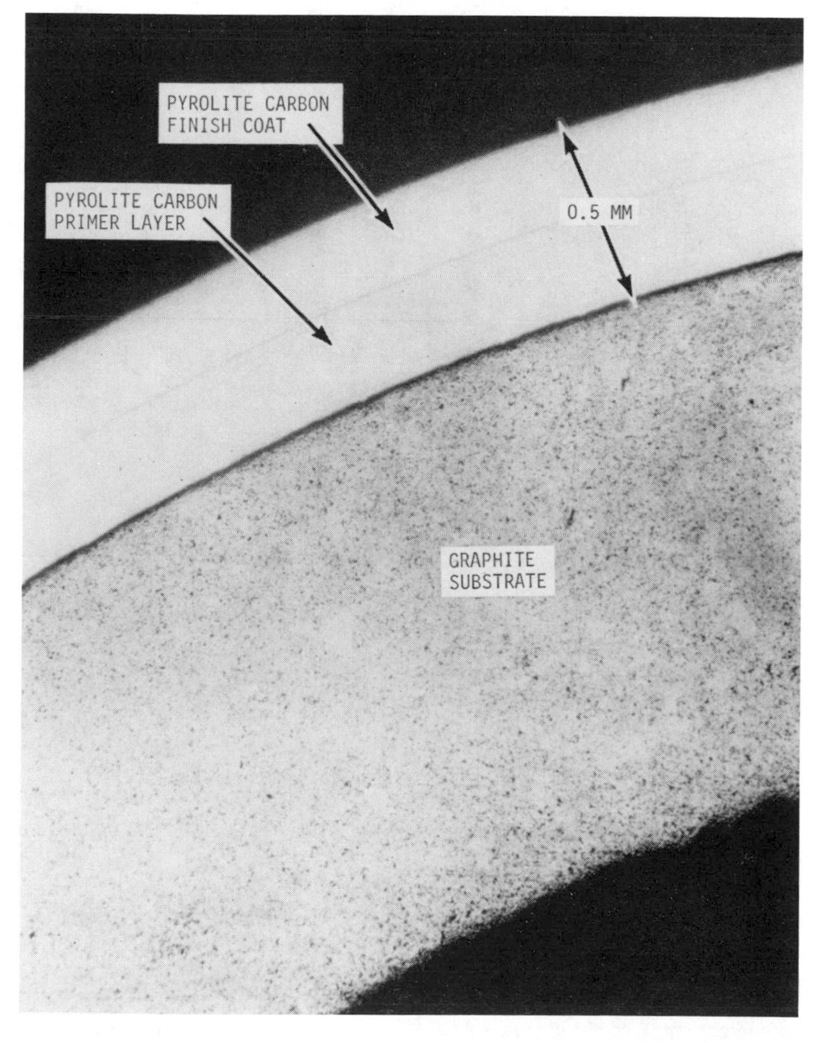

FIG. 19. Cross section of the hollow ball, coated with LTI Pyrolite carbon, DeBakey-Surgitool aortic valve.

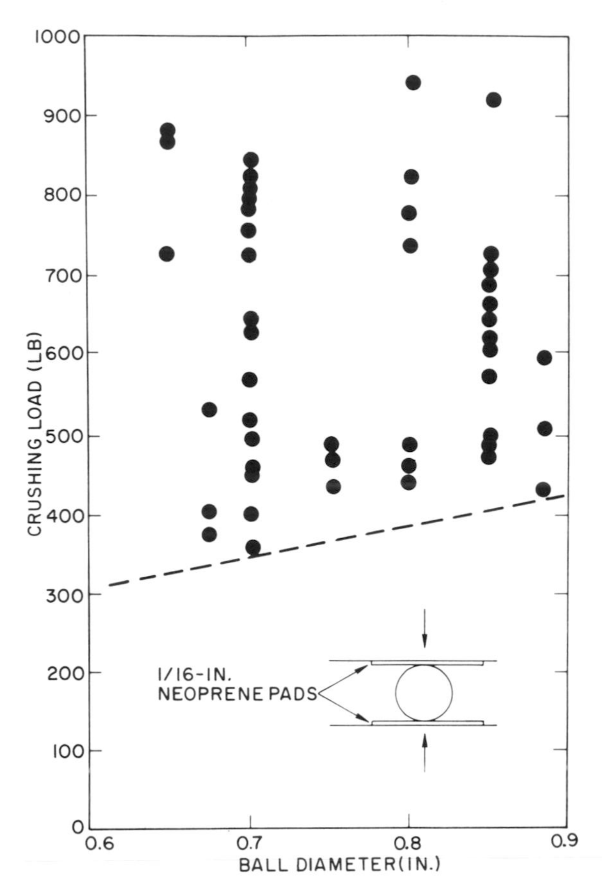

FIG. 20. Crushing loads versus ball diameter for hollow balls used in the DeBakey-Surgitool aortic valve.

higher than any load applied in practice. Since the fatigue limit for carbon is nearly equal to the fracture stress [100], the data in Fig. 20 suggest that fatigue problems should be minimal.

Since prosthetic valves must open and close approximately 40 million times a year, the wear characteristics of a material can be as important a consideration as its biocompatibility. Accordingly extensive wear tests of carbon components are being carried out [60]. Data that describe the wear rate of a silicon-alloyed LTI Pyrolite ball in a valve with a titanium cage and a cloth-covered seat are plotted in Fig. 21. The data established (by weight change) a maximum wear rate of 4×10^{-5} in. of carbon per year. This rate indicates that it would take at least 200 years to wear 50% of the distance through the LTI carbon coating.

Accelerated tests using 500 pulses/min, 140/100 torr pressure (ratio of downstream pressure when valve is open to that when it is closed), and pumping 10 l/min for a total of 50×10^6 cycles yielded an equivalent maximum wear rate of 4×10^{-4} in. year, that is, an amount per cycle 10 times higher than that observed when a rate of 100 cycles/min was used in the test. The surfaces of the tested ball and a control surface (compared in Fig. 22) showed no evidence of wear (i.e., the wear was uniform). A test conducted at

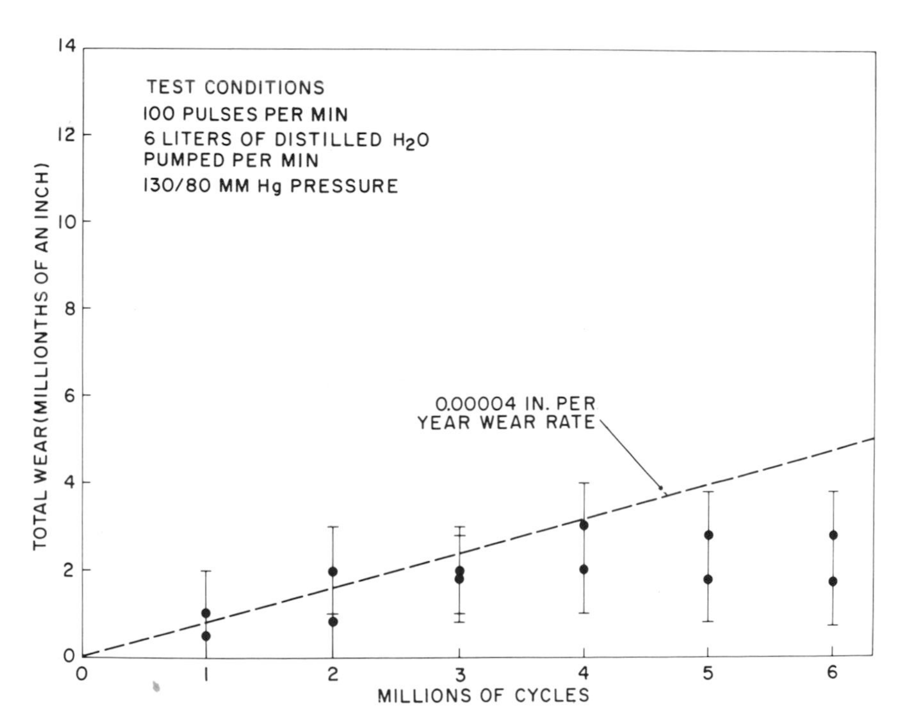

FIG. 21. Wear rate versus number of cycles for a ball coated with silicon-alloyed LTI Pyrolite carbon in a valve with a titanium cage and a cloth-covered seat. Test conditions: 100 pulses/min; distilled water pumped at a rate of 6 l/min; 130/80 torr pressure.

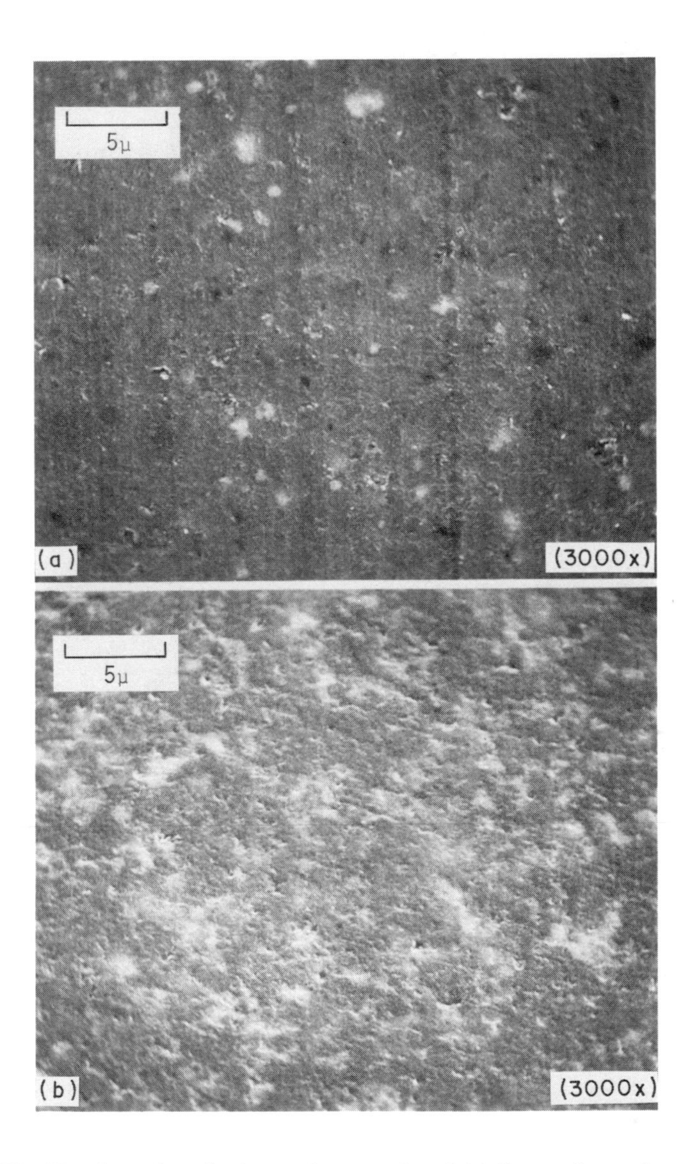

FIG. 22. Scanning electron micrographs of (a) the surface of an untested ball coated with silicon-alloyed LTI Pyrolite carbon and (b) the surface of a similar ball tested at 500 cycles/min, 140/100 torr pressure, and pumping 10 l/min of distilled water to a total of 50 x 10^6 cycles. There is no visible change. The total wear was 5 x 10^{-4} in.

Cutter Laboratories at 2400 cycles/min gave an equivalent wear rate of about 0.1 in./year. This value is several thousand times the rate observed when the test rate was 100 pulses/min. The test run at Cutter was to about 7×10^6 cycles, at which time the wear was 0.02 in. The data in Fig. 21 show that at 100 cycles/min the wear in 7×10^6 cycles is less than 4×10^{-6} in. Thus it is clear that accelerated tests can yield misleading high values for the wear per cycle.

A general comment applicable to accelerated tests of all materials cannot be made because the effect of cycle rate on the wear per cycle is likely to be dependent on the physical and mechanical properties of the material being tested as well as on the exact test conditions. The results show, however, that data obtained from accelerated testing, whether they show an encouraging or discouraging result, should be viewed with suspicion until the dependence of the wear per cycle on cycle rate is established. In most cases it is sounder to establish very sensitive methods of measuring wear and to conduct the tests at physiological, rather than accelerated, rates.

Because of the promising results, a clinical trial of the DeBakey-Surgitool aortic valve was conducted by Dr. M. E. DeBakey in Houston during 1969 and 1970. The valve is now generally available, and to date more than 2000 balls have been supplied for human implant. The valve uses a titanium cage, a carbon-coated seat, and a hollow LTI-carbon-coated ball. So far there have been no contraindications.

2. Beall-Surgitool Mitral Valve

Late in 1968 an effort was initiated to develop an LTI-carbon mitral prosthesis. To facilitate comparison, it was decided to replicate as nearly as possible the Beall-Surgitool mitral prosthesis [101] using silicon-alloyed LTI Pyrolite carbon, instead of Teflon, on all the rigid valve parts. The earliest model, shown in Fig. 23, had a solid-graphite substrate disk and a refractory-metal cage, both coated with silicon-alloyed LTI Pyrolite carbon. The refractory metal chosen for the fabrication of the cage has a slightly higher thermal expansion than the carbon, so that after coating a small compressive stress is generated in the coating during cooling. The cage alloy has a special composition that is not embrittled during the high-temperature coating procedure. A sectional view of a coated metal cage, together with micrographs of the carbon-metal bond, is shown in Fig. 24.

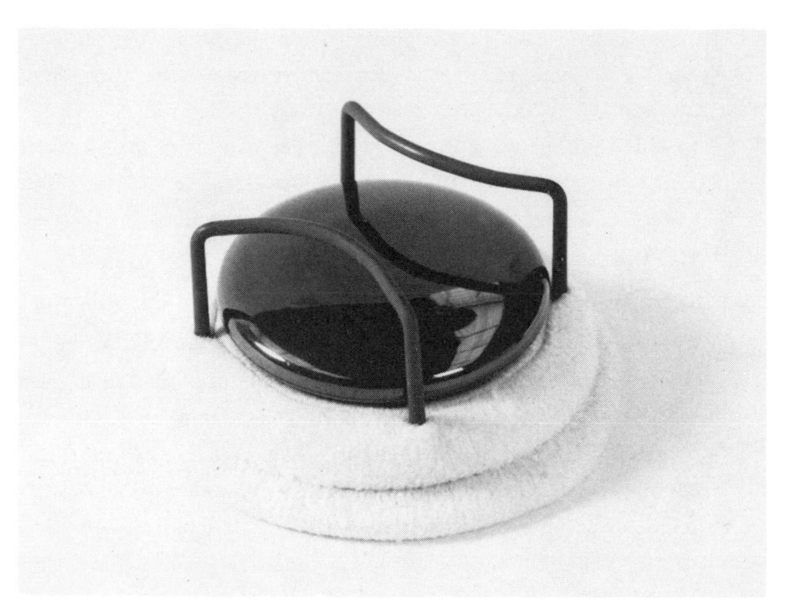

FIG. 23. An experimental version of the Beall-Surgitool mitral valve with silicon-alloyed LTI Pyrolite carbon on the struts and the disk. The seat is cloth-covered. (In the current clinical model, the top of the struts are not curved.)

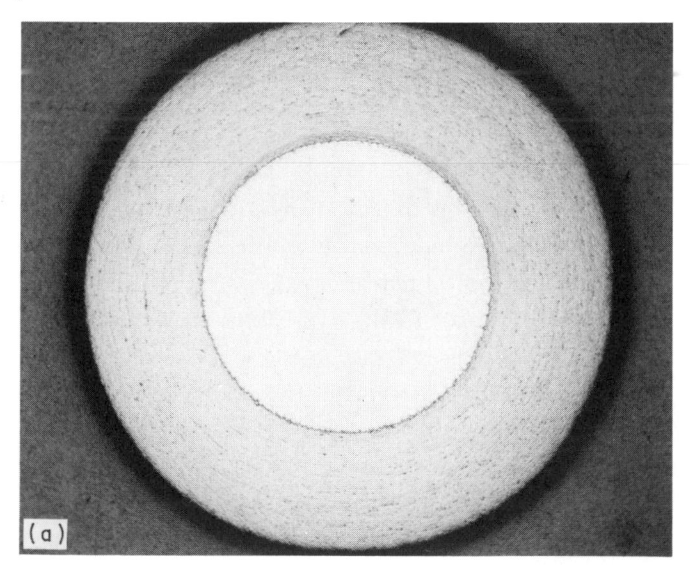

FIG. 24. Cross-sectional photomicrographs of the LTI Pyrolite carbon on a metal cage: (a) unetched 40X; (b) unetched, 340X; (c) etched; illustrates diffusional bond (340X).

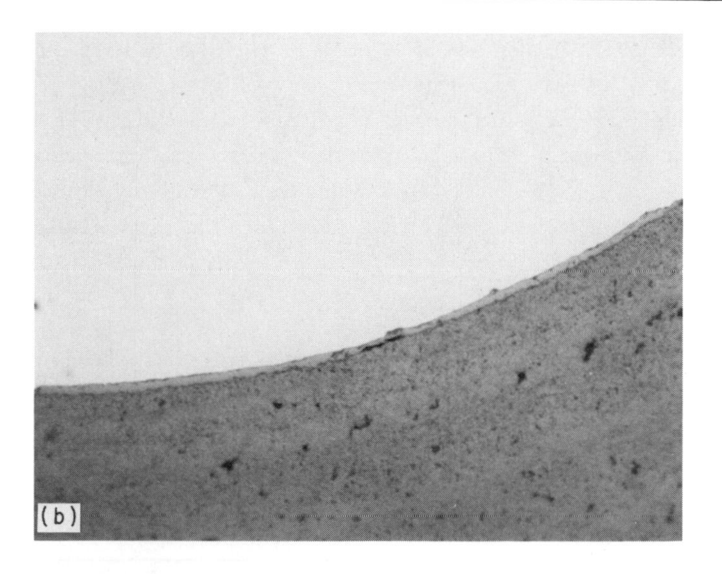

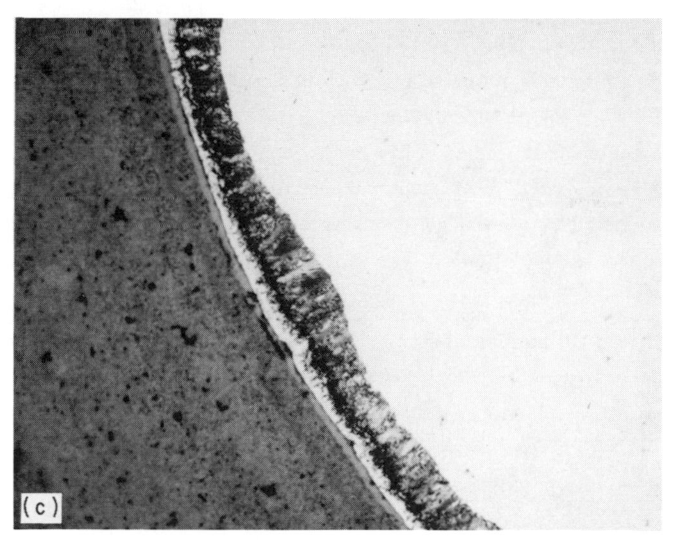

In the unetched sample (Fig. 24b), a thin carbide layer at the interface is visible; etching reveals a diffusion zone about 0.001 in. thick in the metal at the interface (Fig. 24c).

Extensive testing in calves for times extending to a year yielded wear data that project to 140 years as the time to wear through the coating on the struts (Fig. 25). Interestingly, the wear varied from strut to strut. For the 6-month calf, two struts showed no wear at all, one had a barely detectable wear mark, and the fourth had the wear mark shown in Fig. 25a. In all cases the wear at the edge of the disk was too small to be detectable with scanning electron microscopy. More than 3000 all-carbon Beall-Surgitool mitral valves have been introduced for clinical use [101].

3. Pivoting-Disk Valves

During the last 4 years three pivoting-disk valves have been introduced for clinical use. The Lillehei-Kaster valve (Fig. 26) uses a silicon-alloyed LTI Pyrolite carbon disk in a titanium cage. For this particular design a Pyrolite carbon was chosen as much for its mechanical behavior as for its compatibility with blood. Extensive accelerator and animal tests showed that the carbon disks consistently outperformed polypropylene, polycarbonate, Delrin (acetal hopolymer, E. I. duPont de Nemours & Co., Inc.), stainless steel, and titanium [102]. The good results were attributed to the structural and biochemical properties of LTI Pyrolite carbon. No galling of the sort generated when metal rubs against metal [103, 104] was observed when a Pyrolite carbon disk was used in a metal cage. Further details of the Lillehei-Kaster valve have been presented [105]. From the spring of 1971 through 1972, about 5000 disks have been supplied to Medical Incorporated for the Lillehei-Kaster valve and no contraindications have been reported.

In 1969 a tilting disk of somewhat different design was introduced by the Shiley Laboratories for Björk [106-109]. The early clinical version used a Delrin occluder, but a newer modification uses an alloyed LTI Pyrolite carbon disk (Fig. 27). After 4.6 x 10^8 cycles in an accelerator at 1200 cycles/min pumping a water-glycerin mixture, the wear mark appeared as a polished ring, and examination with the scanning electron microscope showed that the depth of the mark was less than 5 μm [106]. More than 5000 Björk-Shiley valves with Pyrolite disks have been supplied by Shiley Laboratories for clinical use since the spring of 1972.

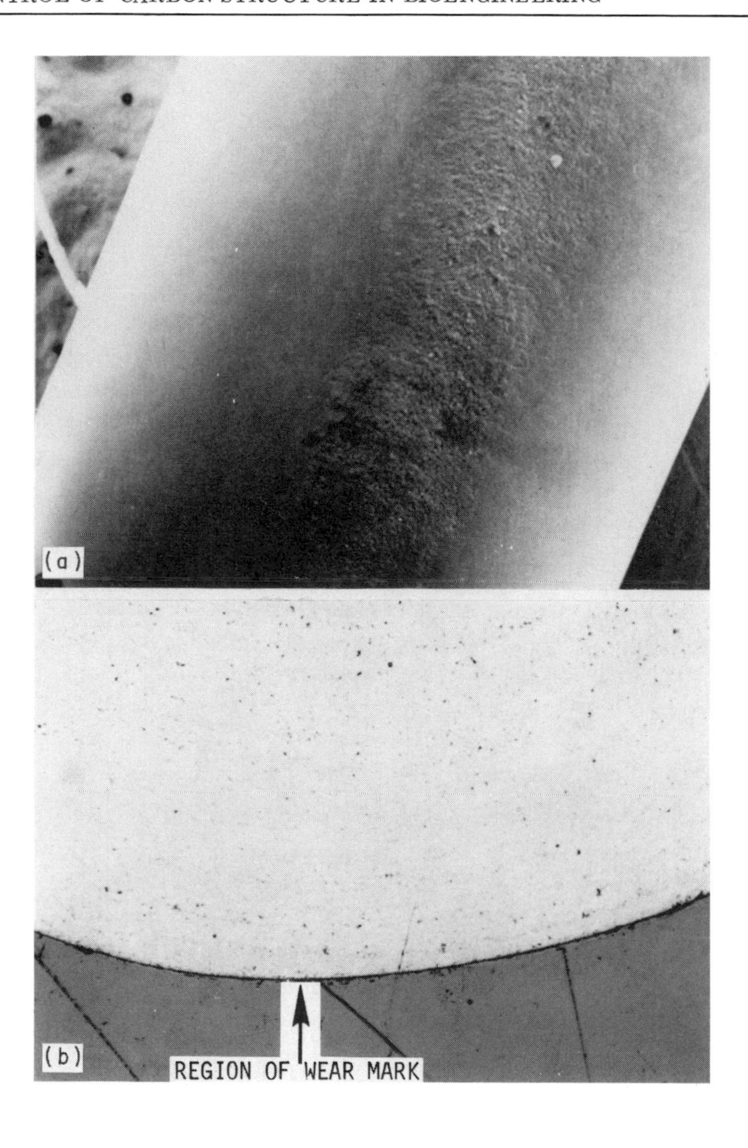

FIG. 25. (a) Scanning electron micrograph (65X) of the wear mark on the most severely worn strut from an experimental Beall valve tested for 6 months in a calf; (b) photomicrograph (1200X) through the region of maximum wear.

FIG. 26. The clinical Lillehei-Kaster pivoting-disk valve.

4. Other Concepts

Two experimental valve designs that utilize pyrolytic carbon in their construction are shown in Fig. 28. The upper valve (a) is a full-orifice disk valve currently under test at Cutter Laboratories. Accelerator tests pumping distilled water and running at 500 to 1500 cycles/min to 2.6×10^8 cycles resulted in no significant wear on the disk. The lower disk valve is an interesting concept developed by Cutter Laboratories for E. B. Diethrich and M. Kaplitt. All exposed surfaces of the valve are Pyrolite carbon; a segment of the patient's vein is used as a sewing ring, thus avoiding the complication of infection from the cloth sewing ring. The concept could prove to be an important one, since surgical excision is often the only chance of cure for an infected valve [110-112].

As a final comment in this section, it is worth mentioning one example of problems that can be encountered when accelerated tests are used in valve

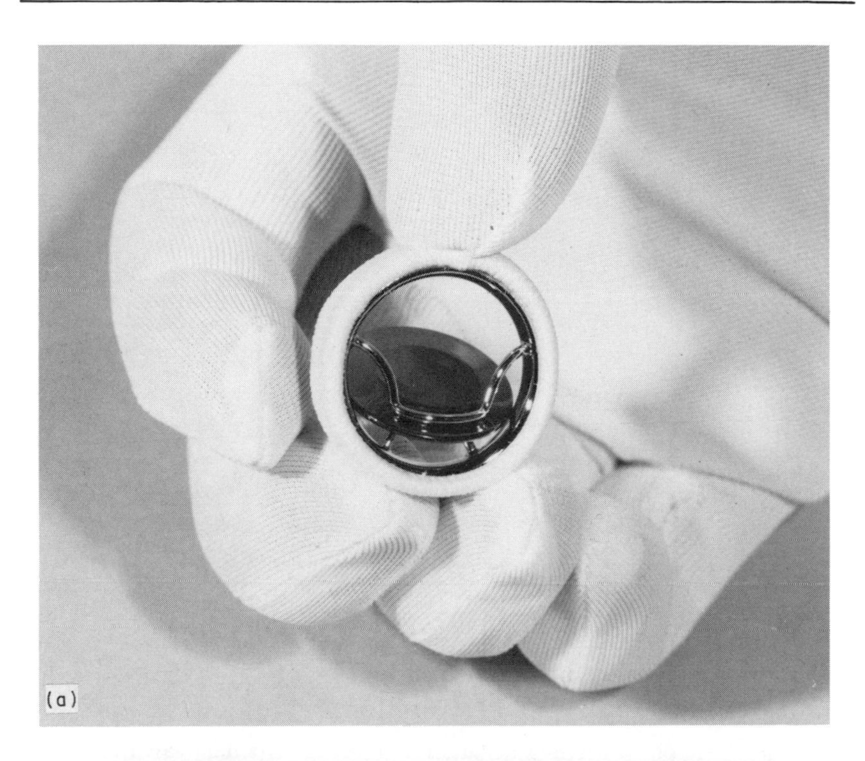

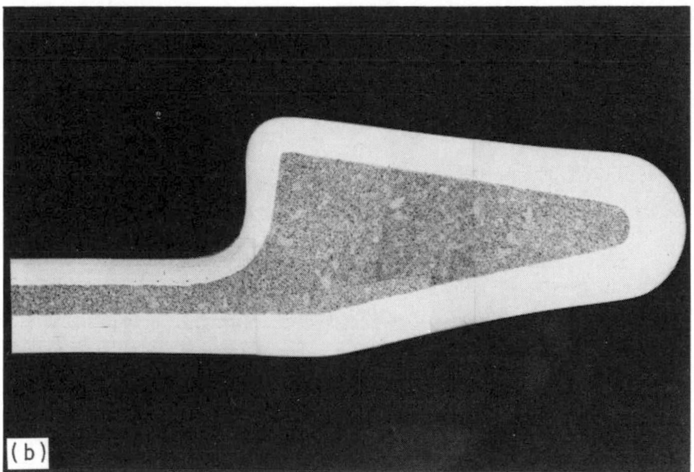

FIG. 27. (a) An experimental Bjork-Shiley valve with an LTI Pyrolite carbon disk; (b) a section through a carbon-coated graphite disk fabricated for the Bjork-Shiley valve.

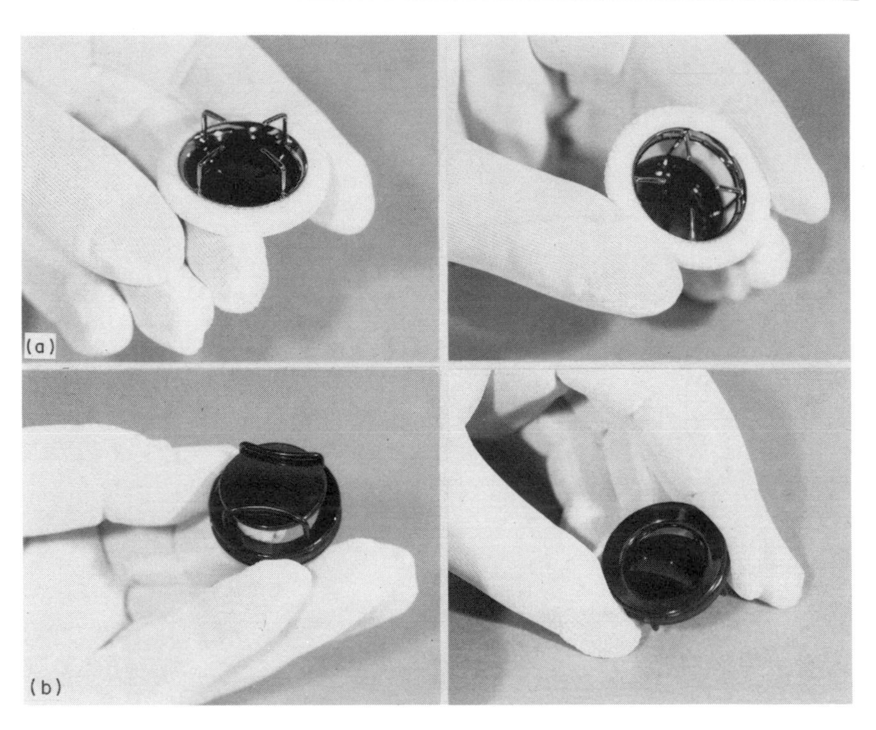

FIG. 28. (a) experimental Cutter full-orifice titanium valve with solid Pyrolite carbon disk; (b) experimental all-Pyrolite carbon valve designed by Cutter Laboratories for E. B. Diethrich and M. Kaplitt.

evaluation. Examination of a valve tested by Kaster in a pulse accelerator at 2500 cycles/min to a total of 110×10^6 cycles showed that the wear on the rubbing surfaces was less than that detectable with a micrometer (i.e., less than 25 μm). However, the interior portions of a disk from such a test, which were not contacted by the titanium housing, were eroded by the liquid itself. At this test rate cavitation was clearly visible with a strobe light, which showed clouds of bubbles generated on both the opening and closing cycles of the valve. The cavitation caused severe localized erosion, an example of which is shown in the scanning electron micrograph in Fig. 29. This well-known mechanism [113, 114] of erosion is most severe for hard materials and has a high probability of occurring, as an artifact, in any high-flow-rate accelerated test. The observation is presented to reemphasize the uncertainty of data obtained through accelerated testing.

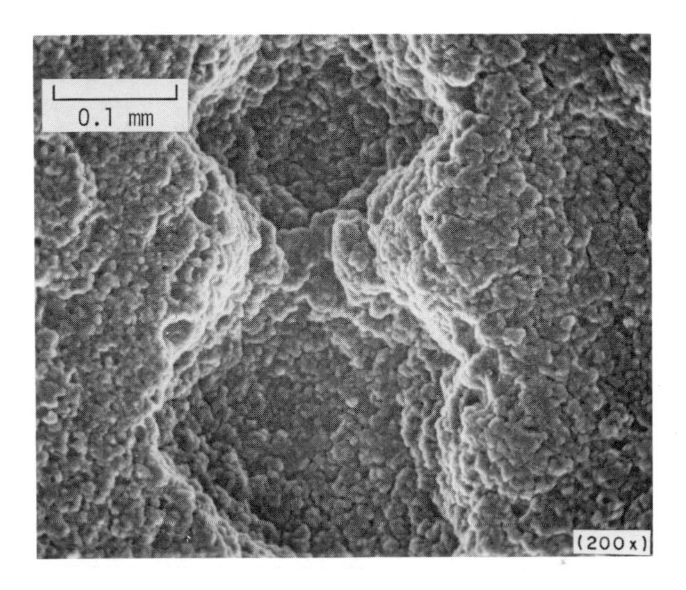

FIG. 29. Cavitation erosion in an accelerated test (2500 cycles/min). Cavitation was visible from the clouds of bubbles emitted on opening and closing of the valve and observed with the aid of a strobe light. This type of erosion is an artifact of the accelerated test [113 and 114].

B. Applications in Artificial Hearts

A variety of approaches have been taken in programs to develop implantable artificial heart and cardiac-assist devices [115]. In a majority of cases the devices have been designed with the anticipation that flexible materials that are nonthrombogenic, able to withstand continual flexing without failure, and resistant to in-vivo degradation will be developed [116-119]. In other approaches the problems associated with flexible components relying on durable polymers are circumvented by using designs with all rigid parts. Examples of applications of Pyrolite carbons in such devices are shown in Fig. 30. The impeller for the Bernstein centrifugal pump (Fig. 30a) was fabricated for use in an experimental version of the pump design described in Ref. 120. Components for the all-carbon version of the nonpulsatile, force-vortex Rafferty-Kletschka design, are coated with silicon-alloyed LTI Pyrolite carbon (Fig. 30b). The Rafferty-Kletschka design uses an

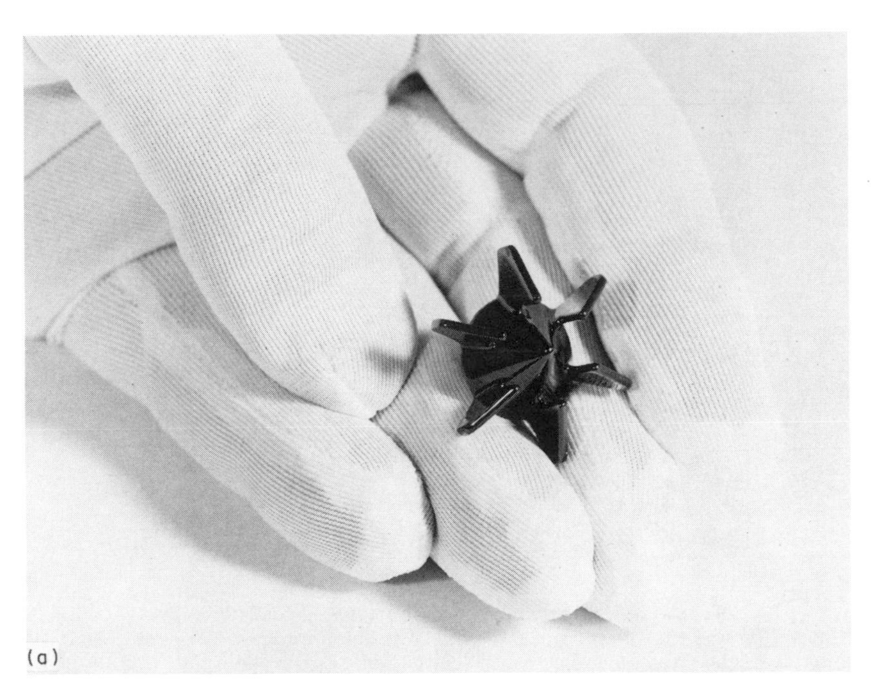

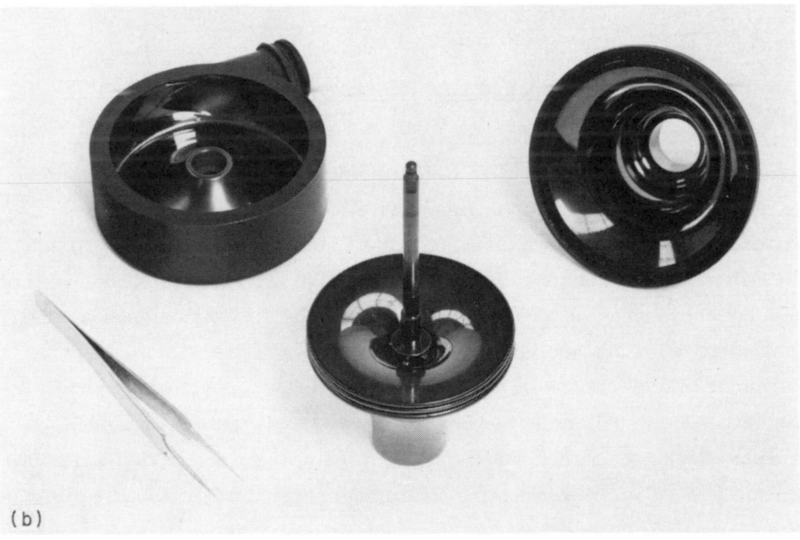

FIG. 30. Examples of components for centrifugal cardiac-assist devices coated with silicon-alloyed LTI Pyrolite carbon: (a) the impeller for the Bernstein pump [120]; (b) components of the all-carbon version of the Rafferty-Kletschka pump [121].

impeller made up of three nested cones spaced about 1 mm apart. The cones rotate about a common axis, thereby imparting centrifugal flow with a minimum of turbulence. In both applications Pyrolite carbon offers resistance to wear and thrombus formation without anticoagulation and biodegradation.

An implantable artificial heart complete with an energy source requires a method by which waste heat can be dissipated. The obvious method is to transfer the heat to the bloodstream and to allow natural processes to expel the heat from the body. Gillis and Walkup [122] have studied the effects of added endogenous heat on Hanford miniature swine. The longitudinally finned lumen extended-area heat exchanger coated with unalloyed LTI Pyrolite carbon (Fig. 31) was implanted in the thoracic aorta of a pig; it dissipated 60 watts of heat for 19 months at which time it was removed. No thrombus was found on the device. The animal at no time received anticoagulants. The thromboresistance demonstrated in this application has occurred in spite of the fact that the surfaces were not polished and that there were regions of stasis in cul-de-sacs oriented perpendicular to the flow between the internal fins.

C. Applications in Orthopedics

Although metals and polymers have been used extensively in orthopedics, these materials have proved to be far from ideal. Degradation of metals caused by the corrosive action of body fluid not only can cause a deterioration of their mechanical properties, but, in addition, the corrosion products are often found to be toxic, leading to allergenic and even carcinogenic responses [123, 126]. The presence of a metal implant can prevent healing when a sepsis develops in the area of the implant [127]. When metal-to-metal contacts are employed, as in a total hip prosthesis, galling is a common occurrence; it produces wear dust, which, because of its large surface area, causes inflammation and pain, and often requires device removal [125, 128].

An additional problem is created by the mismatch between the elastic properties of metals and bone. Whereas bone has moduli that are low ($\sim 3 \times 10^6$ psi) and extremely rate sensitive [56,61], the moduli of metals commonly used in orthopedics are about an order of magnitude higher and more constant. Accordingly a metal-bone composite does not bend uniformly, and stresses are concentrated. This situation leads to bone resorption and contributes to the loosening of a prosthesis that must bear high and fluctuating loads.

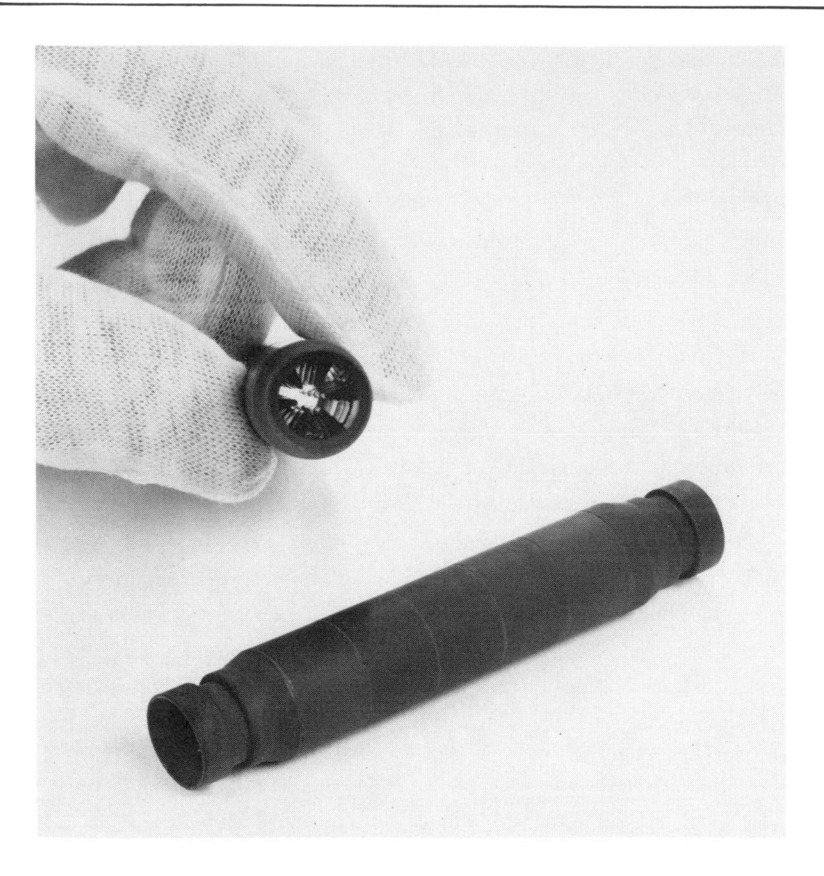

FIG. 31. The Battelle Northwest Laboratory's internally finned LTI-carbon-coated heat exchanger that was implanted in the thoracic aorta of a pig and has been dissipating 60 watts of heat for a year [122]. At no time did the animal receive anticoagulants.

Polymers suffer from shortcomings that are similar to those mentioned for metals [129, 130]. The properties of many polymers are degraded in vivo, and the corrosion products in some cases have led to carcinogenicity, tumors, and excessive inflammation. Of the polymers, only the pure silicone rubbers have demonstrated in-vivo inertness for long-term implants (except the deterioration of silicone rubber that occurs in prosthetic heart valves).

Because of its desirable combination of properties, carbon may overcome many of the shortcomings of metals and polymers in orthopedic applications.

The elastic moduli of all the carbons listed in Table 2 are in the range reported for bone, and the fracture stress of LTI Pyrolite is much higher than that of bone [131-134]. Although the stress response of carbon and bone differs in some respects, it is interesting to note that the strain energy to fracture (a rather crude measure of toughness) reported by Evans [61] for a variety of bones is about 100 in.-lb/in.^{3} or less, that is, substantially less than the values reported by Kaae [8] for LTI Pyrolite carbon (Table 2).

Benson [135] has cited considerable evidence for the good compatibility of a wide variety of carbonaceous materials with tissue and bone. Unlike metals and polymers, carbon appears to be free of the problems associated with corrosion by the in-vivo environment. These findings, together with the extensive intravascular experience and the similarity of the mechanical properties of carbon to those of bone, indicate exciting potentials for carbon in orthopedics.

The LTI-carbon-coated ball shown in Fig. 32 is an example of an application in which the use of carbon circumvents the wear problems

FIG. 32. An LTI-carbon-coated ball for a hip-joint prosthesis. The metal parts are a test fixture.

associated with metal-to-metal wear in a total hip prosthesis. Several prototypes have been implanted in dogs at UCLA by Amstutz. In an alternative approach Charnley [128] reports that metal against high-density polyethylene in vivo eliminates galling of the metal, but that the polymer wears

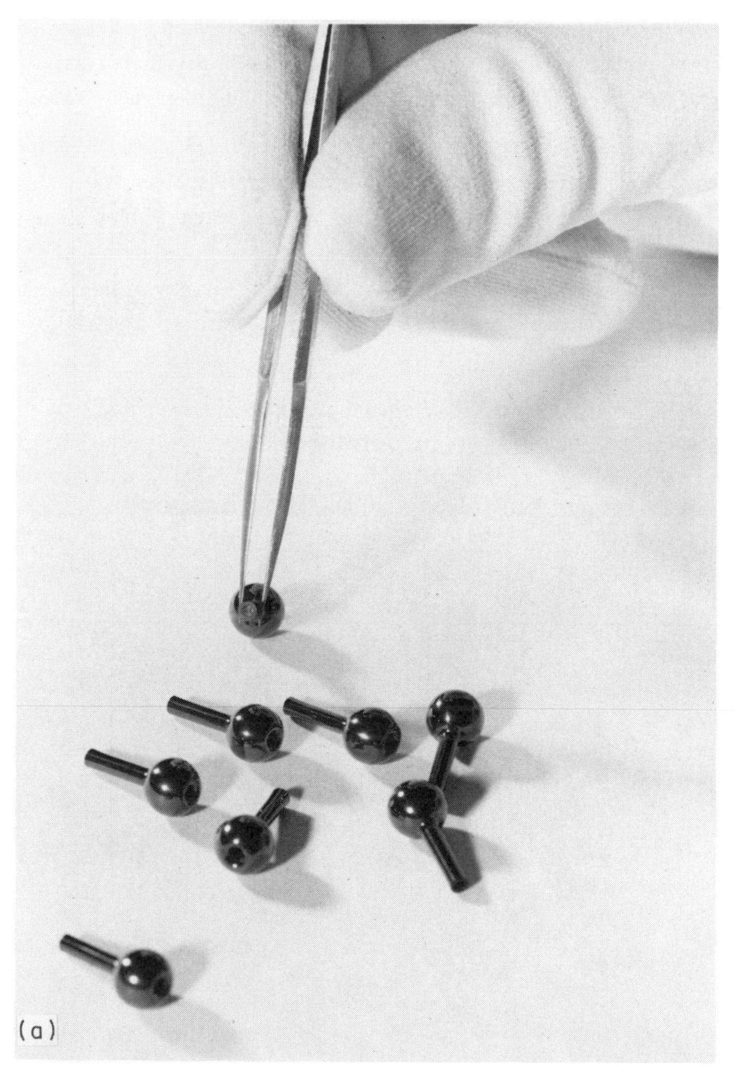

FIG. 33. (a) Pyrolite carbon cannula tips used on cannulae that were inserted through the wall of canine left ventricle to monitor pressure; (b) half-section showing the porous benzalkonium-heparin reservoir in the cannula tip providing for temporary anticoagulation in the stagnant pocket in the cannula tip.

at the rate of 0.5 mm in 3 to 3.5 years, producing 5-μm polyethylene particles. Histology revealed foreign-body giant cells in the synovial lining of these joints [128]. Preliminary tests indicate that metal-to-carbon rubbing surfaces will be more durable than metal-to-polyethylene interfaces and will eliminate problems associated with long-term in-vivo deterioration of the polymer.

D. Other Applications

There are, of course, many potential applications for carbon other than those cited for heart valves, blood pumps, and orthopedics. The carbon cannula tips shown in Fig. 33a are presented to illustrate a method by which heparin can be administered in vivo in a localized area. The cannula tips were fabricated for Boucher [136] for use in experiments designed to monitor the instantaneous pressure within the canine heart. The construction is shown schematically at the bottom of Fig. 33b. Within the LTI Pyrolite carbon tip a porous region of polycrystalline graphite was provided. In practice the porous reservoir is treated with benzalkonium chloride and heparin, using the same procedures that are employed in the preparation of a GBH surface. The insoluble benzalkonium-heparin complex is formed in the pores of the reservoir and, in the subsequent in-vivo situation, the heparin

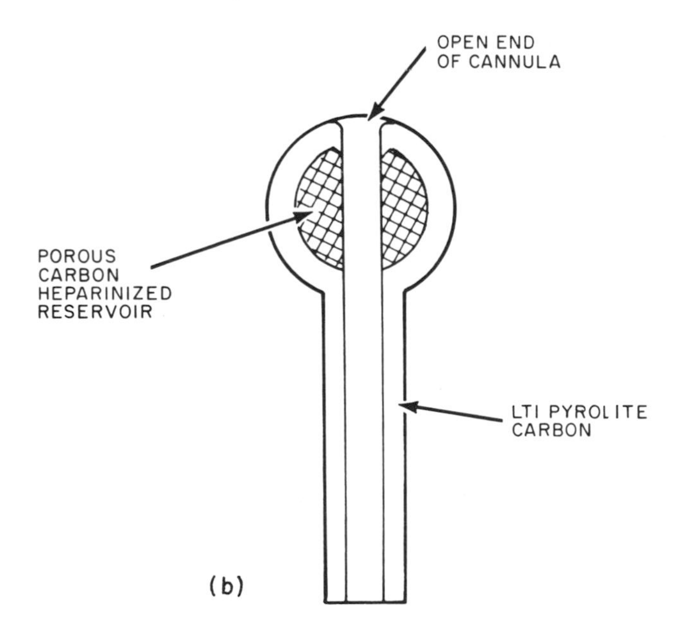

FIG. 33. — continued.

complex is very slowly elutriated, thus providing anticoagulation activity in the stagnant pocket in the cannula tip.

Boucher [136] inserted the cannula tips through the wall of the left ventricle and measured blood pressures for 2-week periods. In the two experiments the cannula tips remained free of thrombus for the period of the experiments. Surprisingly, in the second experiment, the reservoir was not heparinized and the tip remained clean for the full 2-week period. Boucher's experiments further illustrate both the thromboresistance of LTI Pyrolite carbon and a method by which heparin can be administered in vivo at a localized troublesome position, allowing time for the surfaces of a nearby intravascular prosthesis to be "conditioned" by the adsorption of protein from the blood [85, 89-92]. The in-vivo elutriation data in Table 4 of Ref. 41 for the porous Poco AXZ graphite suggest that the reservoir would be useful in vivo for as long as a month.

Lee and co-workers [137] implanted four LTI carbon and three silicon-alloyed LTI carbon test samples of the sort shown in Fig. 34 in rhesus monkeys so that the wide perforated flange was subcutaneous and the small button protruded through the skin. After 7 months, Amromin [138] made a detailed anatomic study of the implant sites. The outstanding feature of all the implants was the thin collagenous wall surrounding them and the paucity of chronic infection. The stratified squamous epithelium in most cases interfaced with collagen, rather than inflammatory or granulation tissue, as has been common with percutaneous implants. Amromin concluded from the thin collagenous wall surrounding the implants and the presence of little to no tissue response in the surrounding dermi that an adequate bacterial seal had formed. Benson [135] and others [139] have reported good results from similar tests of vitreous carbon in the percutaneous situation.

Because of the very encouraging results obtained so far with Pyrolite carbons, an expanded effort is directed toward other applications in bioengineering. Among these are developments of prosthetic knee joints, bone pins, and dental implants. The latter have been implanted for almost a year in dogs and baboons [140] with excellent results. It is anticipated that the new carbons will be instrumental in advancing the science and technology of prosthetics.

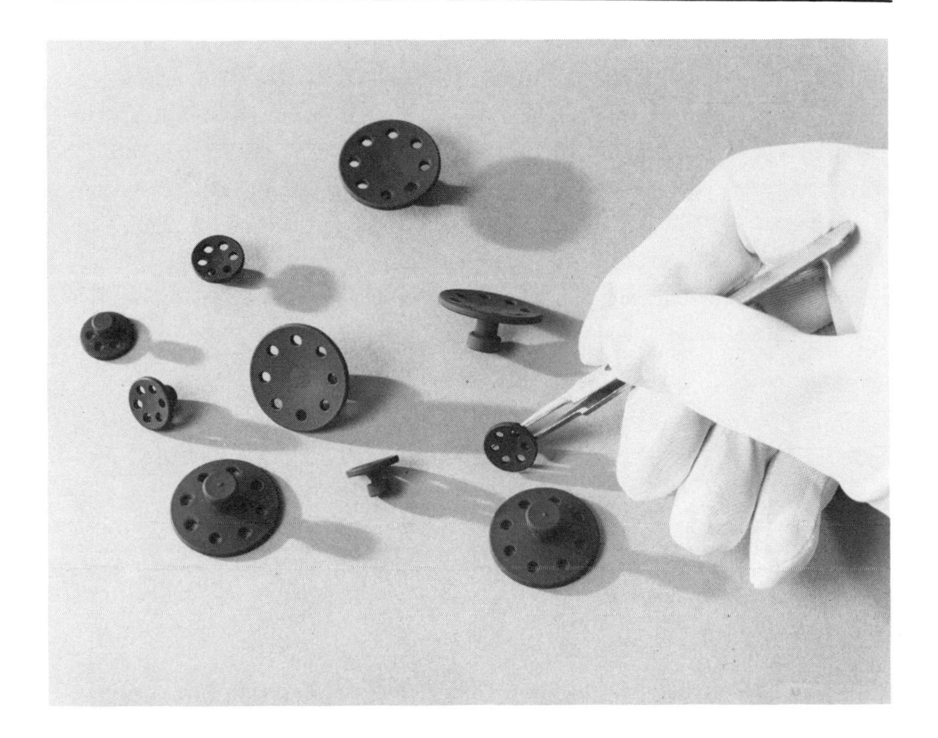

FIG. 34. Through-the-skin Pyrolite carbon test equipment.

ACKNOWLEDGMENTS

We gratefully acknowledge the experimental assistance of W. H. Ellis, F. J. Gagnon, L. J. Noble, and V. Slivenko. Much of the data reported resulted from collaborative efforts. We express our appreciation for the cooperation of Mr. R. G. Averill of Meditec, Inc; Dr. R. E. Baier of the Cornell Aeronautical Laboratories, Inc.; Dr. A. C. Beall of Baylor University; Dr. S. D. Bruck of the National Institutes of Health; Dr. V. O. Björk of the Thoracic Surgical Clinic, Sweden; Mr. F. W. Child of Medical, Inc.; Mr. H. Cromie of Surgitool, Inc.; Dr. M. E. DeBakey of Baylor University; Mr. B. E. Fettel of Shiley Laboratories, Inc.; Dr. M. F. Gillis of the Battelle Northwest Laboratory; Mr. S. H. Goodenough of Cutter Laboratories; Dr. V. L. Gott of the Johns Hopkins University; Dr. S. P. Halbert of Cordis Laboratories; Dr. J. E. Hamner, III, of the National Institutes of Health; Dr. R. I. E. Johnsson-Hegyeli of BMI-Columbus (now with the AHP of NHI); Mr. R. L. Kaster of Cornell University; Dr. H. L.

Lee of the Epoxylite Corporation; Dr. C. W. Lillehei of Cornell University; Mr. M. Kriesel of Medical, Inc.; Dr. R. G. Mason of the University of North Carolina; Dr. O. M. Reed of the Southwest Foundation for Research and Education; and Mr. E. H. Rafferty of Biomedicus, Inc. Parts of the program were coordinated and sponsored by the Artificial Heart Program (Contract PH-43-67-1411), National Heart Institute, National Institutes of Health, under the direction of Dr. F. W. Hastings, Chief, and Dr. L. T. Harmison, Assistant Chief. The scanning electron microscopy was carried out at Applied Space Products, Anaheim, California, by Mr. Roland Marti.

REFERENCES

1. B. E. Warren, J. Chem. Phys., 2, 551 (1934).
2. J. C. Bokros, in Chemistry and Physics of Carbon, Vol. 5 (P. L. Walker, Jr., ed.), Dekker, New York, 1969, pp. 70-81.
3. J. C. Bokros and R. J. Price, Carbon, 3, 503 (1966).
4. J. C. Bokros, Carbon, 3, 201 (1965).
5. J. C. Bokros, in Chemistry and Physics of Carbon, Vol. 5 (P. L. Walker, Jr., ed.), Dekker, New York, 1969, pp. 8-23.
6. J. D. Bernal, Proc. Roy. Soc. (London), A106, 749 (1924).
7. J. Biscoe and B. E. Warren, J. Appl. Phys., 13, 364 (1942).
8. J. L. Kane, J. Nucl. Mater., 38, 42 (1971).
9. A. R. Ubbelohde, Endeavour, 24, 63 (1965).
10. W. P. Eatherly and E. L. Piper, in Nuclear Graphite (R. E. Nightingale, ed.), Academic Press, New York, 1962, Chapter 2.
11. R. W. Cahn and B. Harris, Nature, 221, 132 (1969).
12. J. B. Donnet, Carbon, 6, 161 (1967).
13. H. P. Boehm, Angew. Chem., 78, 617 (1966).
14. H. P. Boehm, E. Diehl, and W. Heck, Rev. Gen. Caoutchouc, 41, 461 (1964).
15. V. L. Gott, D. E. Koepke, R. L. Daggett, W. Zarnstorff, and W. P. Young, Surgery, 50, 382 (1961).
16. V. L. Gott, J. D. Whiffen, and R. C. Dutton, Science, 142, 1297 (1963).

17. J. D. Whiffen, R. C. Dutton, W. P. Young, and V. L. Gott, Surgery, 56, 404 (1964).

18. V. L. Gott, J. D. Whiffen, R. L. Daggett, W. C. Boake, and W. P. Young, Trans. Amer. Soc. Artif. Intern. Organs, 10, 213 (1964).

19. J. D. Whiffen, W. P. Young, and V. L. Gott, J. Thorac. Cardiovasc. Surg., 48, 317 (1964).

20. J. D. Whiffen and V. L. Gott, Proc. Soc. Exptl. Biol. Med., 116, 314 (1964).

21. V. L. Gott, J. D. Whiffen, R. C. Dutton, R. I. Leininger, and W. P. Young, in Biophysical Mechanisms in Vascular Homeostasis and Intravascular Thrombosis (P. N. Sawyer, ed.), Appleton-Century-Crofts, New York, 1965, p. 297.

22. J. D. Whiffen and V. L. Gott, Surg. Gynecol. Obstet., 121, 287 (1965); J. Surg. Res., 5, 51 (1965).

23. J. D. Whiffen and D. C. Beeckler, J. Thorac. Cardiovasc. Surg., 51, 121 (1966).

24. R. I. Leininger, C. W. Cooper, R. D. Falb, and G. A. Grode, Science, 152, 1625 (1966).

25. R. E. Leininger, M. M. Epstein, R. D. Falb, and G. A. Grode, Trans. Amer. Soc. Artif. Intern. Organs, 12, 151 (1966).

26. L. Fourt, A. M. Schwartz, A. Quasius, and R. L. Bowman, Trans. Amer. Soc. Artif. Intern. Organs, 12, 155 (1966).

27. R. D. Falb, G. A. Grode, M. L. Luttinger, and R. I. Leininger, in Proceedings, Polymeric Materials for Biomedical Engineering, 19th Annual Conference on Engineering in Medicine and Biology, 1966, p. 261.

28. E. W. Merrill, E. W. Saltzman, B. J. Lipps, E. R. Gilliland, W. C. Austen, and J. Joison, Trans. Amer. Soc. Artif. Intern. Organs, 12, 139 (1966).

29. E. W. Saltzman, W. G. Austen, B. J. Lipps, E. W. Merrill, E. R. Gilliland, and J. Joison, Surgery, 61, 1 (1967).

30. J. C. Eriksson, G. Gillberg, and H. Lagergren, J. Biomed. Mater. Res., 1, 301 (1967).

31. R. D. Falb, M. T. Takahashi, G. A. Grode, and R. I. Leininger, J. Biomed. Mater. Res., 1, 239 (1967).

32. R. I. Leininger, R. D. Falb, and G. A. Grode, Ann. N.Y. Acad. Sci., 146, 11 (1968).

33. L. W. Hersh, H. H. Weetall, and I. W. Brown, Jr., J. Biomed. Mater. Res., 3, 403 (1969).

34. S. Uy and K. Kammermeyer, J. Biomed. Mater. Res., 3, 587 (1969).

35. G. A. Grode, R. D. Falb, and S. J. Anderson, in Proceedings, Artificial Heart Program Conference, Washington, D.C., June 9-13, 1969, p. 19.

36. R. L. Merker, L. J. Elyash, S. H. Mayhew, and J. Y. C. Wang, in Proceedings, Artificial Heart Program Conference, Washington, D.C., June 9-13, 1969, p. 29.

37. J. C. Bokros and L. D. LaGrange, The Compatibility of Carbon and Blood, Annual Report, Contract PH-43-67-1411, National Heart Institute, Gulf General Atomic Report GA-8770, October 1, 1968.

38. J. C. Bokros, V. L. Gott, L. D. LaGrange, A. M. Fadali, K. D. Vos, and M. D. Ramos, J. Biomed. Mater. Res., 3, 497 (1969).

39. J. C. Bokros, V. L. Gott, L. D. LaGrange, M. M. Fadali, K. D. Vos, and M. D. Ramos, J. Biomed. Mater. Res., 4, 145 (1970).

40. L. D. LaGrange, V. L. Gott, J. C. Bokros, and M. D. Ramos, in Proceedings, Artificial Heart Program Conference, Washington, D.C., June 9-13, 1969, p. 47.

41. J. C. Bokros and L. D. LaGrange, The Compatibility of Carbon and Blood, Annual Report, Contract PH-43-67-1411, National Heart Institute, Gulf General Atomic Report GA-9777, September 20, 1969.

42. P. C. Frei and A. Fleish, Vox Sang., 6, 489 (1961).

43. G. Duc, Vox Sang., 7, 63 (1962).

44. V. T. Mirkovitch, T. Akutsu, and W. J. Kolff, J. Appl. Physiol., 16, 381 (1961).

45. M. Silberberg, Physiol. Rev., 18, 197 (1938).

46. S. E. Moolten, L. Vroman, G. M. S. Vroman, and B. Goodwin, Arch. Intern. Med., 84, 667 (1949).

47. W. V. Sharp, A. F. Finelli, W. H. Falov, and J. W. Ferraro, Circulation, 29 (Suppl.), 165 (1964).

48. J. S. Wright, J. B. Johnston, and L. J. Laber, J. Thorac. Cardiovasc. Surg., 52, 740 (1966).

49. R. S. Kramer, J. S. Vasko, and A. G. Morrow, J. Thorac. Cardiovasc. Surg., 53, 130 (1967).

50. J. S. Wright and J. B. Johnston, Surgery, 60, 1036 (1966).

51. R. A. Indeglia, F. D. Dorman, A. R. Castaneda, R. L. Varco, and E. F. Bernstein, Trans. Amer. Soc. Artif. Intern. Organs, 12, 166 (1966).

52. R. B. Farb and H. M. Kaplan, J. Biomed. Mater. Res., 1, 427 (1967).

53. H. L. Milligan, J. Davis, and K. W. Edmark, J. Biomed. Mater. Res., 2, 51 (1968).

54. H. L. Milligan, J. Davis, and K. W. Edmark, J. Biomed. Mater. Res., 4, 121 (1970).

55. J. C. Bokros, Carbon, 3, 17 (1965).

56. J. D. Currey, Clin. Orthoped. Relat. Res., 73, 210 (1970).

57. S. Yamada, A Review of Glasslike Carbons, DCIC Report 68-2, Battelle Memorial Institute, April 1968; also available as AD668465, U. S. Department of Commerce, Office of Technical Services, Washington, D. C.

58. J. J. Gebhardt and J. M. Berry, Astronaut. Aeronaut., 3, 302 (1965).

59. R. F. Wehrman, Poco Graphite Company, Decater, Texas, private communication.

60. R. J. Akins, J. L. Kaae, R. J. Price, K. Koyama, F. J. Schoen, and J. C. Bokros, Gulf Energy and Environmental Systems Report Gulf-EL-A12250, p. 41 (August 25, 1972).

61. H. L. Leichter and E. Robinson, J. Amer. Ceram. Soc., 53, 197 (1970).

62. J. L. Kaae and T. D. Gulden, J. Amer. Ceram. Soc., 54, 605 (1971).

63. P. J. Hart, F. J. Vasola, and P. L. Walker, Jr., Carbon, 5, 363 (1967).

64. V. R. Usdin and L. Fourt, J. Biomed. Mater. Res., 3, 107 (1969).

65. S. P. Halbert, M. Anken, and A. E. Ushakoff, in Proceedings, Artificial Heart Program Conference, Washington, D. C., June 9-13, 1969, p. 223.

66. S. P. Halbert, M. Anken, and A. E. Ushakoff, Compatibility of Blood with Materials Useful in the Fabrication of Artificial Organs, Annual Report, Contract PH-43-66-980, National Heart and Lung Institute, 1969.

67. T. Akutsu and A. Kantrowitz, J. Biomed. Mater. Res., 1, 33 (1967).

68. H. G. Clark, R. G. Mason, and N. F. Rodman, in Proceedings, Clemson Symposium on the Use of Ceramics in Surgical Implants, Clemson, South Carolina, January 31-February 1, 1969.

69. R. G. Mason, N. F. Rodman, D. E. Scarborough, and I. D. Ikenberry, in Proceedings, Artificial Heart Program Conference, Washington, D.C., June 9-13, 1969, p. 193.

70. R. G. Mason, D. E. Scarborough, S. R. Saba, K. M. Brinkhous, L. D. Ikenberry, J. J. Kearney, and H. G. Clark, J. Biomed. Mater. Res., 3, 615 (1969).

71. R. D. Landell, R. H. Wagner, and K. M. Brinkhous, J. Lab. Clin. Med., 41, 637 (1953).

72. N. F. Rodman, E. M. Barrow, and J. B. Graham, Amer. J. Clin. Pathol., 29, 525 (1958).

73. J. B. Miale, Laboratory Medicine-Hematology, 2nd ed., Mosby, St. Louis, 1962.

74. R. M. Hardisty and R. A. Hutton, Brit. J. Haematol., 12, 764 (1966).

75. G. L. Hody and J. A. Breeden, Amer. J. Med. Technol., 31, 367 (1965).

76. R. J. Hegyeli, R. Gallagher, and L. K. Peterson, in Proceedings, Artificial Heart Program Conference, Washington, D.C., June 9-13, 1969, p. 203

77. R. I. E. Johnsson-Hegyeli and A. F. Hegyeli, J. Biomed. Mater. Res., 3, 115 (1969).

78. R. I. E. Johnsson-Hegyeli and A. F. Hegyeli, Trans. Amer. Soc. Artif. Intern. Organs, 14, 48 (1968).

79. C. J. Pennington, Jr., A. C. Peters, R. A. Gallagher, and L. L. Peterson, Biological Testing of Prosthetic Materials, Final Report, Contract PH-43-67-1404, National Heart Institute, October 31, 1969.

80. D. J. Lyman, W. M. Muir, and I. J. Lee, Trans. Amer. Soc. Artif. Intern. Organs, 11, 301 (1965).

81. I. J. Lee, W. M. Muir, and D. J. Lyman, J. Phys. Chem., 69, 3220 (1965).

82. D. J. Lyman, Chem. Eng. News, 47(4), 37 (1969).

83. D. J. Lyman, J. L. Brash, S. W. Chaikin, K. G. Klein, and M. Carini, Trans. Amer. Soc. Artif. Intern. Organs, 14, 250 (1968).

84. K. B. Bischoff, J. Biomed. Mater. Res., 2, 89 (1968).

85. R. E. Baier, Trans. Amer. Soc. Artif. Intern. Organs, 16, in press.

86. E. G. Shafrin, in Polymer Handbook (J. Brandrup and E. H. Immergut, eds.), Interscience, New York, 1967, pp. 111-113.

87. R. E. Baier and W. A. Zisman, paper presented at American Chemical Society 153rd National Meeting, Miami Beach, Fla., 1967.

88. R. E. Baier and W. A. Zisman, Wettability and MAIR Infra-Red Spectroscopy of Solvent Cast Thin Films of Polyamides and Polypeptides, Naval Research Laboratory Report No. 6755, 1968.

89. R. E. Baier and R. C. Dutton, J. Biomed. Mater. Res., 3, 191 (1969).

90. R. C. Dutton, T. J. Webber, S. A. Johnson, and R. E. Baier, J. Biomed. Mater. Res., 3, 13 (1969).

91. D. E. Scarborough, R. G. Mason, R. G. Dalldorf, and K. M. Brinkhous, Lab. Invest., 20, 164 (1969).

92. L. Vroman and A. L. Adams, J. Biomed. Mater. Res., 3, 43 (1969).

93. L. A. Brewer III (ed.), Prosthetic Heart Valves, Thomas, Springfield, Ill., 1969.

94. Anon., J. Amer. Med. Assoc., 205(3), 28 (1968).

95. R. L. Reis, D. L. Glancy, K. O. O'Brien, S. E. Epstein, and A. G. Morrow, J. Thorac. Cardiovasc. Surg., 59, 84 (1970).

96. W. C. Roberts and A. G. Morrow, Amer. J. Cardiol., 22, 614 (1968).

97. A. Starr, W. R. Pierie, D. A. Raible, M. L. Edwards, G. G. Siposs, and W. D. Handcock, Circulation, 33 (Suppl. I), 115 (1966).

98. E. W. Saltzman, discussion in Surgery, 63, 67 (1968).

99. J. C. Bokros and R. J. Akins, in Proceedings of the 4th International Conference on Materials, Pittsburgh, November 16-18, 1971 (M. C. Shaw, ed.), Carnegie Press, Pittsburgh, 1971.

100. F. J. Schoen, Gulf Energy and Environmental Systems Report GA-10156, June 1, 1970.

101. A. C. Beall, Jr., Presented at the 21st Annual Scientific Sessions of the American College of Cardiology, Chicago, Illinois, March 4, 1972.

102. R. L. Kaster, C. W. Lillehei, and P. J. K. Starek, Trans. Amer. Soc. Artif. Intern. Organs, 16, 233 (1970).

103. R. L. Reis, D. L. Glancy, K. O'Brien, and A. G. Morrow, J. Thorac. Cardiovasc. Surg., 59, 84 (1970).

104. R. D. Jones, discussion in J. Thorac. Cardiovasc. Surg., 59, 116 (1970).

105. C. W. Lillehei, R. L. Kaster, P. J. Starek, J. H. Block, J. R. Rees, and F. J. Schoen, paper presented at the 20th Annual Scientific Session, American College of Cardiology, Washington, D.C., February 5, 1971.

106. V. O. Björk, Scand. J. Thorac. Cardiovasc. Surg., 6, 109 (1972).

107. V. O. Björk, Thorax, 25, 439 (1970).

108. V. O. Björk, Scand. J. Thorac. Cardiovasc. Surg., 4, 15 (1970).

109. V. O. Björk and C. Olin, Scand. J. Thorac. Cardiovasc. Surg., 4, 31 (1970).

110. W. C. Roberts and A. G. Morrow, Arch. Pathol., 82, 164 (1966).

111. B. A. Braniff, N. E. Shumway, and D. C. Harrison, New Eng. J. Med., 276, 1464 (1967).

112. D. E. Detmer, A. G. Morrow, H. H. Marsh III, and N. W. Braunwald, J. Thorac. Cardiovasc. Surg., 60, 46 (1970).

113. R. T. Knapp, Trans. ASME, October 1955, p. 1045.

114. M. S. Plesset and A. T. Ellis, Trans. ASME, October 1955, p. 1055.

115. Ad Hoc Task Force on Cardiac Replacement, Cardiac Replacement, October 1969; available from Superintendent of Documents, U.S. Government Printing Office, Washington, D.C.

116. T. Akutsu, H. Takagi, H. Takano, and C. Farish, in Proceedings, Artificial Heart Program Conference, Washington, D.C., June 9-13, 1969, p. 529.

117. F. Giron, W. Birtwell, and H. Soroff, ibid., p. 541.

118. W. Bernard, C. LaFarge, S. Kitrilakis, and T. Robinson, ibid., p. 559.

119. T. Akutsu, H. Takagi, and C. Farish, ibid., p. 581.

120. F. Dorman, E. F. Bernstein, P. L. Blackshear, R. Sovilj, and D. R. Scott, Trans. Amer. Soc. Artif. Intern. Organs, 15, 441 (1969).

121. E. H. Rafferty, discussion in Trans. Amer. Soc. Artif. Intern. Organs, 15, 460 (1969).

122. M. F. Gillis and P. C. Walkup, in Proceedings, Artificial Heart Program Conference, Washington, D. C., June 9-13, 1969, p. 883.

123. A. B. Ferguson, Jr., in Metals and Engineering in Bones and Joint Surgery, Williams and Wilkins, Baltimore, 1959.

124. B. S. Oppenheimer, E. T. Oppenheimer, I. Danishefsky, and A. P. Stout, Cancer Res., 16, 439 (1956).

125. M. A. R. Freeman, S. A. V. Swanson, and J. C. Heath (H. D. Ritchie), Brit. J. Surg., 56, 701 (1969).

126. J. C. Heath, Nature, 173, 822 (1956).

127. P. G. Laing, in Metals and Engineering in Bone and Joint Surgery, Williams and Wilkins, Baltimore, 1959.

128. J. Charnley, in Reconstruction Surgery and Traumatology, Vol. II, Phiebig, New York, 1969, p. 9.

129. S. F. Hulbert, J. J. Klawitter, C. D. Talbert, and C. T. Fitts, in Research in Dental and Medical Materials, Plenum Press, New York, 1969, p. 19.

130. S. C. Woodward, in Plastics in Surgical Implants, ASTM STP 386, symposium presented by Committee F-4 on Surgical Implant Materials, Indianapolis, Ind., 1964.

131. F. G. Evans and M. Lebow, Amer. J. Surg., 83, 326 (1952).

132. G. H. Bell, in The Biochemistry and Physiology of Bone (G. H. Bourne, ed.), Academic Press, New York, 1956.

133. F. R. Morral, J. Mater., 1, 384 (1966).

134. G. H. Bell and J. B. deV. Weir, Med. Res. Council Memo., No. 22, 85 (1949).

135. J. Benson, Presurvey on Biomedical Applications of Carbons, North American-Rockwell, Rocketdyne Report R-7855, May 9, 1969.

136. J. M. Boucher, U. S. Army Medical Research Institute of Infectious Diseases, Fort Detrick, Frederick, Md., private communication.

137. H. L. Lee, D. E. Ocumpaugh, G. W. Culp, and A. L. Cupples, in Proceedings, Artificial Heart Program Conference, Washington, D. C., June 9-13, 1969, p. 793.

138. G. D. Amromin, Division of Pathology, City of Hope National Medical Center, Duarte, Calif., private communication.

139. R. Kadefors, J. B. Reswick, and R. L. Martin, Med. and Biol. Eng., 8, 129 (1970).

140. J. E. Hamner, III, and O. M. Reed, Private communication, January, 1973.

DEPOSITION OF PYROLYTIC CARBON IN POROUS SOLIDS

W. V. Kotlensky

Super-Temp Company
Santa Fe Springs, California

I. INTRODUCTION

The past decade has seen significant advances in the research and technology of carbon fibers and pyrolytic carbon. Carbon fibers and pyrolytic carbon have evolved during this time period from laboratory research materials into major industries today. Carbon fibers are commercially available today with strengths in the 200,000-psi range and over. Free-standing pyrolytic graphite plates have been deposited in thicknesses of over 1 in. and in closed shapes up to 5 ft in height and 2 ft in diameter. Many thousands of technical reports and technical publications have described their processing, structure, property characteristics, and application performance. The growth of these materials from a laboratory viable stage to a multifaceted application stage has been phenomenal. The current decade will see further significant growth in the use of these materials for a larger variety of applications.

The terms "pyrolytic carbon," "pyrolytic graphite," and "chemically vapor deposited carbon" (CVD carbon) have been widely used by different investigators in the carbon and graphite field. In specific cases these terms can all refer to the same material. "Pyrolytic carbon" and "CVD carbon" are general generic terms relating to the carbon material that is deposited on a substrate by the thermal pyrolysis of a carbon-bearing vapor. The term "CVD carbon" describes the processing used, whereas "pyrolytic carbon" refers more to the type of carbon material that is deposited. "Pyrolytic graphite," on the other hand, is a trade name given to carbon deposited from a hydrocarbon gas over the temperature range 1750 to 2250°C. It is a specific high-temperature form of CVD carbon. It is sometimes also referred to as PG and by some investigators by the trade name of "pyrographite." The usage of these terms is generally a matter of preference, with "pyrolytic carbon" and "CVD carbon" being preferred by most investigators.

The scope of this chapter is not to review the fields of carbon fibers or pyrolytic carbon, nor is the purpose to list the many uses that have been found for these materials. A number of excellent review papers deal with these two classes of materials, and the reader is referred to them. Recently Ezekiel [1] has reviewed the preparation and properties of high-strength, high-modulus graphite fibers. Palmer and Cullis [2] have provided a fundamental background on the phenomenology, kinetics, and mechanism of carbon formation from gases. Bokros [3] has provided a comprehensive overall review of the deposition, structure, and properties of pyrolytic carbon and, in particular, the deposition of pyrolytic carbon in fluid beds. Smith and Leeds [4] have reviewed the manufacturing, properties, and uses of pyrolytic graphite, with emphasis on the latest property data, methods of quality control, and applications in the aerospace and commercial markets.

The scope of this chapter is to review the mechanism, structure, and properties of pyrolytic carbon deposited in porous solids. The porous solids considered include fibrous carbon substrates along with porous carbons and graphites. The process of depositing pyrolytic carbon in porous substrates is generally referred to as infiltration in the industry.

Terms like "CVD carbon infiltration" and "pyrolytic carbon infiltration" have been used to describe the carbon densification processing of porous fibrous and particulate substrates. The difference in depositing pyrolytic graphite by chemical vapor deposition (CVD process) and infiltrating pyrolytic carbon by the CVD process is due to the different processing conditions employed and the substrate. In the case of depositing pyrolytic graphite much higher temperatures are employed, the substrate is usually nonpermeable, and a dense surface coating is produced. The substrate can be removed to give a free-standing pyrolytic graphite structure, or the coating may be left intact and the substrate and pyrolytic graphite used as a coherent body. In the case of infiltration with pyrolytic carbon the porous substrate is the body that is densified. The deposition temperature is much lower. The porous substrate then becomes an integral part of the infiltrated structure.

The processes used for infiltrating porous solids with pyrolytic carbon and for coating particles with pyrolytic carbon in a fluid bed are somewhat similar. The fluid-bed process presents a dynamic surface for pyrolytic carbon deposition, whereas infiltration requires a rigid surface. Compacting particles in a column, heating to a suitable temperature, and passing a hydrocarbon gas through the column is an infiltration process that combines both pyrolytic carbon coating of particles and pyrolytic carbon deposition in the pores of a solid. With appreciable densification, a useful load-bearing structure is produced. The structure and properties of the pyrolytic carbon deposited in these two processes are similar on a macroscopic scale. On a microscopic scale infiltration with pyrolytic carbon leads to a multiphase composite whose properties are a weighted average of the phases present.

II. PHENOMENOLOGY OF PYROLYTIC CARBON DEPOSITION IN POROUS SOLIDS

"Phenomenology" is defined in Webster's New Collegiate Dictionary as a scientific description of actual phenomena with avoidance of all interpretation, explanation, and evaluation. As intended in this section, phenomenology refers to a brief review of the mechanism, kinetics, and growth of pyrolytic carbon as translated to porous solids.

A. Mechanism

Numerous qualitative theories have been proposed to describe the mechanism of pyrolytic carbon formation. Gaydon and Wolfhard [5] present a general description of carbon formation in flames and in the gas phase, with Palmer and Cullis [2] reviewing the more popular theories of carbon formation. Schwind [6] groups the proposed mechanisms of carbon formation into six principal theories. The following is a brief summary of this classification:

1. Condensation theory. Studies by Smith [7] on the effect of pressure on carbon formation in premixed flames suggested that the carbon was formed by the polymerization of C_2 species.

2. Atomic carbon theory. Gaydon and Wolfhard [5] suggest that monatomic carbon is the principal species in the nucleation of carbon particles in premixed flames.

3. C_3 Theory. Cabannes [8] observed C_3 molecules in flames and theorized that C_3 species are the main constituent in the formation of solid carbon.

4. Acetylene theory. Porter [9, 10] and Anderson [11] hold that carbon comes directly from acetylene by simultaneous polymerization and dehydrogenation. These and other investigators [12, 13] have observed that acetylene and diacetylene are the last stable products to appear before carbon formation.

5. Hydrocarbon-polymerization theory. Numerous investigators have detected high-molecular-weight hydrocarbons during the pyrolysis of simple carbon-bearing source gases [12-17]. Grisdale et al [14, 15] have proposed that droplets nucleate in the gas phase, and as these droplets come in contact

with a hot surface, they can, depending on the conditions, dehydrogenate and carbonize in situ, or they can rupture and flow over the hot surface prior to dehydrogenation and solidification.

6. Surface decomposition theory. Film carbon was first produced in controlled experiments for lamp filaments during the late 1800s [18, 19]. The work of Gomer and Meyer [20] and Meyer [21] demonstrated that methane and ethane do not decompose on clean graphite substrates at temperatures of up to 2300° C as long as the mean free path of the molecules exceeds the dimensions of the reaction cell. Tesner [22] studied the competitive processes for forming dispersed carbon particles in the gas phase and forming carbon surface deposits without homogeneous nucleation during the pyrolysis of a hydrocarbon flowing through a heated tube. Near the deposition surface under the flow conditions used, diffusion of active species and condensation on the surface were reported to be more favorable than nucleation and growth of large carbon particles. Earlier work of Iley and Riley [23], Grisdale [15], and Kinney and Walker [16] in depositing carbon by passing a hydrocarbon gas through a heated tube showed that the deposition of a film carbon was favored when the tube was packed with silica and porcelain rods. In these experiments an unpacked tube led to the formation of powdery, low-density, amorphous carbons.

Diefendorf's studies [24-26] along with the thermodynamic work of Duff and Bauer [27] give a systematic picture of pyrolytic carbon formation from low-molecular-weight hydrocarbon gases. Diefendorf's results [28] as applied to pyrolytic carbon infiltration, showing the sum of all species from methane in equilibrium with solid carbon, are shown in Fig. 1. At low temperatures all the gas is methane.

Thermal decomposition at temperatures below nominally 800° C leads to free-radical formation [29] and recombination to more stable hydrocarbons [30, 31]. At 700° C [30] ethane and substituted butadienes were the principal products of the thermal decomposition of substituted cyclohexanes. Short contact times favor the formation of paraffins and olefins, whereas long contact times favor coke formation [31].

At 1700° K the equilibrium concentration of gas-phase species in contact with solid carbon goes through a minimum. At still higher temperatures the carbon content in the gas phase increases as acetylene and acetylenic species

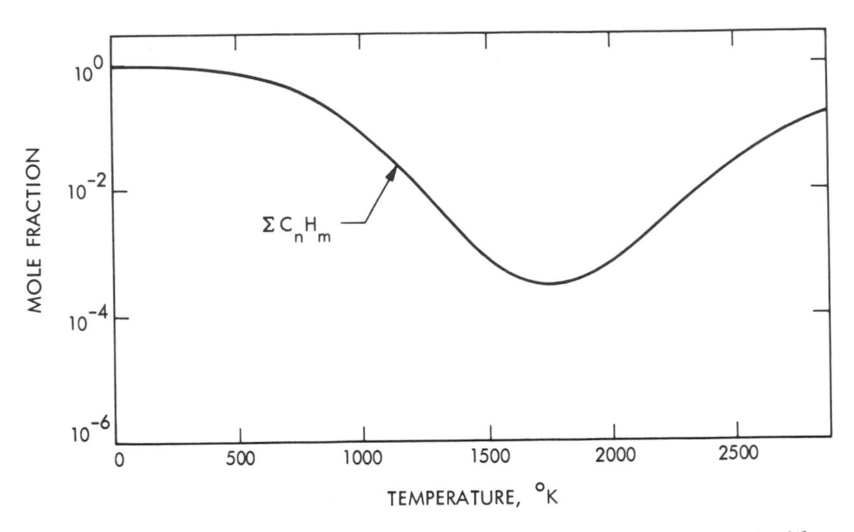

FIG. 1. Equilibrium calculation of gas-phase species in contact with solid carbon at 0.01 atm. From Ref. 28.

become stable. If methane at low pressures is heated to high temperatures very rapidly without decomposing at intermediate temperatures to form high-molecular-weight hydrocarbons, the decomposition products will be essentially acetylene and hydrogen [28]. Figure 2 shows the calculated equilibrium gas-phase hydrocarbon species for a carbon-hydrogen ratio of 1:4, at a pressure of 7.6 torr [26-28].

Thermodynamic calculations can be used to show the equilibrium gas-phase species, but not the structural characteristics of solid carbon. The nature of solid carbon produced from the gas-phase pyrolysis of methane is shown schematically in Fig. 3a. This figure, which is based on the work of a number of investigators, shows that surface-nucleated, dense pyrolytic graphite is formed at high temperatures and low pressures (see also Fig. 3b). Increasing the pressure results in a continuously nucleated structure, then sooty material, and ultimately a low-density soot. At low pressures and temperatures a high-density layered pyrolytic carbon is produced. As the pressure is increased, large species are formed in the gas phase, and these species dehydrogenate to form a laminar aromatic type of pyrolytic carbon [3, 4, 24, 25, 33].

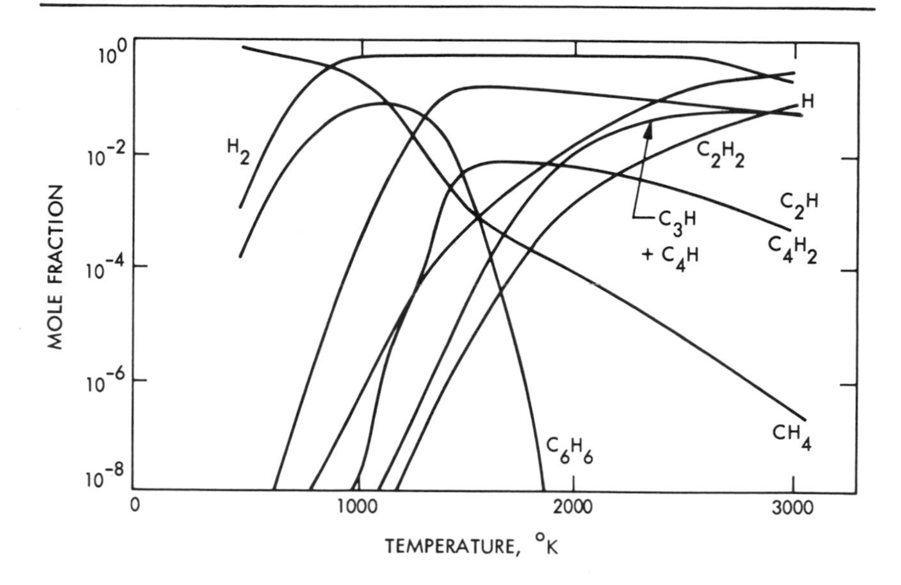

FIG. 2. Calculated equilibrium gas-phase species at 0.01 atm. Carbon-hydrogen ratio is 1:4. From Ref. 28.

B. Kinetics

Numerous investigators have studied the kinetics of the thermal decomposition of methane and other gases to pyrolytic carbon [2, 14, 34-39]. Grisdale, Pfister, and Van Roosbroeck [14] studied the kinetics of pyrolytic carbon deposition on ceramic rods over the temperature range 975 to 1300° C. The rate of decomposition was found to be a first-order one with respect to methane concentration, with an activation energy of 107 kcal/mole. Murphy, Palmer, and Kinney [34] also studied the kinetics by measuring the rate of change of electrical conductance produced by the deposition of pyrolytic carbon on a ceramic rod. These investigators found the rate of pyrolytic carbon deposition to be controlled by a first-order homogeneous reaction and concluded that the rate-determining step appeared to be the rupturing of a C-H bond, since the activation energy of 108 kcal/mole obtained was close to the C-H bond energy.

Hirt and Palmer [35, 36] enumerated the following factors that may affect the kinetics of pyrolytic carbon formation and growth by gas-phase pyrolysis:

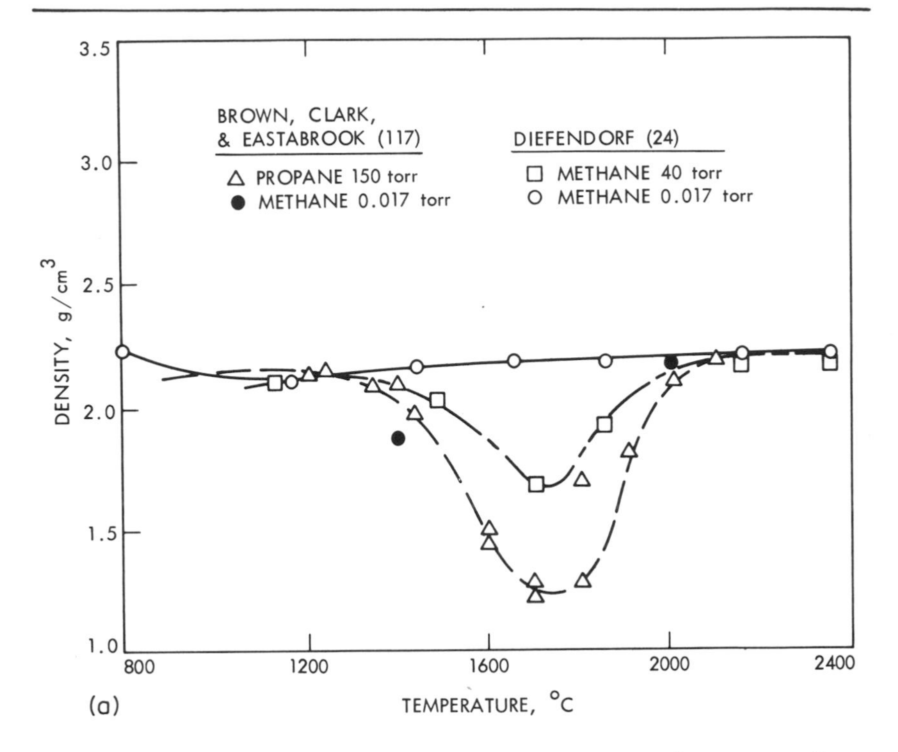

FIG. 3a. Relationship of structure to temperature and pressure. From Ref. 24.

1. Gas-phase decomposition reactions.
2. Gas-phase nucleation of solids.
3. Growth of solid carbonaceous nuclei.
4. Formation of droplets.
5. Carbonization of droplets.
6. Diffusional transport to surfaces.
7. Aerodynamic effects.
8. Decomposition at surfaces.
9. Nucleation on surfaces.
10. Influence of substrates on film growth.
11. Effects of nonisothermal temperature distributions.

Using steady-state conditions and a high surface-to-volume ratio in the reactor, the rates of gas-phase decomposition and decomposition at the

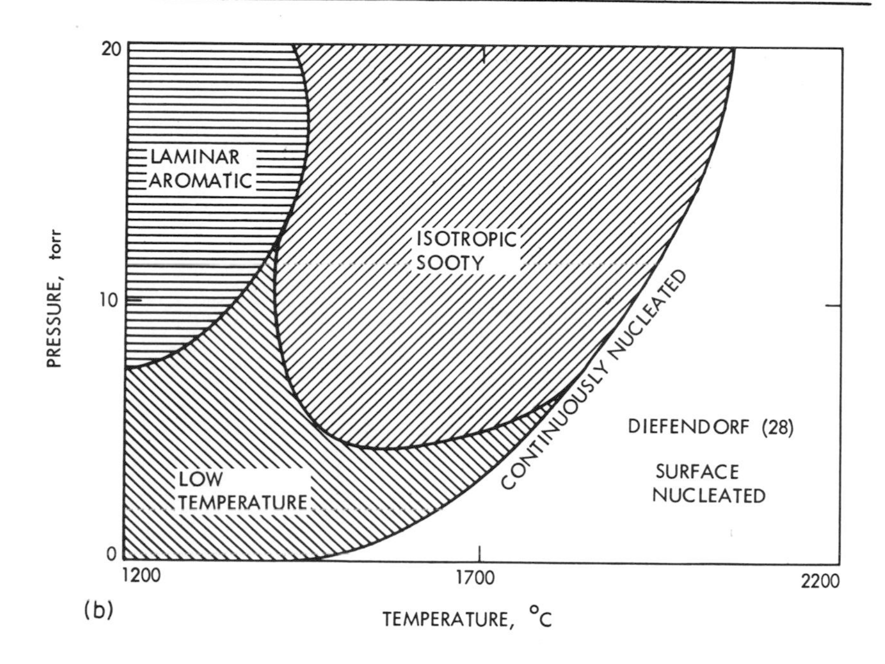

FIG. 3b. Relationship of density to temperature and pressure.

surface were found to be the controlling factors. The average activation energy for carbon formation from carbon suboxide was 59.9 kcal/mole, with the gas-phase reaction being a first-order one in carbon suboxide; from methane, an average value of 103 kcal/mole was found [35]. This work was extended by Palmer [37], who derived kinetic expressions that give carbon-deposition rates for flow systems of different geometries. Figure 4 summarizes the best available rate-constant data for methane decomposition [38]. The best fit through the data gave the following rate expression for the initial step:

$$k_1 = 10^{14.58} \exp(-103 \text{ kcal/RT}) \text{ sec}^{-1}. \tag{1}$$

The frequency factor and activation energy were concluded to be consistent with either or both of the following reactions as the first step in the pyrolysis [38]:

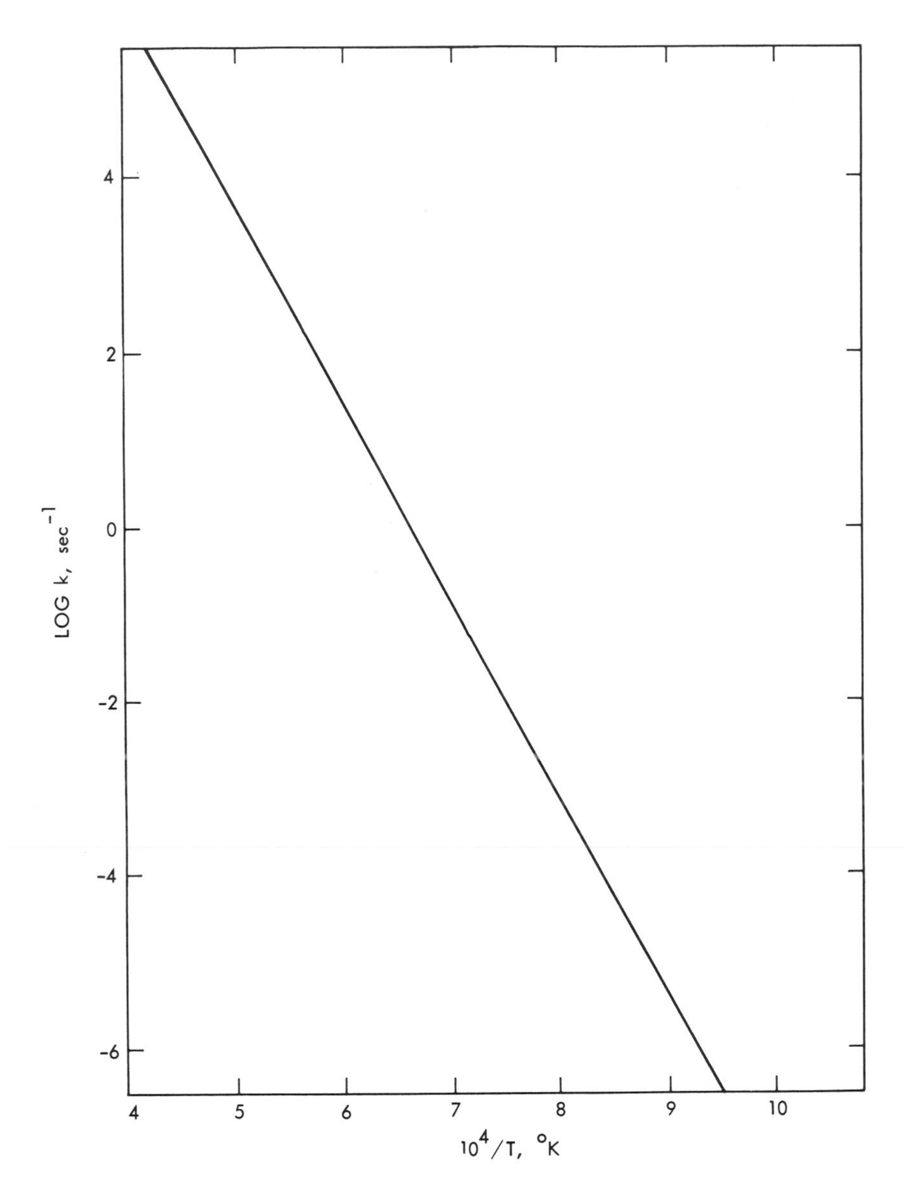

FIG. 4. Rate constants for the decomposition of methane. From Ref. 38.

$$CH_4 \rightarrow CH_2 + H_2$$

$$CH_4 \rightarrow CH_3 + H$$

Eisenberg and Bliss [39] studied the pyrolysis of methane in a flow reactor and reported a first-order reaction for temperatures above 1200° C; in the temperature range 1100 to 1200° C, the reaction was reported to be autocatalytic and not a first-order one. Other kinetic studies have identified C_2H_6, C_2H_4, and C_2H_2 as the most stable intermediate products during the thermal pyrolysis of methane [40, 41]. It was postulated that carbon was produced by the following consecutive first-order reactions:

$$CH_4 \xrightarrow{k_1} C_2H_6 \xrightarrow{k_2} C_2H_4 \xrightarrow{k_3} C_2H_2 \xrightarrow{k_4} C + H_2$$

Reaction k_1 was postulated to be rate controlling at temperatures of up to 1800° K; above this temperature, k_4 became the rate-controlling step.

Hudson et al. [42, 43] have given a generalized description of the theory of carbon formation from the vapor-phase pyrolysis of carbon-bearing gases. It is assumed that the pyrolysis of a carbon-bearing gas gives an active species that can add to carbon particles and that carbon particles containing two or more carbon atoms can grow in the gas phase or deposit on the wall. Active species from the pyrolysis of hydrocarbon gases may be either C_2, C_2H, or C_2H_2, and these propagate particle formation by means of the following reactions:

$$C + C_n \rightarrow C_{n+1}$$

The two treatments describe cases with the active-species concentration remaining constant [42] and varying [43]. The following equations govern C_n in the constant-active-species treatment:

$$n=2 \quad \frac{\partial(C_2)}{\partial t} = D_2\left(\frac{1}{r}\right)\frac{\partial}{\partial r}\left[r\,\frac{\partial(C_2)}{\partial r} + k_1\right](C)^2 - k_2(C)(C_2), \tag{2}$$

$$n=3 \quad \frac{\partial(C_n)}{\partial t} = D_n\left(\frac{1}{r}\right)\frac{\partial}{\partial r}\left[r\,\frac{\partial(C_n)}{\partial r} + k_{n-1}\right](C)(C_{n-1}) - k_n(C)(C_n), \tag{3}$$

where $(C_n) = [C_n(r,t)]$. Equations (2) and (3) are solved for particles reflected from the wall and particles deposited on the wall by using a polynomial expansion. Good agreement between theory and actual deposition-rate results were reported by the authors for high temperatures, high gas flow, and low pressures.

C. Carbon Growth in Pores

Before carbon growth can occur on the walls of a porous solid, carbon source molecules or species must be able to penetrate into the porous structure. Pyrolytic carbon deposition must then occur preferentially within the pores in order to achieve the desired carbon growth. Carbon growth in the pores will not occur or will stop when pyrolytic carbon is deposited on the surface of the porous solid at a much greater rate than within the pores, thereby closing the surface pores, or if bottleneck pores are present within the porous solid and these become closed.

The growth of pyrolytic carbon in the pores of an aggregate was considered by Bickerdike et al. [44] as being analogous "to the oxidation of a block of graphite which occurs throughout the material at a sufficiently low temperature but becomes increasingly concentrated at the outer surface as the temperature is raised." At low temperatures the carbon source gas penetrates the pores by diffusion and deposits pyrolytic carbon on the solid surface. Following the treatment of Thiele [45] for a heterogeneous reaction taking place in a porous catalyst, these authors [44], assuming a first-order reaction and the rate-controlling step being at the solid surface, developed the following expression for the ratio of reactant concentration Y in the pore at depth X to the concentration at the mouth of the pore Y_s at X_s:

$$\frac{Y}{Y_s} = \frac{\cosh h X}{\cosh h X_s}, \qquad (4)$$

where

$$h = C^{1/2} K V,$$

C being the reaction rate in moles/sec-cm^2 of pore surface per mole/cm^3 of reactant, K the diffusion coefficient of reactant, and V the pore cross-sectional area per pore circumference. Calculated values [44] of Y/Y_s for h values between 0.1 and 10 for an open pore 0.6 cm long are shown in Fig. 5. As long as sufficiently low values of h are employed at measurable

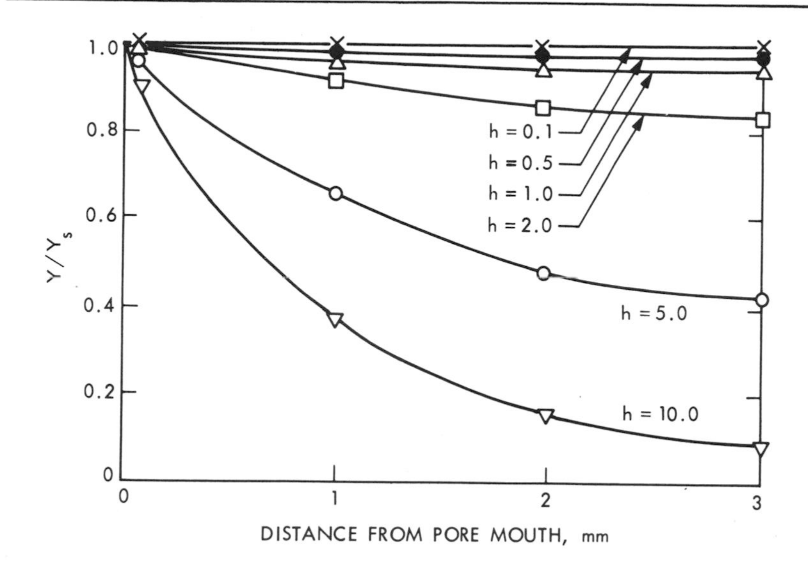

FIG. 5. Ratio Y/Y_s at various distances from mouth of pore for h values between 0.1 and 10. From Ref. 44.

deposition rates, pyrolytic carbon would be expected to grow uniformly along the length of the pore. At high h values pyrolytic carbon would grow more at the pore mouth and in time completely close over the pore.

The work of Vohler et al. [46] also treated the growth of pyrolytic carbon in porous solids in a manner analogous to the treatment of the heterogeneous oxidation of porous graphites with gases. Figure 6 is an Arrhenius plot by Hedden and Wicke [47] describing the general reactions of gases with porous solids. These authors describe the oxidation as follows: In range I, where the temperature is low, the oxidation rate is controlled by the heterogeneous surface chemical reaction on and in the porous solid. With increasing temperature and rate of chemical reaction, as in range II, diffusion through the porous solid influences the oxidation rate. In range III, at high temperatures, the rate constant is independent of temperature and becomes dependent on diffusion through the boundary layer. Vohler et al. [46] consider that pyrolytic carbon will be deposited more preferentially inside the pores the more the experimental conditions correspond to range I of Fig. 6. This is similar to the low h values required in Fig. 5. According

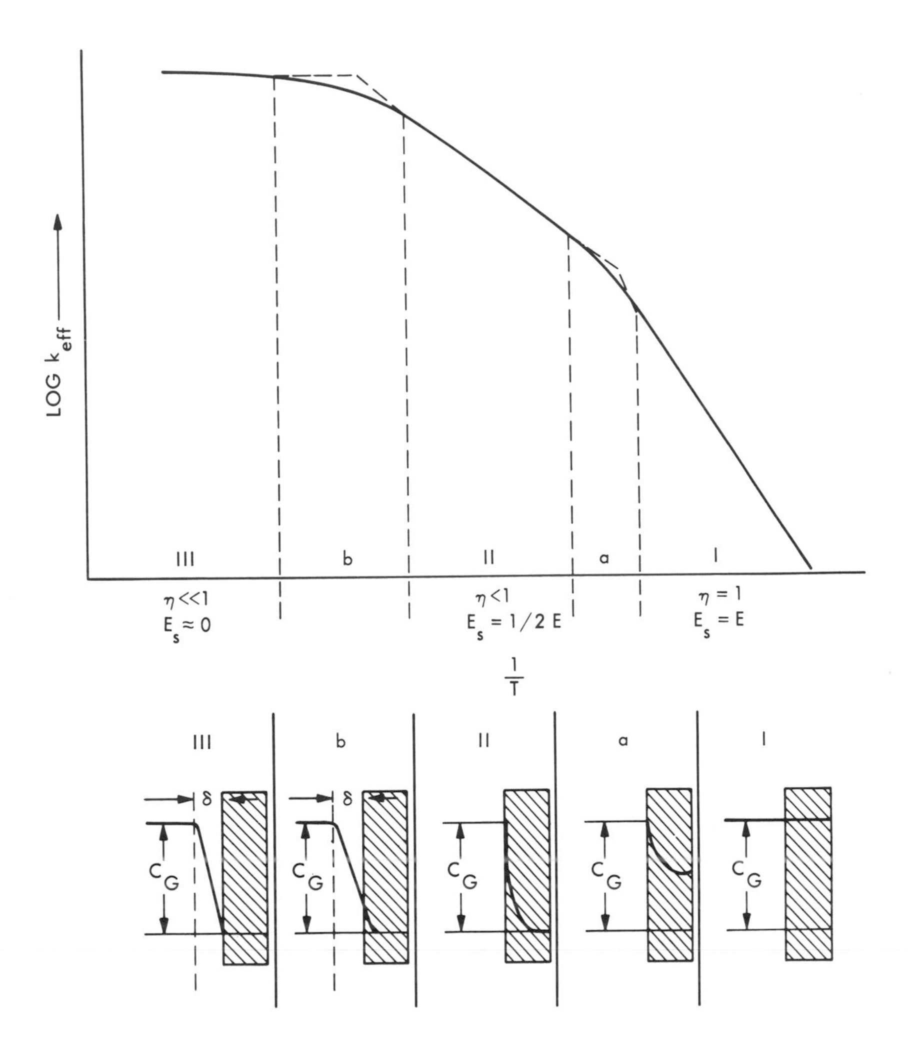

I. THE RATE OF CONVERSION IS DETERMINED ONLY BY THE RATE OF CHEMICAL REACTION

II. THE DIFFUSION THROUGH THE PORES OF THE SOLIDS INFLUENCES THE RATE OF CONVERSION

III. THE CONVERSION RATE IS DETERMINED BY THE DIFFUSION RATE THROUGH A BOUNDARY GAS LAYER

a, b TRANSITION ZONES

FIG. 6. Schematic Arrhenius plot by Hedden and Wicke [47] for reactions of gases with porous solids as applied to infiltration. In region I the rate of conversion is determined only by the rate of chemical reaction; in region II diffusion through the pores of the solid influences the rate of conversion; in region III the conversion rate is determined by the rate of diffusion through a boundary gas layer. The symbols "a" and "b" indicate transition zones.

to these authors [46], carbon growth in the pores will be favored by "low rate of pyrolytic carbon deposition and low hindrance by backdiffusion of the gaseous reaction products. This can be realized by selecting a deposition temperature and a gas pressure as low as possible. Theoretically, the most ideal conditions for pyrolytic carbon deposition in the pores would be given at infinitely small deposition rates. In practice, working conditions that are approximately equal to range a in Fig. 2 [Fig. 6 in this chapter] will be adequate."

A number of investigators [48-50], in studying phenolic ablation materials, observed densification of the char by the deposition of pyrolytic carbon in the pores resulting from the cracking of hydrocarbon gases transpiring through the char.

Weger et al. [51, 52] and Schwind [6], following this work, studied the growth of carbon in the pores of ablative chars at temperatures of up to 3000°K. Growth of carbon in the pores was correlated with the permeability change with time. Using the rate expression for the decomposition of methane after Palmer and Cullis [2] and making a number of assumptions, including the assumption that the carbon-deposition rate is determined by the initial decomposition rate of methane in the gas phase, analytical expressions were developed relating the change in permeability with position and time. Low-temperature results [6] gave larger changes in permeability than those predicted by theory. At temperatures above 2100°K the permeability change with time was less rapid than that predicted by theory. Figure 7 shows a comparison of experimental and theoretical mobilities at 1730°K. Mobility is defined as the ratio of the permeability at any time and position to the initial permeability of the porous material [52].

III. DEPOSITION TECHNIQUES

A. Introduction

Porosity is inevitable, due to the method of manufacturing carbon- and graphite-fiber-reinforced composites as well as polycrystalline graphites.

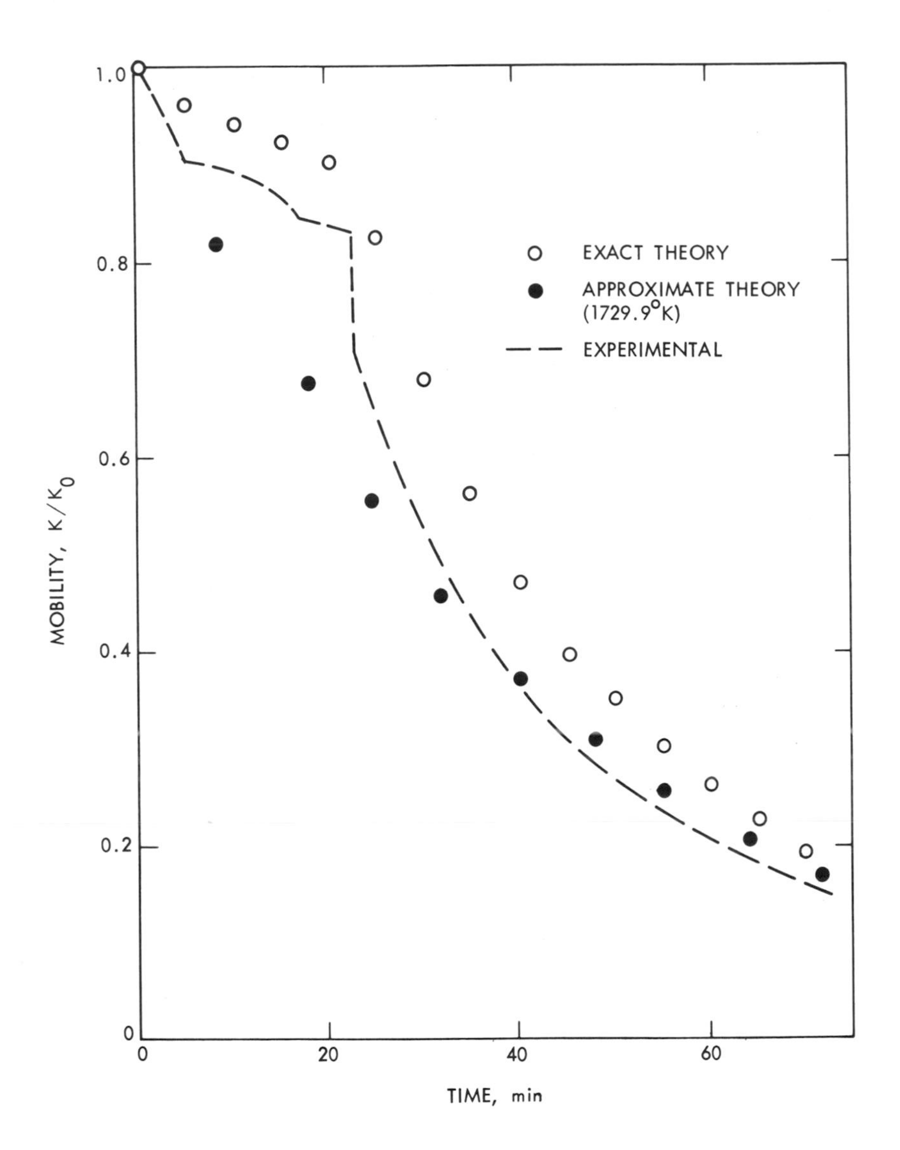

FIG. 7. Mobility of pyrolytic carbon deposited from methane on a porous substrate. Key: ○, exact theory; ●, approximate theory (1729.9°K); the broken curve shows experimental values. From Ref. 6.

Organic resins and coal-tar pitch, which serve as binders in cementing the fibers and coke particles together, may lose up to 50% or more of their weight during pyrolysis to give porous, low-density structures. The density of these bodies can be increased and the gas permeability reduced by impregnating the porous structures with resin and pitch or by infiltration with pyrolytic carbon. A schematic diagram illustrating the pore-filling and pore-blocking mechanism using pitch, resin, and pyrolytic carbon is shown in Fig. 8 [53]. Pitch impregnation has been widely used in the industry for a number of years to improve the density and physical properties of polycrystalline graphites. Resin impregnation can also be used to upgrade the density of porous bodies; however, for high-density bodies, it is difficult to impregnate with resin on other than thin-walled specimens [53]. In both pitch and resin impregnation the pores are initially filled, but during heat treatment volatiles are lost, shrinkage occurs, and a structure with residual porosity is obtained. Multiple impregnations can be used to reduce the residual porosity further.

Infiltration with pyrolytic carbon overcomes a number of the problems associated with pitch and resin impregnation. It is an internal surface-coating process of pores and does not have the drawbacks of loss of volatiles and shrinkage. It has demonstrated significant improvements in physical properties over impregnation techniques [53-55].

Infiltration with pyrolytic carbon is conceptually very simple, though from a fundamental physical and chemical standpoint the process is very complex. The equipment required consists of a metering system for controlling the flow of gases, a substrate that is to be densified, a graphite tube furnace or some means for heating the porous substrate, and throttling valve and vacuum pump for maintaining the desired pressure.

During infiltration a carbon source gas, such as natural gas, methane, or benzene, at a given pressure at room temperature is admitted into the infiltration chamber. Hydrogen and inert gases are also sometimes added. The gases come in contact with hot surfaces within the furnace, are heated and expanded, and diffuse through the porous substrate. Certain active species are formed from the carbon source gas, and as these come in contact with the porous substrate surface, carbon is deposited. As shown in Figs. 5 and 6, deposition rate is very important in controlling the infiltration.

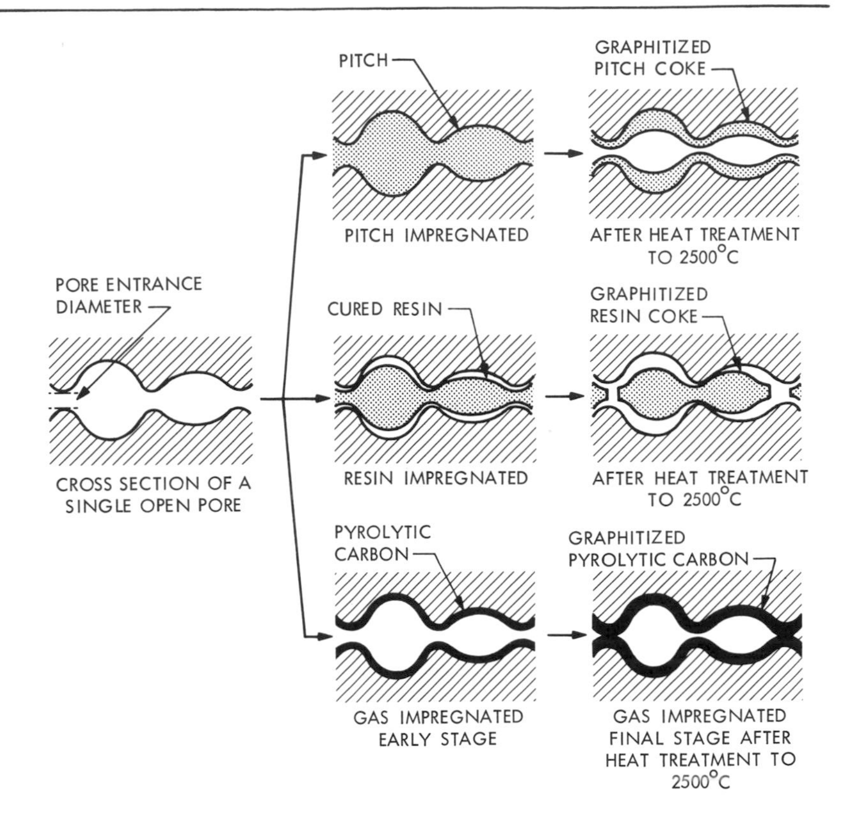

FIG. 8. Schematic illustration of pore-filling and pore-blocking mechanisms. From Ref. 53.

Surface crusting occurs when the pyrolytic-carbon-deposition rate is markedly greater on the exterior surface than on the interior surface. However, by proper selection of temperature, pressure, gas flow, and furnace geometry, the deposition rate on the interior surface can be made to approach that on the exterior surface.

Three principal techniques can be employed in infiltrating porous substrates: isothermal, thermal gradient, and pressure gradient. A combination of these techniques, as well as slight variations in them, is also sometimes employed. A schematic diagram showing the three principal infiltration techniques commonly used as well as impregnation from solution and autoclave impregnation is presented in Fig. 9 [56]. The extent to which porous substrates can be infiltrated is strongly dependent on the compatibility of the starting skeletal structure with the particular processing technique and conditions employed in the infiltration. A description of the principal infiltration techniques is given in this section, which is based on a review of pyrolytic carbon infiltration of porous substrates published by Kotlensky [55]. Additional references to the three principal infiltration techniques, as practiced in the United States, can also be found in two recent technical reports [57, 58].

B. Isothermal Process

The isothermal pyrolytic-carbon-infiltration process is the most widely used of the three infiltration processes for improving the physical properties and reducing the gas permeability of polycrystalline graphites and filamentary carbon composites. A constant temperature is employed. The carbon source gas is passed over the surfaces of the part to be infiltrated. The source gas diffuses into the pores and deposits pyrolytic carbon. Conditions are selected so that the deposition of pyrolytic carbon in the pores is generally carried out in region I described in Fig. 6 and at the low h values cited in Fig. 5. The temperatures employed are below 1400°C, and pressures are kept very low or pressures near atmospheric are used with high dilution of an inert gas, so that the pyrolytic carbon deposited is out of the isotropic, sooty, low-density region defined in Fig. 3a.

1. Graphite Substrates

Pyrolytic carbon infiltration of polycrystalline graphite is principally employed to reduce the gas permeability and oxidation of graphite for nuclear reactors [44, 46, 53]. References 44, 46 and 59 describe some basic experiments conducted in depositing pyrolytic carbon in the pores of graphite.

The experiments of Bickerdike and co-workers [44] were carried out on 1.6-cm-diameter graphite disks, using benzene and methane as the pyrolytic carbon source gases. The graphite disks were 0.16 to 0.64 cm thick, had a density of 1.65 g/cm^3, 15 vol% porosity, and a pore-diameter

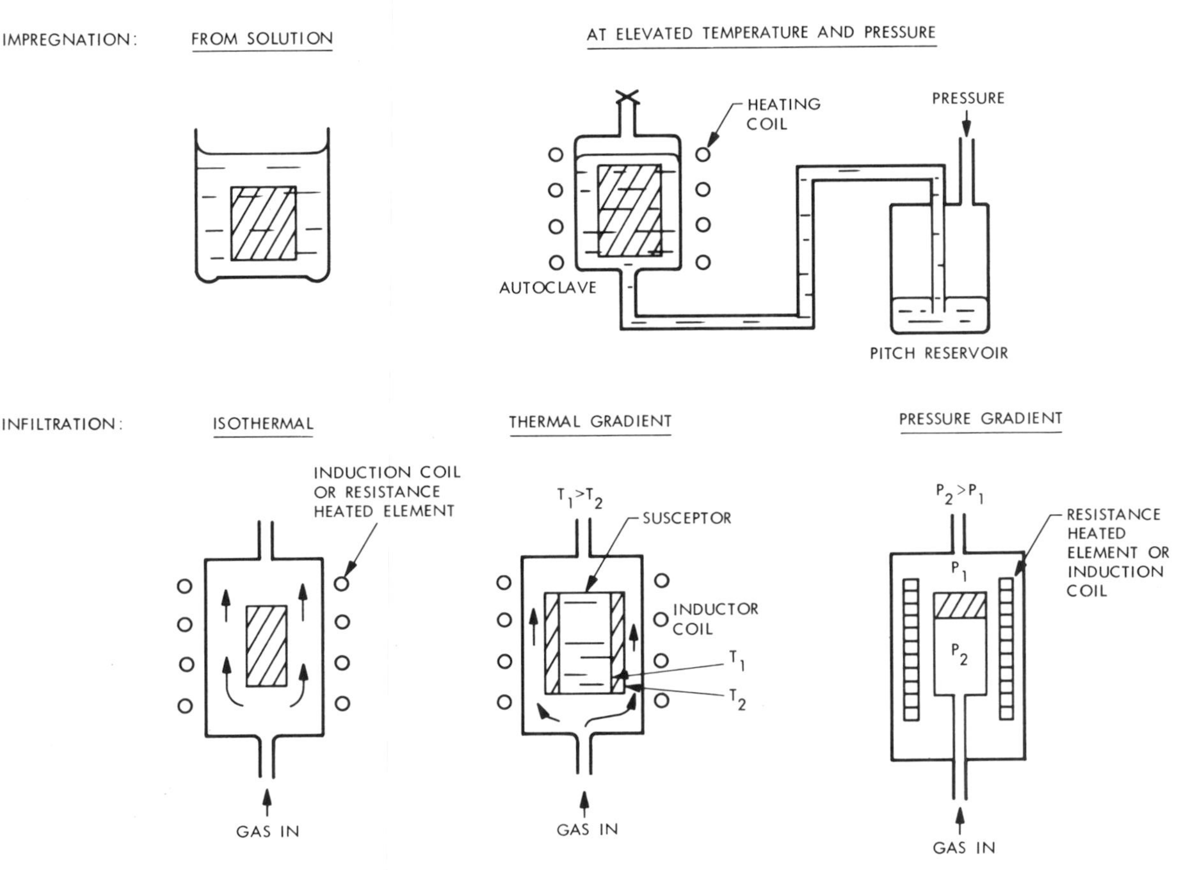

FIG. 9. Impregnation and infiltration techniques. From Ref. 56.

range from 0.3 to 1.0 μm. Using a benzene partial pressure of 80 torr, a weight pickup of 7 to 8% was obtained at 750 °C after 300 h of infiltration; at 780 °C the rate of deposition increased and there was a larger variation of weight pick-up, from 6 to 9%, with disk thickness after 60 h of infiltration; at 820 °C it was possible to increase the weight of the thickest disk by 4%; and at 900 ° C the surfaces of all the disks were sealed with pyrolytic carbon after a few hours. The same trend was found when methane was used in place of benzene, but at temperatures 100° C higher. Small graphite tubes (2.54 cm outside diameter and 1.27 cm inside diameter) were infiltrated at 750° C at 26.5-, 80-, and 210-torr partial pressure of benzene. In all three experiments the weight increase was nominally 8%; however, this was reached after 200 h at the highest partial pressure and after 700 h at the lowest partial pressure.

Vohler, Reiser, and Sperk [46] studied the infiltration of graphite over the temperature range 1100 to 1500° C at pressures of 5, 50, 150, and 300 torr. The graphite used in these experiments had a bulk density of 1.68 g/cm^3 and an open porosity of 15.7 vol%, with 30% of the pores occurring in the 3- to 10-μm diameter range. Methane was used as the pyrolytic carbon source gas. Graphite tubes with a wall thickness of 15 mm were infiltrated for 3.5 h at 5-torr pressure over the temperature range 1100 to 1500° C. At 1200° C the weight pickup in the pores was nominally 1%; at 1500° C most of the pyrolytic carbon was deposited on the surface. The authors explain these results as being in accordance with the transition from range II to range III as described in Fig. 6.

Additional experiments were conducted at 1100° C to investigate the effect of pressure and gas flow on infiltration. Graphite rods (16 mm in diameter) were treated with methane for 17 h at pressures of 50, 150, and 300 torr. The weight pickup was nominally 2% at 50 torr and 4% at 150 torr, with the pyrolytic carbon deposited mostly in the pores. Pickup increased slightly with increase in gas flow. At 300 torr the pyrolytic carbon deposited in the pores was less than that at 150 torr, with approximately one-half pyrolytic carbon deposited on the surface. Infiltration at atmospheric pressure (at 1100° C) required high dilutions with nitrogen to prevent irregular deposition from occurring on the surface.

Isothermal and pressure-gradient infiltration studies on graphite were carried out by Pecik, Makarov, and Tesner [59]. Natural gas was used as the source of pyrolytic carbon, and the infiltration experiments were made at temperatures of 900, 950, and 1000° C. Natural gas with hydrogen dilutions of 0, 15, and 30 vol% was passed into a graphite tube (6 mm internal diameter and 7 mm wall thickness) at an initial flow of 180 ml/min. The graphite tube was sealed to the gas-inlet tube. The graphite used had a density of 1.5 g/cm^3 and a porosity of 25%. The gas residence time was reported to be 0.5 sec. Infiltration was continued until the pressure increased to 1 atm. A weight increase of 13.0% was obtained at 900° C after 58.8 h of infiltration using 30 vol% hydrogen dilution. Without hydrogen dilution, the surface sealed after 9.5 h, showing a weight increase of 8.3%. The same trend with dilution was observed at 950 and 1000° C, but for shorter times to reach atmospheric pressure and corresponding weight increases. The graphite tubes infiltrated at 900° C showed a density gradient through the wall thickness of 1.67 to 1.77 g/cm^3. Larger density gradients were found at higher temperatures.

A unique modification of the isothermal process is the vacuum-pressure pulsing infiltration technique [60]. In this process an inductively heated graphite substrate is cycled between vacuum and 20-psig butadiene for 2 to 30 h at 750 to 950° C. The vacuum and pressure pulse periods used were 0.5 to 1 and 7.5 to 60 sec, respectively. Weight increases of up to 8.0% were achieved in graphite with an original density 1.86 g/cm^3, and helium permeabilities were reduced from about 10^{-2} to less than 10^{-8} cm^2/sec, as required for molten-salt breeder-reactor application.

Application of the infiltration process for improving the properties of graphite has also been reported by other investigators [61-66]. Conventional nuclear graphites [62] were treated in a natural gas atmosphere at 900 to 1000° C, which reduced the open bulk porosity from 15 to about 3%. Bickerdike et al. [66], as well as Carley-Macauly and Mackenzie [67], have also applied the infiltration process for consolidating carbon powders. Table 1 is reproduced from Ref. 67 and describes the results of early consolidation work employing carbon powders and fibers. Carbon-black powders were reported to be densified to 1.8 g/cm^3 and precursor cotton-wool fibers to a density of 1.6 g/cm^3. Permeabilities are seen to be as low as 10^{-10} cm^2/sec.

TABLE 1

Summary of Some Infiltration Experiments [a]

Material	Gas used	Pressure (torr)	Density reached (g/cm^3)	Permeability (cm^2/sec)
Carbon black [b]	Propane	6	1.58	—
	Propane	115	1.80	—
Coke particles [c]	Propane	6	1.78	—
Cotton twill [d]	Propane	6	1.75	10^{-5}
Cotton wool [e]	Methane	6	1.60	10^{-10}

[a] Data from Ref. 67. Infiltration time varied from 4 to 12 h.

[b] Compacted United Carbon Co. Dixitherm M.

[c] Particle size 76 to 150 μm.

[d] Wrapped on element and carbonized.

[e] Wrapped and prebaked.

2. Carbon-Fiber Substrates

Carbon-fiber carbon-matrix composites have attracted wide interest over the past several years for certain load-bearing and high-temperature applications. Carbon fibers are made by the controlled thermal pyrolysis of natural and synthetic organic felts, fibers, yarns, and textiles. A review of their processing is given in Chapter 1 of this volume. Carbon-fiber substrates for infiltration can be used as unidirectional layups, two-dimensional layups, or multidimensional woven structures. The substrates can also be stiffened with resin or pitch and subsequently pyrolyzed prior to densification with pyrolytic carbon. The structural features of carbon-fiber substrates required for effective densification are small pore size, permeable structure, open porosity, and absence of bottleneck pores.

The relationship of initial density and porosity to the theoretical and experimental infiltrated density is shown in Table 2 [55]. The fibers in

TABLE 2

Relation among Initial Bulk Density, Theoretical Infiltrated Density, and Experimental Infiltrated Density[a]

Initial bulk density (g/cm^3)	Initial open porosity (%)	Theoretical infiltrated density (g/cm^3)	Experimental infiltrated density (g/cm^3)
0	100.0	2.1	—
0.2	86.7	2.02	—
0.4	73.3	1.94	1.88
0.6	60.0	1.86	1.79
0.8	46.8	1.78	1.69
1.0	33.3	1.70	1.58
1.2	20.0	1.62	1.48
1.4	6.5	1.54	—

[a] Data from Ref. 55.

this calculation have a density of 1.5 g/cm^3, and it is assumed that there are no closed pores. The theoretical density was calculated on the basis of pyrolytic carbon, with a density 2.1 g/cm^3, completely filling all of the initial open porosity. This is never achieved in practice because of the closing of bottleneck pores and the deposition of a surface coating. Starting with an initial bulk density of 1.2 g/cm^3 and a maximum open porosity of 20%, a theoretical infiltrated density of 1.62 g/cm^3 is calculated. In practice, as shown in the last column, densities of nominally 1.5 g/cm^3 or less are achieved. This table also shows that to reach densities in the range of 1.8 g/cm^3 it is necessary to select a starting density of nominally 0.6 g/cm^3.

The work of Bickerdike et al. [44, 64], following the work of Thiele [45] as shown in Fig. 5, can also be applied to the infiltration of carbon-fiber substrates. Assuming a first-order reaction with no volume change, results obtained on infiltrating graphite could be explained [64] qualitatively in terms of the temperature dependence of a parameter C/D, where C is the rate of reaction per unit area per unit concentration of reactant and D is the gas diffusion coefficient of reactant. For benzene the parameter C was found to be strongly dependent on temperatures over the range 800 to 900°C for pores between 1 and 20 μm. Under the conditions employed the gas diffusion coefficient was reported to be proportional to $T^{1.75}$. Increasing the furnace temperature raised the value of C/D, resulting in a steeper concentration profile of reactant along the length of a pore [65].

In infiltrating carbon-fiber substrates three characteristic flow regions may be considered [68, 69]:

1. Viscous flow. The mean free path is small relative to the pore diameter, and the diffusing species will collide more times with each other than with the pore wall. Kinetic gas theory predicts that the bulk diffusion coefficient is inversely proportional to gas pressure and directly proportional to the 3/2 power of temperature.

2. Slip flow. The mean free path is approximately the same as the pore diameter. The gas flow rate in this regime is higher than that predicted by Poiseuille's law.

3. Molecular flow. The mean free path is many times larger than the pore diameter, and the gaseous species will collide with a pore wall far more often than with each other. The diffusion coefficient should be directly proportional to temperature to the 1/2 power.

Calculated mean free paths for methane and hydrogen over a range of temperatures and pressures are given in Table 3. For typical fabric composites with densities of 1.2 g/cm^3 the average pore size between the fibers within the rovings is 1 to 4 and 10 to 30 μm between rovings. For resin-stiffened and unstiffened low-density felt composites the average pore size would be in the neighborhood of 10 to 50 μm. Considering Fig. 5 and Table 3 along with the three flow regions cited here, the infiltration conditions must be selected to optimize the diffusion of the carbon source gas through the porous substrate and to minimize the deposition of a surface coating.

Isothermal infiltration processing at the Super-Temp Company is generally conducted over a temperature range 950 to 1150°C at pressures from 1 to 150 torr using methane or natural gas [70]. Infiltration continues as a function of time until pores are narrowed to a point where surface overcoating occurs. Infiltration is usually accomplished in multiple cycles, with

TABLE 3

Mean Free Path of Methane and Hydrogen under Various Conditions

Pressure (torr)	Mean free path (μm)					
	20° C		1000° C		1500° C	
	CH_4	H_2	CH_4	H_2	CH_4	H_2
2.5	22	56	148	320	220	460
5	11	28	74	160	110	230
10	5.5	14	36	80	54	120
100	0.55	1.4	3.6	8.0	5.4	11
200	0.27	0.70	1.8	3.9	2.7	5.6

machining in between to clean away surface coatings which block porosity. Optimum times per cycle are usually 60 to 120 h, depending on furnace size. Bulk densities of 1.92 g/cm^3 were obtained for rayon-precursor felt substrates with a starting density of 0.1 g/cm^3, and densities of up to 1.6 g/cm^3 were obtained for wool-precursor substrates with 0.7-g/cm^3 starting densities [57, 71, 72]. Filament-wound cylinders and fabric layup substrates have similarly been infiltrated to densities of up to 1.8 g/cm^3 [73, 74].

The average fiber diameter for carbonized rayon felt is 10 μm; the separation between fibers varies from 10 to 50 μm, with an average separation of 30 μm [75]. For carbonized wool felt the fiber diameter varies from 13 to 20 μm, with an average separation between fibers of 20 μm [76].

C. Thermal-Gradient Process

The thermal-gradient process is similar to the isothermal process in that the infiltration is also diffusion controlled. It differs in that a thermal gradient is established across the thickness of the part to be infiltrated, so that the surface temperature is kept below the threshold pyrolysis temperature of the carbon-bearing gas or at a temperature where the deposition rate at the surface is below the deposition rate of carbon in the pores, thereby minimizing the deposition of a surface coating. Minimizing the deposition of a surface coating can lead to shorter infiltration times required to reach a desired density. Its main drawback, as seen from Fig. 3a, is that, if the process is carried out at atmospheric pressure with the high-temperature side near 1400°C, a sooty, low-density, nongraphitizable pyrolytic carbon could be deposited. The thermal-gradient infiltration process has been utilized mostly with carbon-fiber substrates [58, 75-81].

A thermal gradient is achieved in this process by several means. One surface of the porous material to be infiltrated is in contact with the susceptor and is heated while the other surface radiates to the water-cooled induction coil. The porous substrate is not induction-heated due to its low density and poor coupling. The thermal conductivity and structure of the substrate have a large effect on the gradient. Carbon felt is a better thermal insulator and therefore shows larger gradients than woven substrates made from high-thermal-conductivity fibers. The cooling effect of the high-velocity gas flowing over the substrate surface is the principal means for achieving and controlling the thermal gradient. For nitrogen flow rates greater than 45 standard cubic feet per hour (SCFH) a gradient of 500°C was reported for a susceptor temperature of 1500°C and a gradient of 570°C for the susceptor at 1650°C [79]. The graphite susceptor was in the shape of a hexagon, 10 in. across the diagonal and 12 in. high. Figure 10 shows a correlation between sooting temperature and gas velocity, and Fig. 11 shows deposition profiles obtained in carbon felt in the same study [79]. Deposition rates are strongly dependent on temperature. A plot of deposition temperature against the time required to infiltrate a 0.5-in.-thick carbon felt to a density of 1.8 g/cm^3 is shown in Fig. 12 [77]. Under the conditions employed, no surface crusting was reported for temperatures of up to 1500°C.

As already seen from Figs. 3a and 3b, pressure and temperature have a marked effect on the structure and density of the deposited pyrolytic carbon. At temperatures of 1000 and 2100°C the density is 2.1 and 2.2 g/cm^3, respectively. At 1700°C and high pressure, where isotropic sooty deposits are formed, the density may be as low as 1.2 g/cm^3. The deposition rate for depositing pyrolytic graphite at 2100°C and 10-torr pressure varies from 5 to 10 mils/h [4]. At 1000°C the deposition rate is typically 0.01 mils/h for the isothermal process [55]. In the thermal-gradient process a deposition-rate gradient from 0.1 to 0.01 mils across the thickness was reported for a sample 0.5 in. thick [79]. Infiltrated densities attainable by the two processes, however, are similar.

D. Pressure-Gradient Process

In the pressure-gradient process the part to be infiltrated is cemented to a mandrel attached to the source-gas feed tube [82]; alternatively the part can be sealed directly to the feed tube [59]. A virtually gastight assembly is required, and the differential pressure that is established serves as the

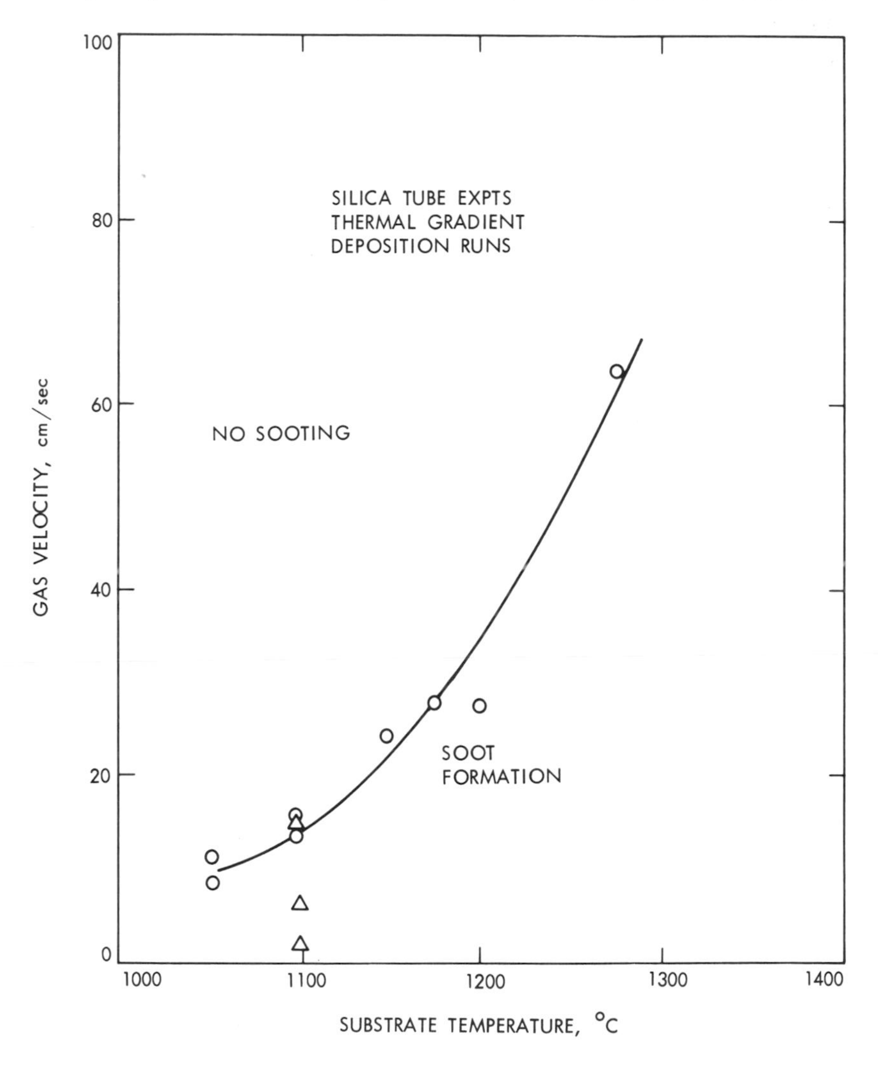

FIG. 10. Correlation between sooting temperature and gas velocity for thermal-gradient infiltration. From Ref. 79.

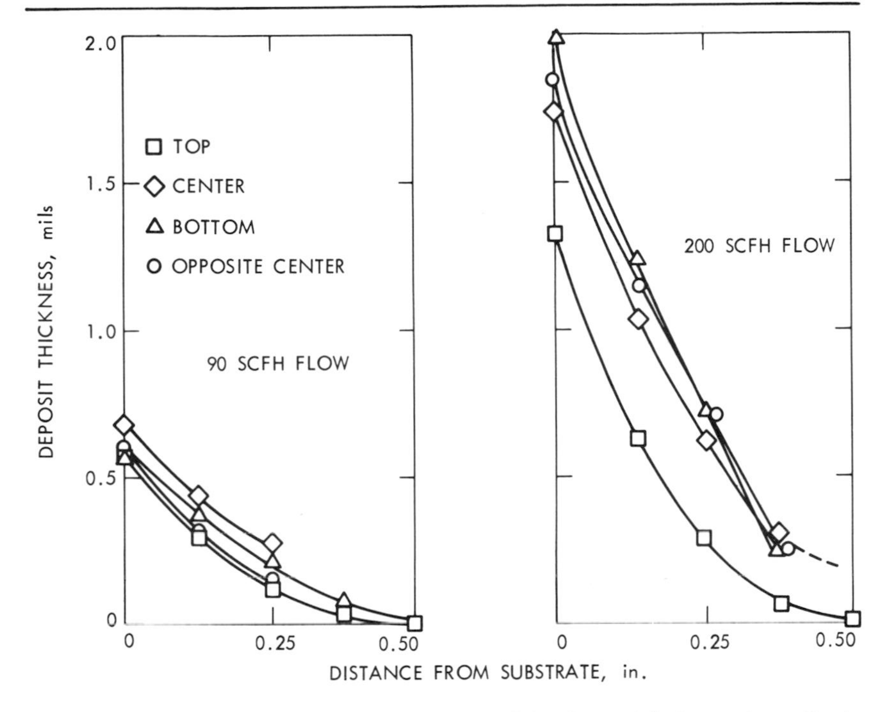

FIG. 11. Thickness profiles in carbon felt after 24-h thermal-gradient infiltration at 1 atm and 90 and 200 standard cubic feet per hour gas flow. Key: □, top; ◇, center; △, bottom; ○, opposite center. From Ref. 79.

driving force for infiltration. This process has not gained wide acceptance because of the obvious difficulties that are encountered. Vohler et al. [46] have stated that pyrolytic carbon deposition in the pores "can be substantially improved by using a pressure gradient forcing the carbonaceous gas (methane or a similar gas) through the pore system. However, a process using forced gas-flow is restricted in its application, not easy to carry out technically, and relatively difficult to reproduce."

The work of Pecik, Makarov, and Tesner [59] with the pressure-gradient process has been described in Section III. B. 1. Graphite with a density of 1.5 g/cm^3 and a porosity of 25% was infiltrated to a density of 1.7 g/cm^3 in typically 60 h. Rayon- and wool-precursor felt substrates have been infiltrated to densities of nominally 1.7 g/cm^3 [54]. For fabric layup substrates stiffened with pyrolyzed resin and having starting densities of about 1.0 g/cm^3 the upper density limit after infiltration appears to be 1.5 g/cm^3 [82]. Closing of bottleneck pores and the deposition of a surface coating limited the extent of infiltration.

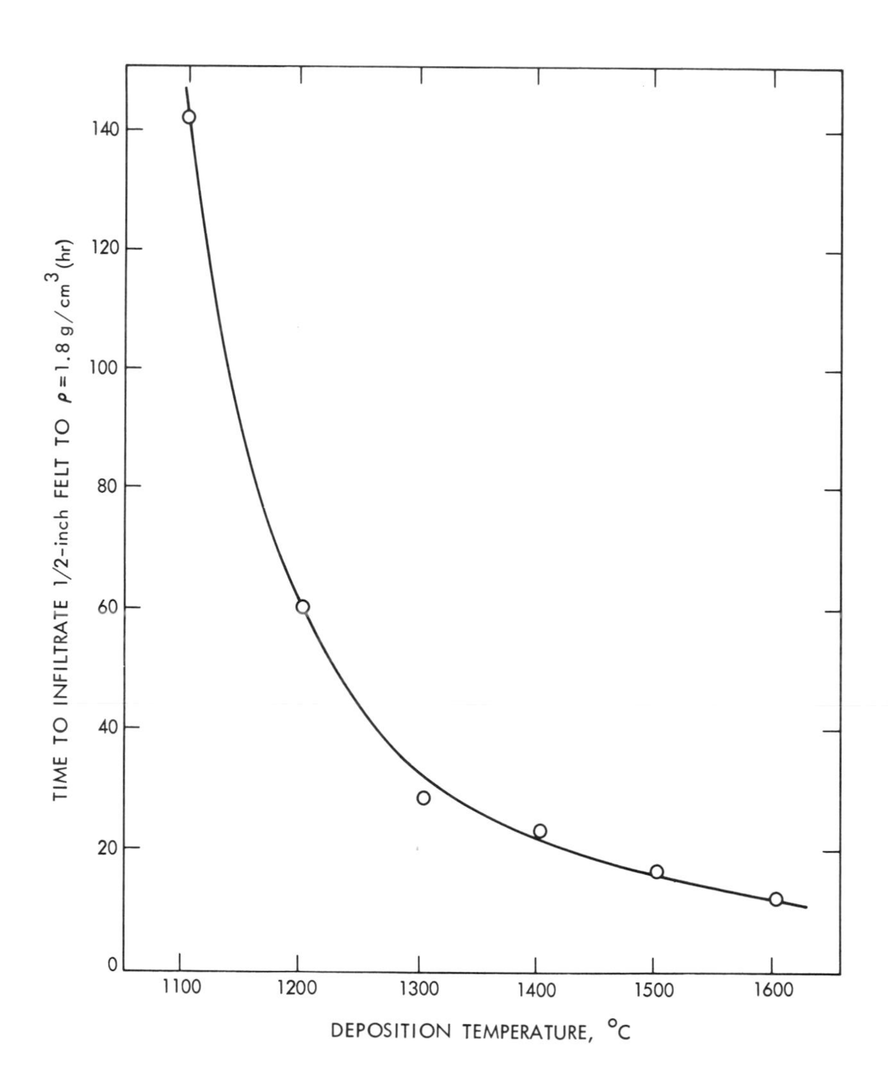

FIG. 12. Dependence of deposition rate on deposition temperature. From Ref. 77.

E. Other Processes

An additional infiltration process that is particularly suitable for densifying particles in a packed bed is the vapor-consolidation technique [83, 84]. This technique combines the characteristics of the thermal-gradient and pressure-gradient processes. Pfeifer and co-workers [84] compressed the packed bed of coke particles and chopped fibers between layers of coke or graphite particles. In a typical consolidation run at 1100°C the reactant gas is passed through the packed bed and the flow is periodically reversed to promote uniform deposition through the substrate. As consolidation occurs and the available porosity decreases, the pressure drop across the packed bed increases. The extent of porosity desired in the final structure is obtained from established porosity-permeability relationships. Typical infiltrated densities of 1.8 g/cm^3 have been reported [84].

A combination of the pressure-gradient and thermal-gradient process has been described by Papalegis and Bourdeau [85] for infiltrating felt and fabric layered substrates. Using methane as the carbon source gas, 0.25-in.-thick layers were densified at 1700°C for 6 h. The felt and fabric substrates increased in density from 0.075 to 1 and from 0.5 to 1 g/cm^3, respectively.

One additional method that could have merit in infiltrating porous substrates is worthwhile mentioning. Studies by Jenkins and Medwell [86] in striking an ac arc between a copper and an electrographite electrode in light contact in transformer oil produced pyrolytic carbon grown on nuclei of amorphous carbon. Pyrolytic carbon was reported only at the interface, and details were not given whether or not any in-depth deposition of pyrolytic carbon was achieved.

IV. RELATION BETWEEN STRUCTURE AND PROCESSING

A. Effect of Temperature and Pressure on Microstructure

Bokros has provided a comprehensive treatise [3] on the deposition, structure, and properties of pyrolytic carbon, with emphasis on the fluid-bed processing technique. Table 4, reproduced from Bokros' work, summarizes the conditions favoring the deposition of three characteristic pyrolytic carbon structures observed in fluid beds. Deposition of pyrolytic carbon on a rigid

TABLE 4

Summary of Conditions Favoring Deposition of Laminar, Granular, and Isotropic Carbon Structures in Fluid Beds [a]

Microstructure	Deposited when	Favored by
Low-temperature laminar	Planar complexes form in the gas and deposit directly on the particle surfaces	Low temperatures, intermediate to high hydrocarbon concentrations, and large bed surface areas
Isotropic	Supersaturation occurs in the gas; gas-borne particles form and are incorporated into the carbon deposit	Long contact times, small bed surface areas, high hydrocarbon concentrations, and low to intermediate temperatures
Granular and columnar	Conditions favor orderly crystal growth	High temperatures, low hydrocarbon partial pressures, and small bed surface areas

[a] From Ref. 3.

substrate can also lead to these three characteristic structures, as seen in Fig. 3a. In the fluid-bed process at low temperatures (900 to 1400°C) and low hydrocarbon concentrations, characteristic growth features are observed and fairly dense deposits are produced. The density of the pyrolytic carbon decreases and the growth features disappear as the hydrocarbon partial pressure is increased at a constant temperature within the low-temperature range. At intermediate temperatures (1400 to 1900°C) and at a hydrocarbon partial pressure favoring gas-phase particle growth, the deposited pyrolytic carbon is isotropic in structure, showing no characteristic growth features. High temperatures, low hydrocarbon partial pressure, and low fluid-bed surface areas favor the deposition of granular and columnar pyrolytic carbon. In the infiltration processing low furnace pressure is analogous to the low hydrocarbon partial pressure in the fluid-bed process, and the characteristic pyrolytic carbon structures as a function of temperature and pressure are found to be similar.

The effect of infiltration temperature and pressure on pyrolytic carbon structure has been investigated in detail at the Sandia Laboratories. A brief summary of their results is presented to illustrate the characteristics of the pyrolytic carbon deposited.

Microstructures of pyrolytic-carbon-infiltrated felt substances at 1100 and 2000°C are shown in Figs. 13 and 14 [75]. At 1100°C the pyrolytic carbon microstructure is seen to be laminar and to coat the fibers uniformly through the thickness. At 2000°C the characteristic columnar growth cone structure is seen at the surface of the felt substrate, with less indepth coating of the carbon fibers. The change in microstructure appearance with infiltration over the temperature range 1100 to 1600°C is shown in Fig. 15 [77]. Infiltration was carried out by the thermal-gradient process, using a 1-atm pressure, a low methane partial pressure, and a felt substrate. At 1300°C and below the microstructure is layered or laminar. Circumferential delaminations occur at 1200°C and become more pronounced at 1300°C, with many delaminations extending all around the fiber. At 1400°C and above the microstructure becomes granular. No delaminations are observed; however, radial cracks appear and increase in number with temperature.

The effect of pressure on the microstructure of pyrolytic carbon deposited on a felt substrate at 1100°C is shown in Fig. 16 [58]. The rough laminar pyrolytic carbon structure deposited at 35 torr is free of microcracks. The smooth laminar pyrolytic carbon structure deposited at 100 and 760 torr is seen to contain numerous circumferential delaminations. Processing conditions determine the structure that is deposited and similarly influence the graphitizability, thermal conductivity, strength, and electrical resistivity of the carbon-carbon composite [58].

Infiltration of graphite substrates is more sensitive to temperature, pressure, and gas velocity than infiltration of fibrous substrates because of the finer pore network. The work of Vohler and co-workers [46] is cited here to illustrate this relationship. Figures 17 and 18 are reproduced from the published work of these authors. The graphite used in this work had a bulk density of 1.68 g/cm^3 and an open porosity of 15.7%. From Fig. 17 it is seen that at an infiltration pressure of 5 torr pyrolytic carbon is increasingly deposited on the surface as the temperature is raised. The microstructure of the coating at this low pressure is columnar. At a constant temperature the growth of pyrolytic carbon in the pores is affected by both pressure and gas velocity. Figure 18 shows in-depth growth of pyrolytic carbon in the pores at a pressure of 150 torr and a methane velocity of 8000 cm^3/h. The microstructure has the laminar appearance.

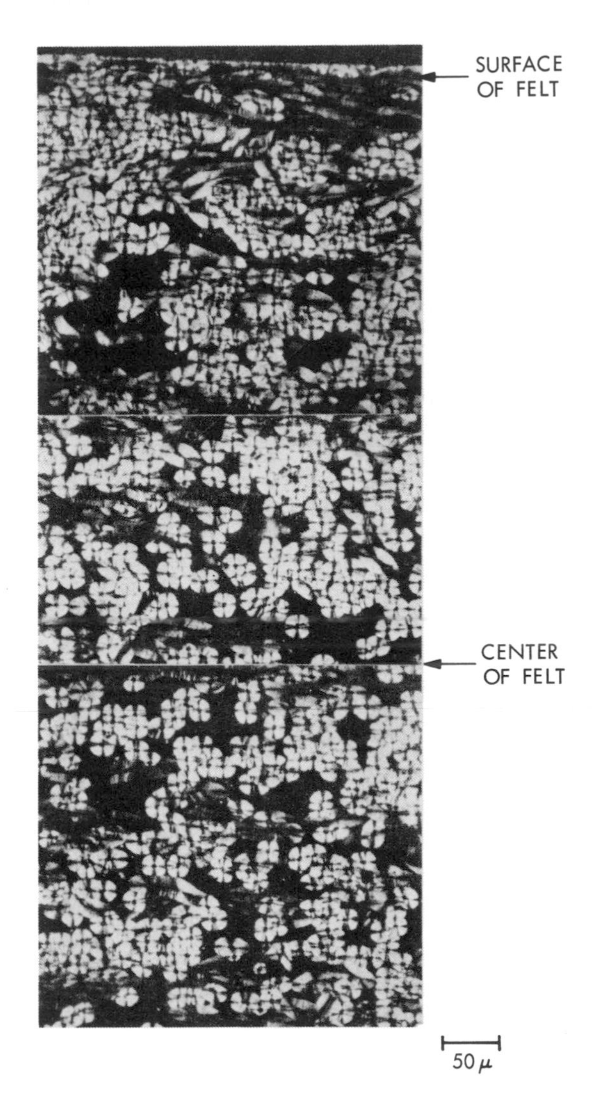

FIG. 13. Microstructure of infiltrated carbon felt at 1100°C, showing in-depth coating of fibers. From Ref. 75.

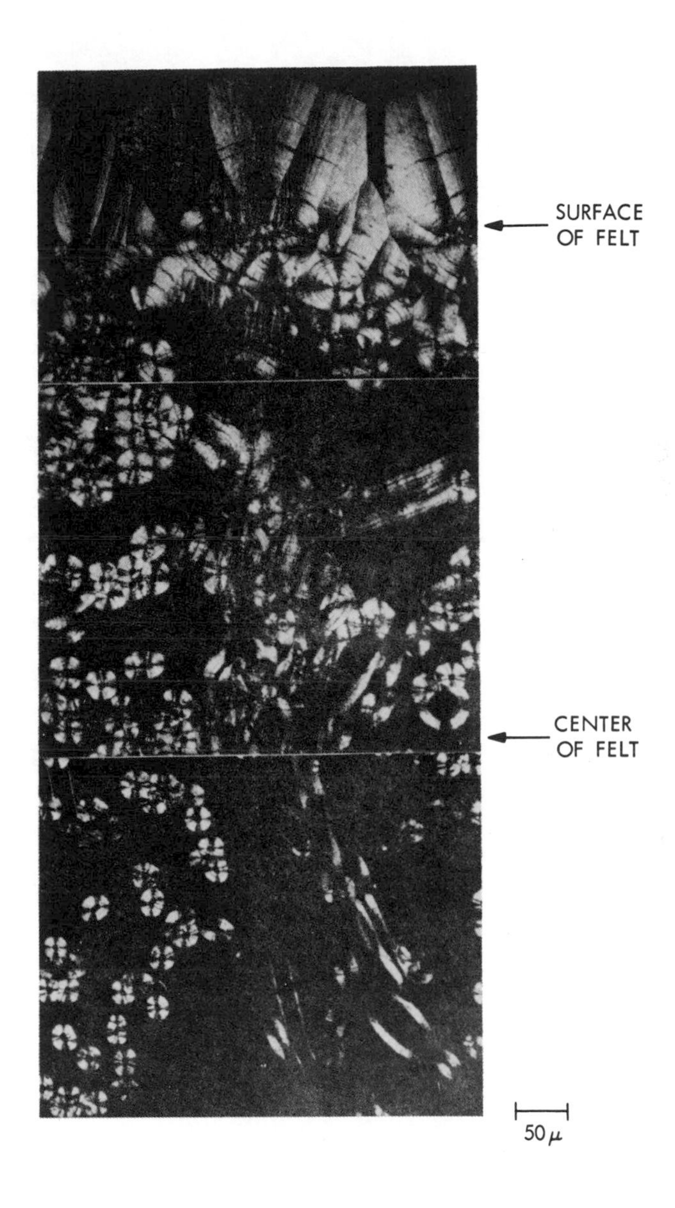

FIG. 14. Microstructure of infiltrated carbon felt at 2000° C, showing dense surface coating. From Ref. 75.

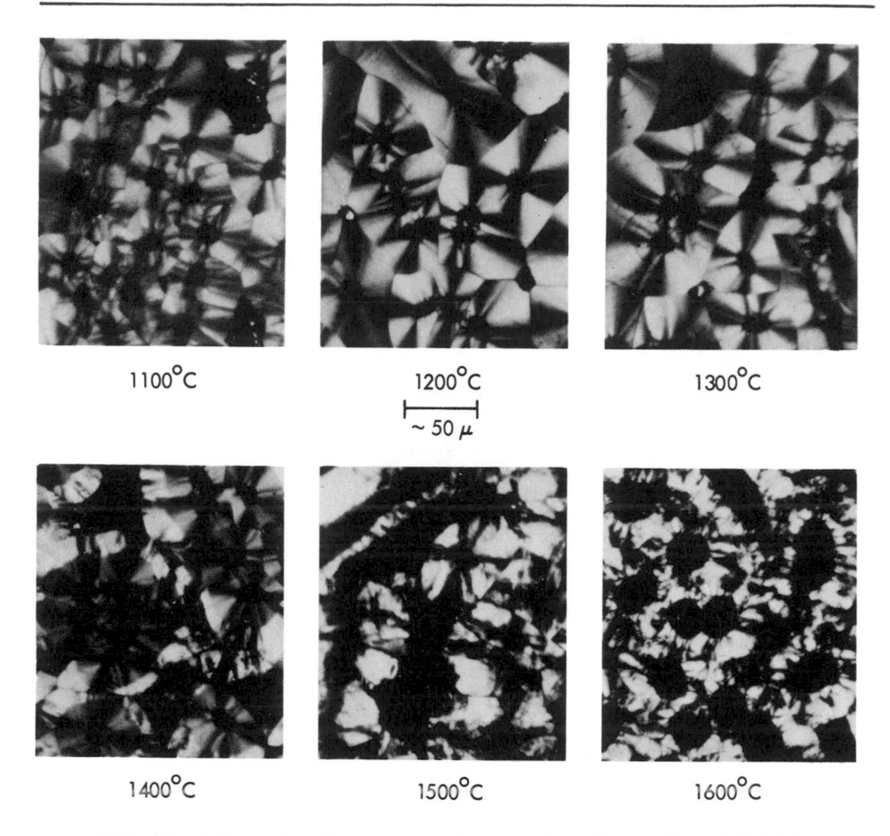

FIG. 15. Microstructure versus temperature for carbon felt infiltrated by the thermal-gradient process. From Ref. 77.

Figure 19 [84] shows the rough laminar appearance of pyrolytic carbon deposited at 1100° C by the vapor-consolidation process. Good bonding is seen between the particles and pyrolytic carbon, with no visual evidence of delaminations.

These and many other investigators working on pyrolytic carbon infiltration have identified these characteristic microstructures [44, 46, 55-61, 64, 66, 70-90].

B. Effect of Initial Density and Porosity

The relationship of the initial density and porosity of fibrous substrates to the theoretical and infiltrated density has been described in Section III. B. 2.

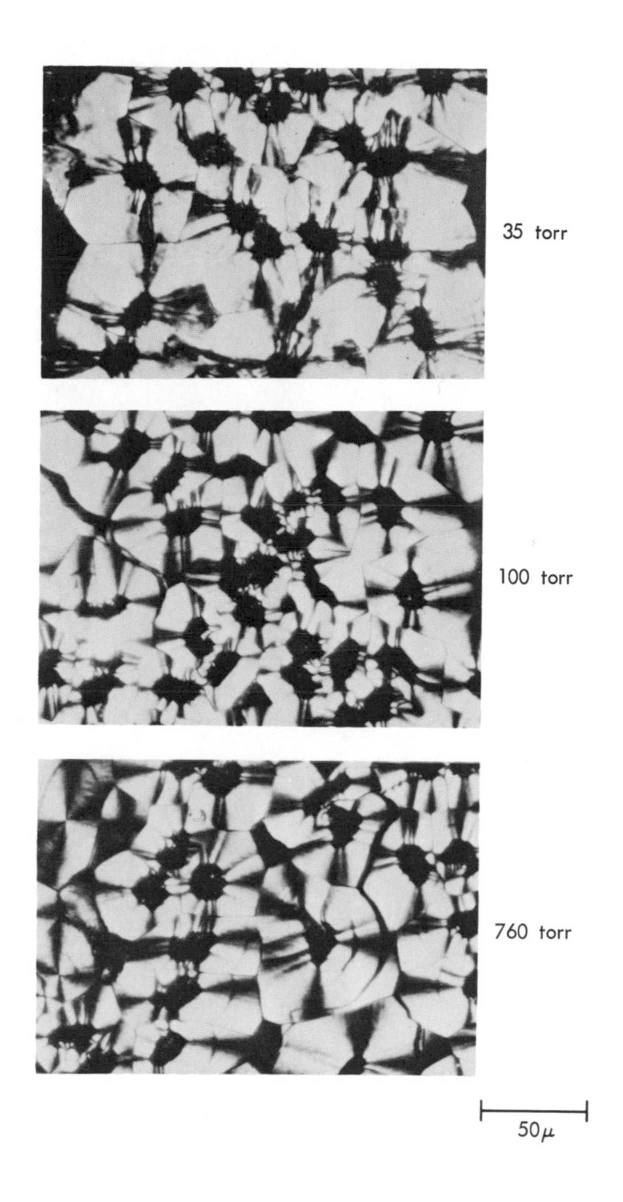

FIG. 16. Microstructure versus pressure for carbon felt infiltrated at 1100° C by the thermal gradient process. From Ref. 58.

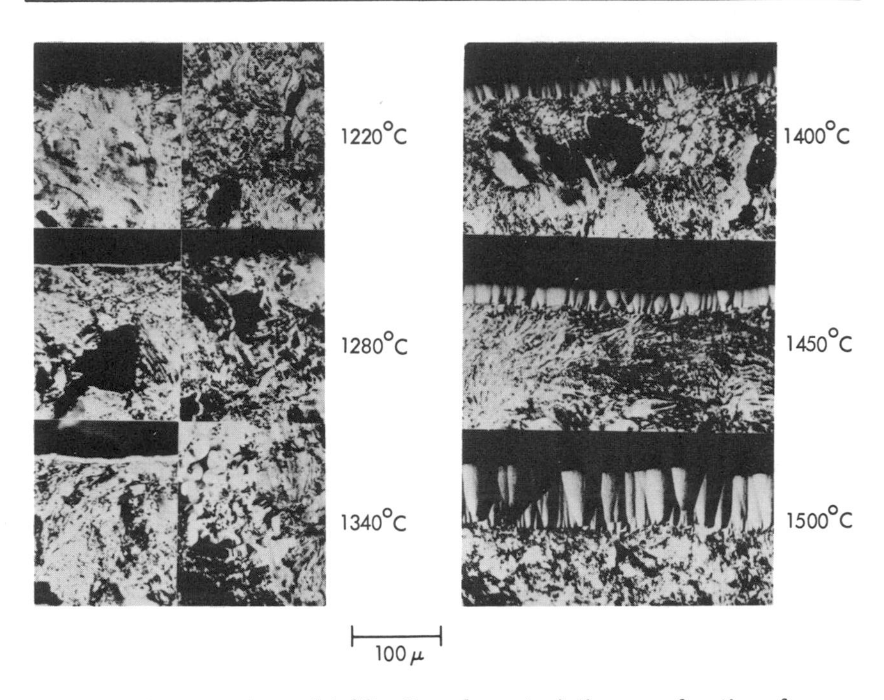

FIG. 17. Coating and infiltration characteristics as a function of temperature. After 3.5-h infiltration at 5 torr, using methane. From Ref. 46.

This relationship for a fiber density of 1.5 g/cm^3 is shown in Table 2. The theoretical density was calculated on the basis of 2.1-g/cm^3 pyrolytic carbon completely filling all of the initial open porosity. Starting with an initial bulk density of 1.2 g/cm^3 and a maximum open porosity of 20%, filling all of the pores with pyrolytic carbon would give an infiltrated density of 1.62 g/cm^3. This is never achieved in practice because of bottleneck pores and the deposition of a surface coating. A similar relationship could be defined for substrates with an initial apparent density greater than 1.5 g/cm^3. In calculating this relationship care must be taken in selecting the proper density for pyrolytic carbon since this is strongly dependent on the processing conditions employed.

Gas-permeability measurements can be used to describe and to follow the growth of pyrolytic carbon in porous solids. This is especially important for the pressure-gradient technique. Typical permeability curves

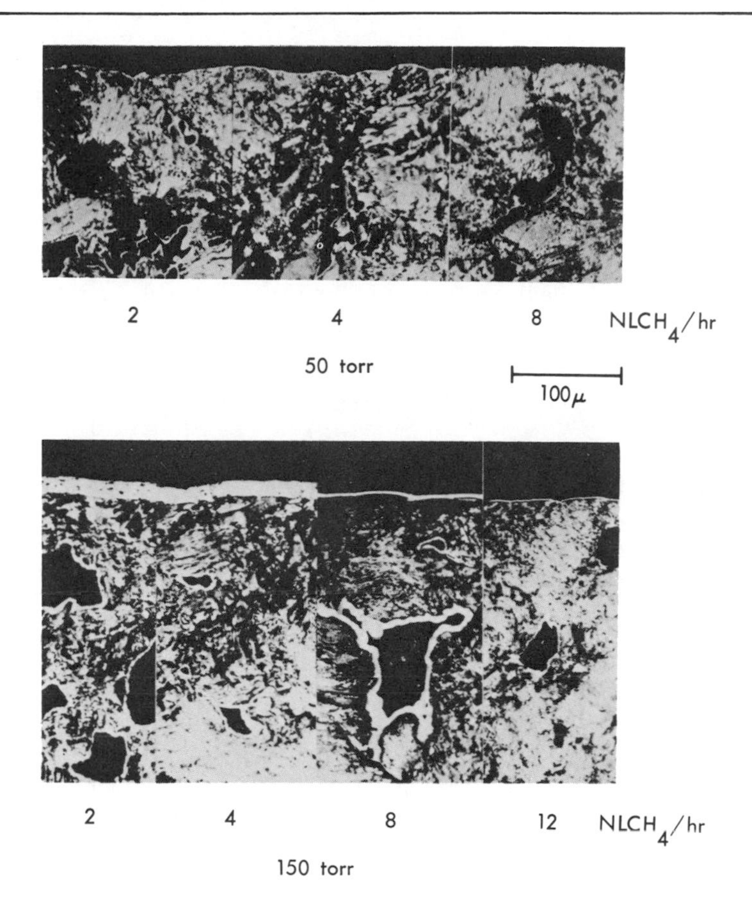

FIG. 18. Coating and infiltration characteristics as a function of pressure and gas velocity. After 17-h infiltration at 1100°C, using methane. Top: 50-torr pressure; bottom: 150-torr pressure. The methane velocity (in cubic centimeters per hour) is indicated by the numbers below the micrographs. From Ref. 46.

plotted as a function of average pressure for fibrous composites with bulk densities of 0.76 to 1.26 g/cm^3 are shown in Fig. 20 [82]. The permeability measurements were made by forcing methane gas through the wall of the porous composite. At a steady-state gas flow the pressure drop across the wall was measured and the methane admittance factor AF was calculated from the following expression [91]:

$$AF = \frac{Qt}{a} \frac{\langle P \rangle}{\Delta P}, \tag{5}$$

where Q is the methane flow, $\langle P \rangle$ is the average pressure, ΔP is the pressure drop across the wall, t is the wall thickness, and a is the mean area through which gas flows.

Cylinders B-3 and F-132 (Fig. 20) had admittance factors at $\langle P \rangle$ = 100 torr of 2×10^{-2} cm^2/sec after infiltration to densities of 1.43 and 1.55 g/cm^3, respectively [82]. At this same average pressure, ATJ and AGR graphites had admittance factors of 5×10^{-2} and 2×10^{-1} cm^2/sec, respectively. The permeability of infiltrated graphite reported by Ford [53] decreased from 3.5×10^{-1} to 10^{-5} cm^2/sec, with a density change from 1.71 to 1.85 g/cm^3. In the work of Beatty and Kiplinger [60] helium permeabilities were reported to be reduced from more than 10^{-2} to less than 10^{-8} cm^2/sec, corresponding to an 8% weight increase after infiltration, from an original density of 1.86 g/cm^3. Similarly a radial permeability after infiltration and graphitization in the region of 10^{-10} cm^2/sec was reported by Carley-Macauly and Mackenzie [67]. Investigators other than those cited have also reported significant decreases in permeability with infiltration.

C. Effect of Initial Pore Size

Recent work by Kotlensky et al. [55, 73, 74, 88] illustrates the relationship between pore size and processing. This relationship for graphite has been described in Section III. Results are presented here for fibrous substrates. Figures 21 and 22 show typical microstructures of filament-wound cones infiltrated by means of the thermal-gradient and isothermal processes. The density after infiltration in both processes was nominally 1.5 g/cm^3. The void size at the yarn crossover points is seen to vary from 750 to 1250 μm. The upper micrographs in Fig. 21 show the principal difference between the thermal-gradient and isothermal processes. Large pores are preferentially filled in the thermal-gradient process; in the isothermal process it is the small pores that are preferentially filled. The upper micrographs in Fig. 22 show an in-depth infiltration between and within plies for the isothermal process and a significantly lower infiltration within plies for the thermal-gradient process. The lower micrographs show the characteristic growth cone structure of pyrolytic carbon obtained by the two processes. The greater extent of pyrolytic carbon coating around the fibers within the ply for the isothermal process is also seen.

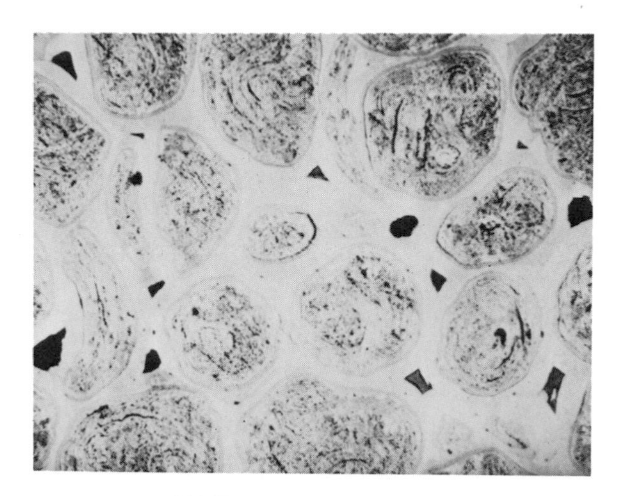

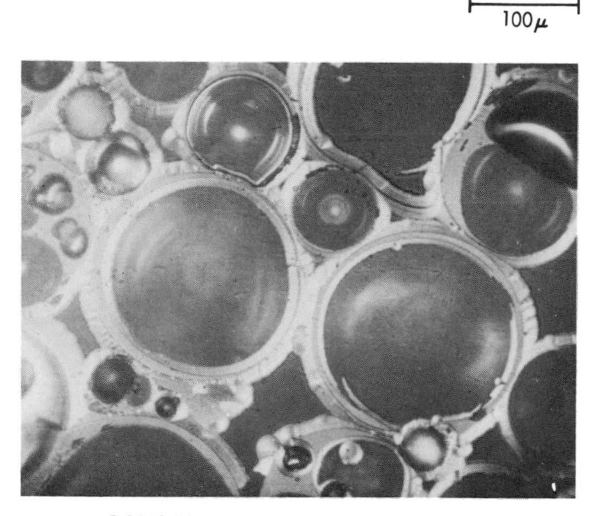

FIG. 19. Microstructure of particles consolidated with pyrolytic carbon. Top: consolidated particles; bottom: consolidated hollow carbon particles. From Ref. 84.

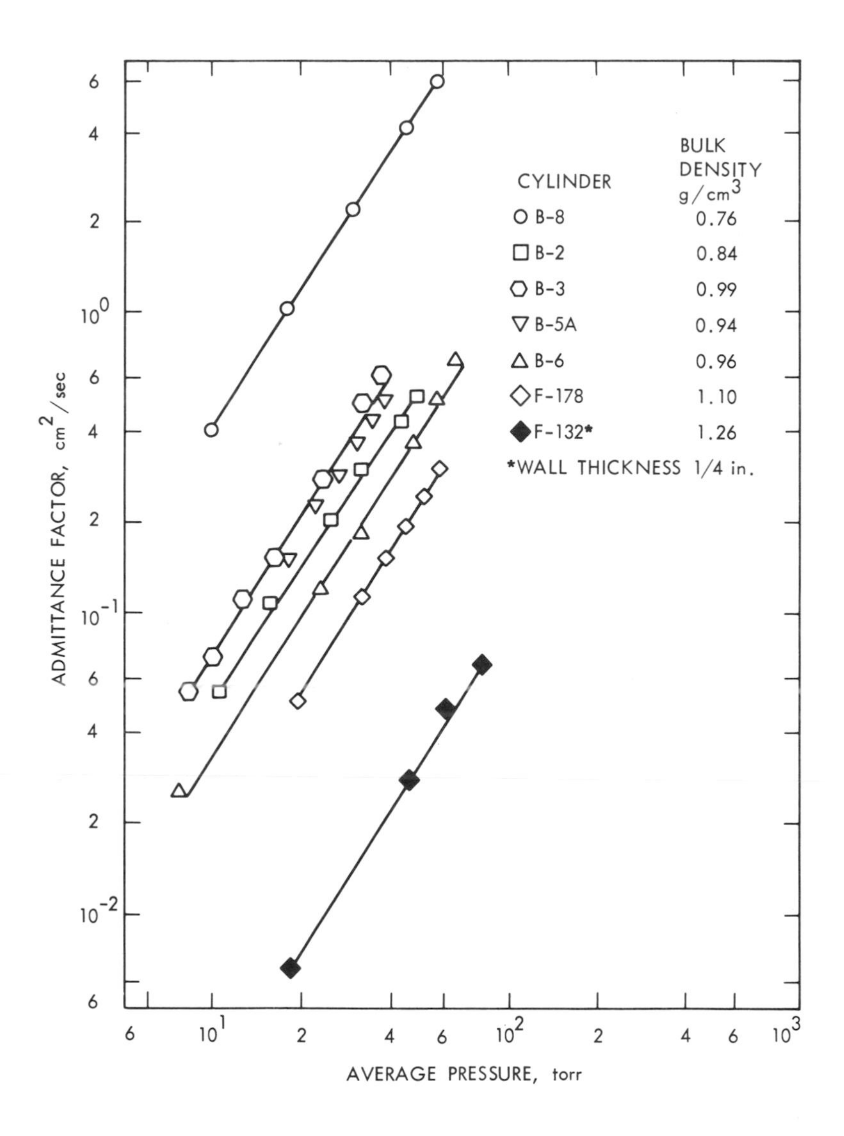

FIG. 20. Relationship between methane permeability and bulk density. All cylinders had 6-in. diameters and a wall thickness of 0.5 in., except for cylinder F-132, which had a wall thickness of 0.25 in. From Ref. 82.

THERMAL GRADIENT

ISOTHERMAL

THERMAL GRADIENT

ISOTHERMAL

1000 μ

FIG. 21. Micrographs of infiltrated filament-wound cones. Top: transverse view; bottom: lower circumferential view. From Ref. 55.

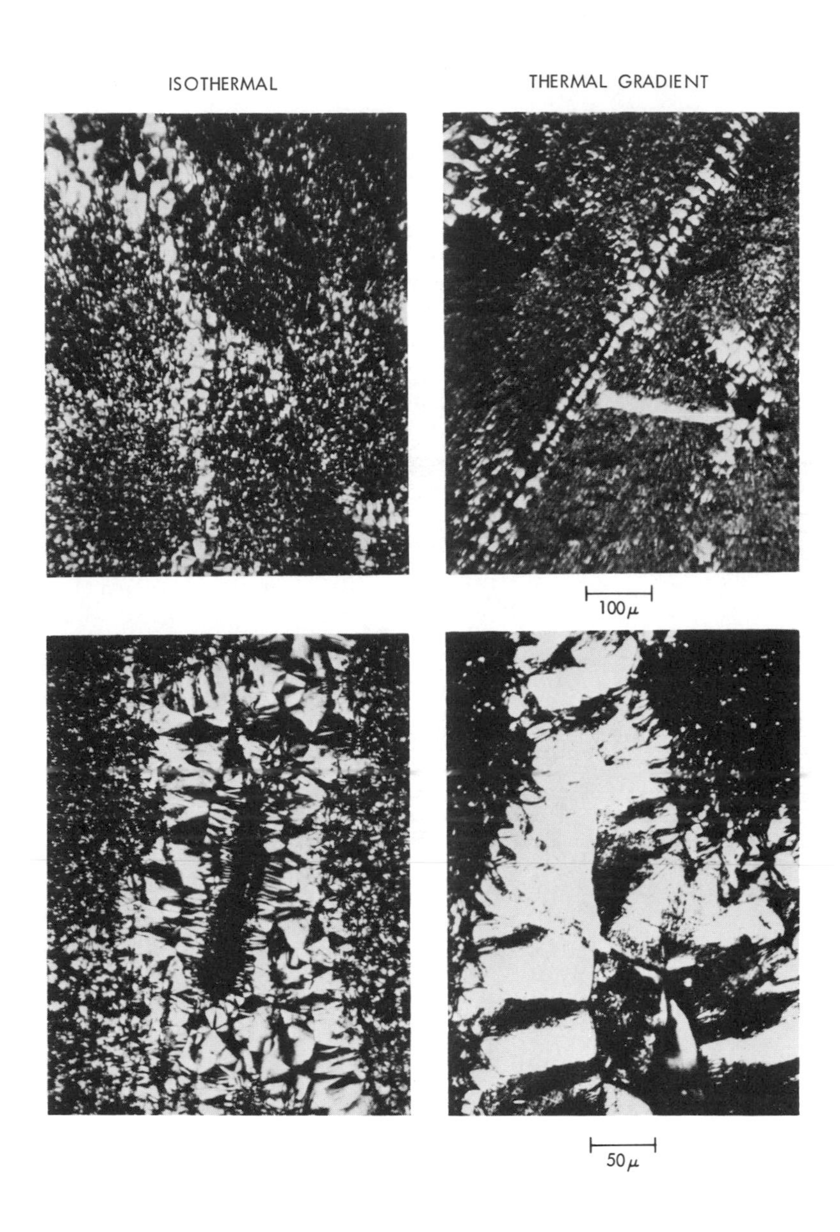

FIG. 22. Micrographs of infiltrated filament-wound cones. From Ref. 55.

The progressive decrease in observable pore structure with isothermal infiltration is illustrated in Fig. 23 [55, 74]. The substrate was a fabric layup construction. The initial density prior to infiltration was 0.5 g/cm^3. On infiltration to a density of 0.87 g/cm^3, an average large-pore size of 431 μm was measured. At a density of 1.59 g/cm^3 the average large-pore size was reduced to 228 μm and then to 128 μm at an infiltrated density of 1.72 g/cm^3. The pore network shown in these micrographs was found to be uniformly distributed throughout the structure. The average large-pore sizes were measured from the photomicrographs shown.

The effect of initial fiber volume on infiltration is shown in Fig. 24 [55, 74]. For the lowest fiber-volume material (30.4%) the fabric layers

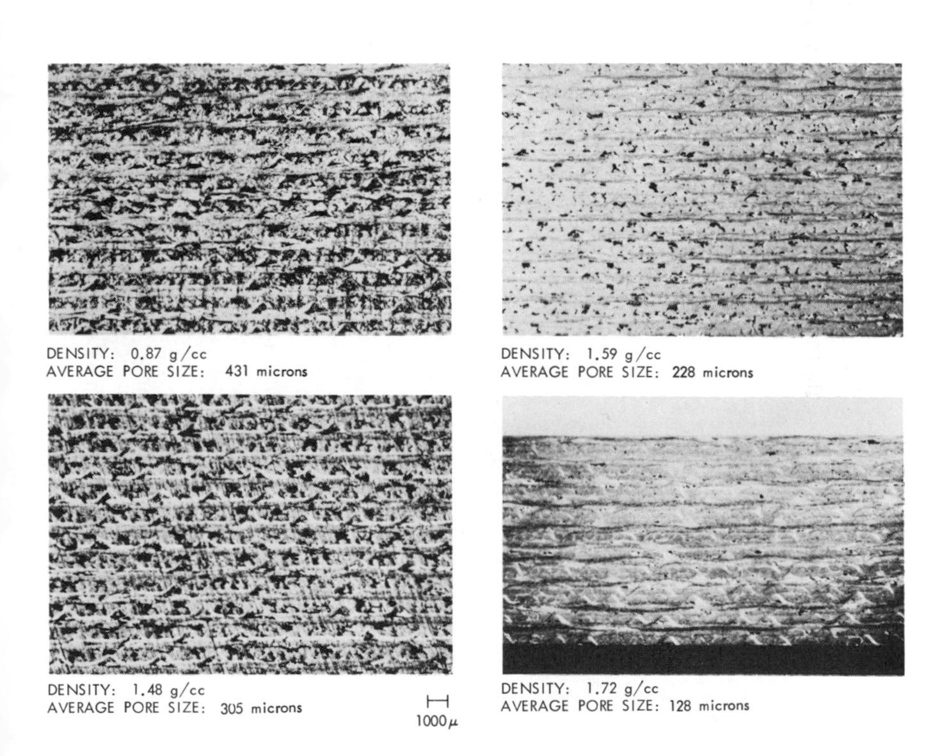

FIG. 23. Micrographs illustrating various stages of infiltration. The average large-pore size decreases from (a) 431 μm at a density of 0.87 g/cm^3 to (b) 305 μm at 1.48 g/cm^3 to (c) 228 μm at 1.59 g/cm^3 to (d) 128 μm at 1.72 g/cm^3. From Refs. 55 and 74.

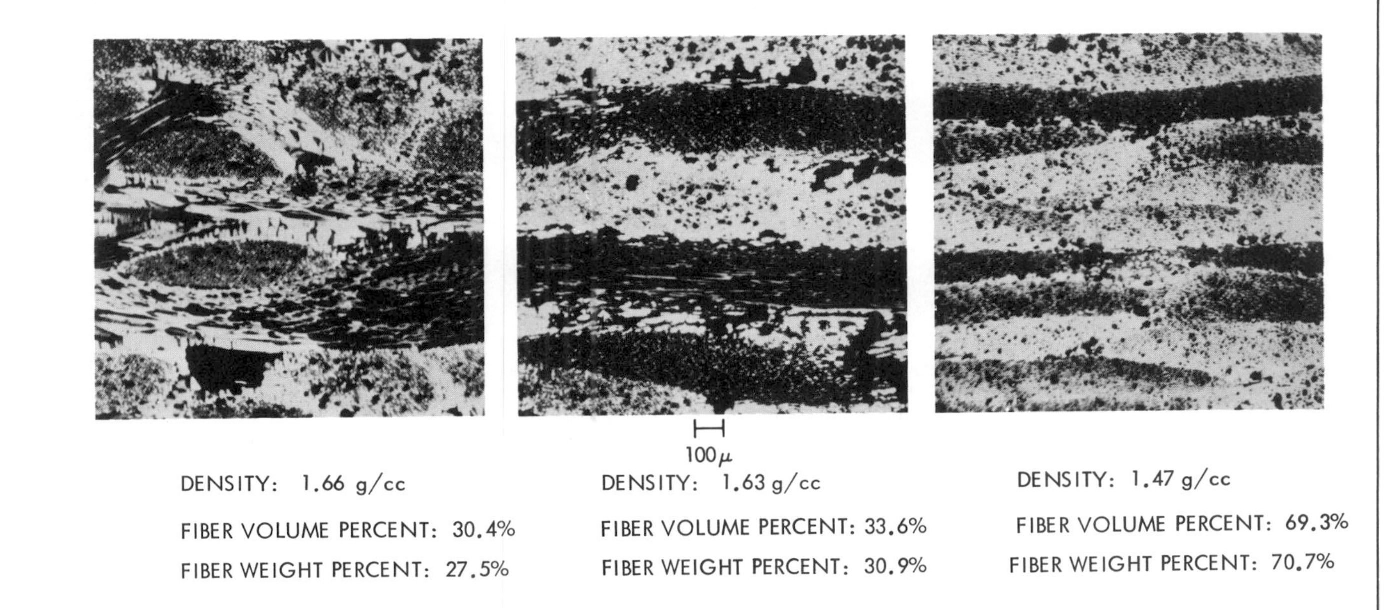

FIG. 24. Structural detail versus fiber volume for isothermal infiltration of fabric layup composites: (a) density 1.66 g/cm^3, fiber volume 30.4%, fiber weight 27.5%; (b) density 1.63 g/cm^3, fiber volume 33.6%, fiber weight 30.9%; (c) density 1.47 g/cm^3, fiber volume 69.3%, fiber weight 70.7%. From Refs. 55 and 74.

remained unflattened, and a coarse-pore structure was produced. At a fiber volume of 69.3% the fabric layers are flattened out appreciably, producing a much smaller pore size for infiltration. The three structures shown in this figure were all infiltrated by the isothermal process for the same length of time. Because of the finer pore structure, which lowered the gas permeability and contributed to the closing of bottleneck pores and the deposition of a surface coating, the 69.3% fiber-volume substrate could be infiltrated to a density of only 1.47 g/cm^3. This substrate had an initial density of approximately 1.1 g/cm^3. The 30.4% fiber-volume substrate with an initial density of 0.5 g/cm^3 was infiltrated to a density of 1.66 g/cm^3. As seen from Table 2, the lower the starting density, the higher the infiltrated density. Similar structures at the 0.5-g/cm^3 starting density range have been infiltrated to a density of 1.75 g/cm^3, which is consistent with the data in Table 2.

The pore distribution for fibrous composites at various densities and after infiltration is shown in Fig. 25 [92]. The open pore volume for the chopped-fiber and fabric layup composites decreases with infiltration. This curve has been included to illustrate this trend.

V. CHARACTERIZATION AND PROPERTIES

A. Crystallographic Structure

A brief review of the early work of Kinney and co-workers at the Pennsylvania State University [12, 13, 16, 93, 94] will be helpful in understanding the crystallographic structure of pyrolytic carbon deposited in porous solids. Pyrolytic carbon was deposited in packed and unpacked ceramic tubes over the temperature range 900 to 1600°C from a variety of hydrocarbons. Helium was used as a diluent as well as a carrier gas. Depositions were carried out at atmospheric pressure over a range of hydrocarbon partial pressures. Four physically different pyrolytic carbon structures were defined [95, 96]:

Type A: flexible, black, shiny pyrolytic carbon film, tightly adhering to smooth surfaces of quartz and Vycor. Rough, gray, metallic looking pyrolytic carbon deposited on unglazed porcelain and at higher temperatures.

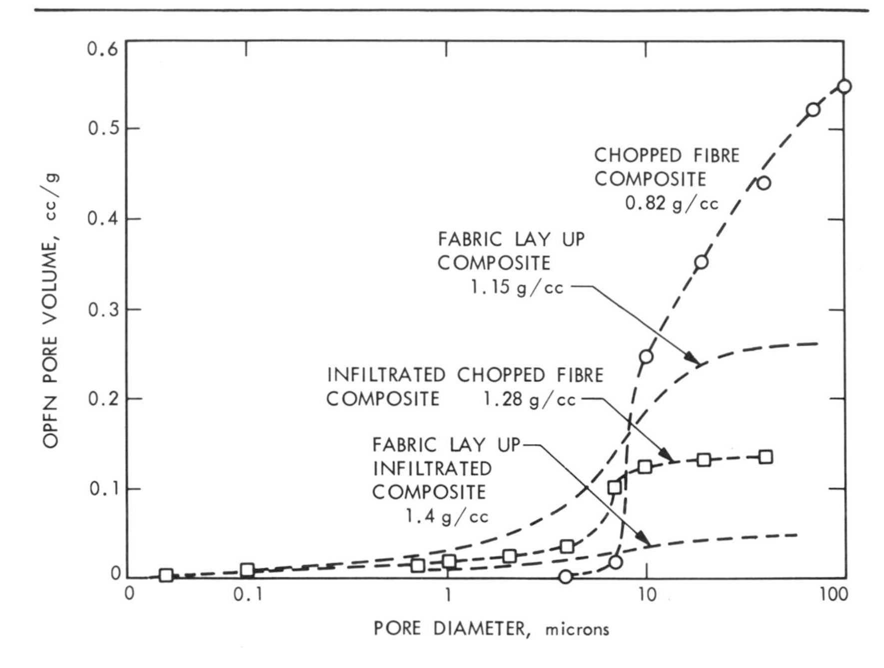

FIG. 25. Open pore volume versus pore diameter for chopped-fiber and fabric layup composites. From Ref. 92.

Type B: spherical particles similar to carbon blacks formed in unpacked tubes.

Type C: mixture of other types.

Type D: also found in unpacked tubes having a hard, gray appearance. Some variations showed a featherlike appearance growing into the gas phase.

Type E: springy, porous, brownish-black material with little strength.

These types of pyrolytic carbon structure correspond in many respects to those described by Iley and Riley [23]. Type A shiny pyrolytic carbon was favored at 900° C using a packed tube. The hard, gray type A variety, which was like the columnar carbon of Iley and Riley [23], was favored at 1200 and 1400° C at the beginning of the hot zone and as an overcoating of the shiny type A [96]. Types B, D, and E were favored by unpacked tubes and high gas concentrations. X-Ray parameters for these four types of pyrolytic carbon structure produced from a variety of hydrocarbons are given in

Table 5 [93, 95, 96]. Type A graphitizes completely from all source hydrocarbons. Types B, C, and D were more turbostratic than type A in the as-deposited state and showed a significantly lower degree of graphitization. (Types C and D in Table 5 are identical with types D and E, respectively, of Refs. 95 and 96.)

Franklin [97] has proposed the term "graphitic carbons" to define carbons that, when heated between 1700 and 3000°C, develop three-dimensional (hkl) graphite reflections and approach the 3.354-Å $d_{(002)}$ spacing of perfect graphite; carbons that show no trace of three-dimensional graphite structure are called nongraphitic carbons. On the basis of Franklins definitions, the type A carbons are graphitic carbons, while the other carbon types described here are nongraphitic carbons. It has been established [97, 98] that nongraphitic carbons are formed with considerably crosslinked structures between layer planes and have a finely porous structure; graphitic carbons, on the other hand, are produced with much less crosslinking between planes and have a more compact mass.

It is difficult to characterize rigorously the crystallographic structure of pyrolytic carbon deposited in porous substrates since a multiphase composite is produced. The crystallographic structure associated with the deposited pyrolytic carbon and its change with processing conditions and heat treatment needs to be separated from the diffraction pattern produced by the substrate. Relatively little work, however, has been done in this area. On graphite substances, since the amount of pyrolytic carbon is generally less than 10%, the diffraction pattern is due principally to the filler particles and binder of the original substrate. On fibrous substrates, where the amount of pyrolytic carbon may vary from 20 to 92%, the diffraction pattern will be composed of the contributions from the fiber, pyrolyzed binder if present, and pyrolytic carbon. For infiltrated rayon-precursor felt composites, where the amount of pyrolytic carbon generally exceeds 90%, the diffraction pattern will be principally that of pyrolytic carbon.

Results obtained at Sandia Laboratories for infiltrated felt composites are given in Tables 6, 7, and 8. Table 6 [58] shows the interlayer spacing to decrease and the layer stacking to increase with increasing mandrel temperature. The same trend is observed in the data given in Table 7 [58] at a constant temperature with decrease in pressure. Heat treatment at

TABLE 5

X-Ray Parameters of Carbons Prepared at 1200°C and Heat-Treated at 2500 and 3000°C[a]

Compound	Carbons prepared at 1200°C			Carbons heat-treated at 2500°C			Carbons heat-treated at 3000°C		
	d-Spacing	L_c	L_a	d-Spacing	L_c	L_a	d-Spacing	L_c	L_a
A. Thin-film carbons (packed tube)									
Benzene	3.49	33	56	3.358	1950	>2000	3.356	>2000	>2000
Toluene	3.48	29		3.363	1440	>2000	3.356	2200	>2000
Biphenyl	3.48	29	56	3.362	1300	>2000	3.358	1600	>2000
Naphthalene	3.50	28	51	3.367	1160	>2000	3.359	>2000	>2000
Anthracene	3.52	20	56	3.361	1580	>2000	3.356	>2000	>2000
Pyridine	3.46	25		3.361	810	>2000	3.357	2110	>2000
Thiophene	3.50	30		3.360	970	1550	3.358	1160	>2000
Methane	3.48	32	47	3.356	1800	1550	3.357	1850	>2000
B. Carbon blacks (empty tube)									
Benzene	3.66	11	48	3.390	190	300	3.392	210	370
Toluene	3.61	14		3.401	200	270	3.391	260	410
Biphenyl	3.65	12	42	3.401	160	250	3.389	210	310
Naphthalene	3.68	13	40	3.395	180	330	3.385	270	440
Anthracene	3.70	12	36	3.398	170	350	3.388	240	400
Pyridine	3.64	12		3.398	200	330	3.389	240	330
Thiophene	3.67	11		3.393	220	310	3.390	270	370
Methane	3.61	14		3.394	190	310	3.392	200	340

C. Thick-film carbons (empty tube)									
Benzene	3.52	29	45	3.401	130	390	3.379	320	990
Biphenyl	3.52	21	38	3.386	250	450	3.376	340	740
Naphthalene	3.52	21	43	3.380	540	620	3.360	930	1560
Anthracene	3.54	24	50	3.376	330	790	3.371	420	860
D. Spongy carbons (empty tube)									
Benzene	3.51	31	49	3.407	120	200	3.383	240	350
Toluene	3.54	23		3.392	170	260	3.384	290	330
Biphenyl	3.52	25	42	3.405	130	370	3.383	270	460
Naphthalene	3.55	26	50				3.384	230	410
Anthracene	3.55	27	41				3.380	330	710
Pyridine	3.49	26		3.395	170	320	3.385	220	330
Thiophene	3.61	14		3.402	190	250	3.387	290	370
Methane	3.51	23		3.388	190	330	3.387	180	390

[a] Data from Refs. 93, 95, and 96. All measurements are in angstroms.

3000°C (Table 8) shows that the CVD/felt A composite is graphitic, whereas the CVD/FW A composite is nongraphitic [99]. The reader is referred to the original reports for more details.

It should be recognized that there is a very limited amount of X-ray data on pyrolytic-carbon-infiltrated composites. However, it is nonetheless possible to draw the obvious observation that, after heat treatment, the infiltrated pyrolytic carbon falls between the type A and type C carbons described in Table 5. Some of this difference may be due to the contribution to the diffraction pattern by the nongraphitic felt and filament-wound components.

B. Mechanical Properties

1. Introduction

The potential tensile strength for highly oriented, single-crystal graphite fibers at 10% offset elongation is estimated to be 14.5×10^6 psi [100]. This estimation is based on a theoretical Young's modulus for graphite whiskers of 145×10^6 psi [101]. At 1% offset elongation the fiber strength would be 1.45×10^6 psi. The latter value is thought to represent more of an upper limit for the practical tensile strength that can be achieved for graphite single-crystal fibers. The dependence of whisker strength on size, crystalline stacking faults, and surface defects has been fairly well documented. A number of recent reports have also described similar relationships between strength and structure.

Strengths as high as 7×10^5 psi have been reported for pyrolytic graphite whiskers [102]. Corresponding modulus values were as high as 90×10^6 psi. These properties are associated with small-diameter, short-length whiskers. In another report by the same investigators a marked effect of whisker diameter on strength and modulus was reported [103]. Whiskers 3 to 6 μm in diameter had tensile strengths of up to 4×10^5 psi, with a modulus of 29×10^6 psi; whiskers in the 20- to 30-μm-diameter range had strengths and modulus values of nominally 2×10^5 and 13×10^6 psi [103]. Because of their short lengths, such graphite whiskers are not very useful in fabricating panels that require careful and precise orientation of the reinforcing media over large dimensions.

TABLE 6

Effect of Deposition Temperature on the Crystallographic Parameters of CVD Carbon on Felt Substrate [a]

Mandrel temperature (°C)	d_{002} (Å)	L_c (Å)	Density (g/cm^3)
1100	3.448	38	1.84
1200	3.449	35	1.80
1300	3.439	33	1.67
1350	3.436	37	1.40
1400	3.427	79	1.62
1500	3.431	51	1.40

[a] From Ref. 58. Material source: H. O. Pierson, Sandia Laboratories; infiltration technique: thermal gradient; deposition pressure 630 torr; methane source gas, flow rate 6500 cm^3/min; argon carrier gas, flow rate 500 cm^3/min; fiber volume 9%.

TABLE 7

Effect of Deposition Pressure on the Properties of CVD Carbon on Felt Substrate [a]

Pressure (torr)	d_{002} (Å)	L_c (Å)	Thermal conductivity at 600°C (cal/cm-sec-°C)	Resistance (ohm-cm x 10^6)	Density (g/cm^3)
35	3.449	50	0.039	2900	1.79
100	3.450	37	0.028	3000	1.80
760	3.459	32	0.024	3200	1.80

[a] From Ref. 58. Material source: Super-Temp Co.; infiltration technique: thermal gradient; deposition temperature 1100°C; fiber volume 9%.

TABLE 8

Crystallographic Properties, Density, and Porosity of Infiltrated Felt and Filament-Wound (FW) Composites[a]

Parameter	CVD/Felt A		CVD/Felt B		CVD/FW A	
	As deposited	3000°C heat-treated	As deposited	2750°C heat-treated	As deposited	3000°C heat-treated
d_{002} (Å)	3.450	3.365	3.452	3.378	3.440	3.420
L_c (Å)	50	338	52	245	30	55
Graphitization factor (%)	0	76	0	52	0	6
Density (g/cm^3)	1.79	1.79	1.75	1.75	1.45	1.55
Porosity (%)	13[b]	13[b]	15[b]	15[b]	4[c]	5[c]

[a] From Ref. 99.

[b] Obtained by helium pycnometry.

[c] Obtained from water-immersion measurements.

Investigations aimed at coating filaments, yarns, and fabrics with pyrolytic graphite to improve their mechanical strength have had limited success [102-105]. Pyrolytic graphite coatings produced at nominally 2000°C are turbostratic in structure and would not be expected to have a particularly high strength or modulus because of the high concentration of low- and high-angle boundaries in the fiber major-axis direction.

It is significant to note here that Kotlensky [106,107] had demonstrated several years previously that the strength and modulus for pyrolytic graphite could be significantly increased on graphitization under an applied load. A room-temperature modulus of 81 x 10^6 psi was measured for pyrolytic graphite that had been hot-stretched 16% at 2750°C. The as-deposited pyrolytic graphite had a modulus of 2.3 x 10^6 psi. Ultimate strengths of over 100,000 psi were measured for hot-stretched pyrolytic graphite [107]. Although this strength is not too spectacular in comparison with values that are being reported today for graphite fibers, it should be noted that the hot-stretched pyrolytic graphite had a rectangular cross section of 60 by 100 mils. Specimen size and edge effects undoubtedly limited the strength attainable.

Typical high-strength, high-modulus graphite fibers produced today are nominally 0.3 mil in diameter. Strengths as high as 450,000 psi are reported for single filaments in this diameter range [108]. Commercially available high-strength, high-modulus yarns, on the other hand, are 15 to 30 mils in diameter and have tensile strengths that are approximately an order of magnitude less than the values measured for the individual filaments comprising the yarn bundle [109]; for example, the strength for Thornel 50 single filament is listed at 285,000 psi with a breaking strength of 6.4 lb for the 15-mil-diameter, 720-filament-per-ply, two-ply yarn. Normalizing for cross-sectional area, the calculated breaking strength is 36,217 psi. Adding a sizing to the yarn improves the load transfer from fiber to fiber to give a yarn breaking strength of 7.4 lb (41,875 psi). Understanding the mechanism by which load is transferred from fiber to fiber in a composite and seeking ways to improve the coupling between fiber and matrix are areas that have received appreciable attention over the past several years.

Fiber volume, dimensions, and orientation are important characteristics that need to be optimized in order to achieve good structural integrity from infiltrated composites. The interrelationship of these fiber characteristics with a ductile matrix has been well established [110]. In a brittle matrix, such as pyrolytic-carbon-infiltrated composites, the effect of fiber characteristics and interaction with the brittle matrix is less well understood.

In brittle-fiber, ductile-matrix composites the strength of the fiber determines the strength achieved in the composite. Where good coupling is obtained between fiber and matrix, high-strength fibers lead to high-strength composites. The relationship of fiber and matrix strength to composite strength is given by the rule of mixtures [111]:

$$\sigma_C = \gamma_F \sigma_F + \gamma_B \sigma_B, \tag{6}$$

where σ_C is the strength of the composite, σ_B and σ_F are the strengths of the binder and fiber at the same strain, and γ_B and γ_F are the volume fractions of each. A similar relationship can be written to describe the modulus.

Figure 26 shows an idealized rule-of-mixtures plot for brittle-fiber, brittle-matrix composites. The brittle-matrix binders shown at the left include zirconium carbide-PG composites, PC, CVD carbon, PG, BPG, glassy carbon, charred epoxy, and phenolic resins. On the right side of the plot reinforcement brittle fibers are shown. These include high-modulus graphite single filaments as well as intermediate- and low-modulus carbon and graphite yarns and fabric.

The broken lines in Fig. 26 represent an idealized predicted rule-of-mixtures relationship for carbon and graphite fibers with pyrolytic carbon binder. The high-modulus graphite filaments show a strength of up to 400,000 psi. At 50% fiber volume, the upper broken line predicts a composite strength of 200,000 psi. The difficulty, however, in achieving this strength depends on transferring the load from fiber to fiber through the brittle matrix. Some results that have been obtained indicate that the rule of mixtures does hold for carbon fibers in a pyrolytic carbon matrix. For carbon yarns and fabrics whose strengths are lower than that of the pyrolytic carbon binder, increasing the fiber volume percentage can lead to a decrease in composite strength. This is shown by the lower broken line. Currently, the most usable and practical combinations are between such binders as pyrolytic carbon and charred epoxy and phenolic resins with high- and low-modulus carbon and graphite yarns and fabrics.

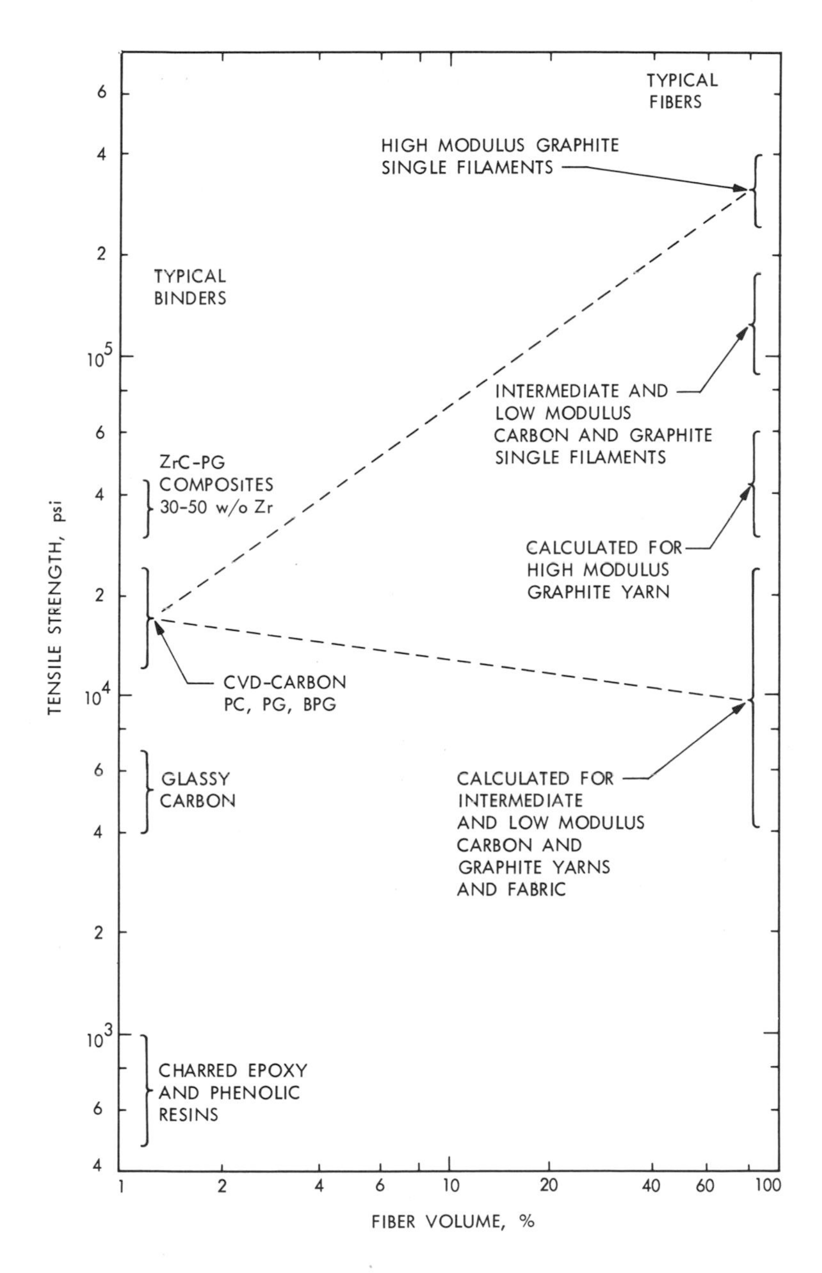

FIG. 26. Idealized rule-of-mixtures relationship for typical binders and typical fibers.

2. Graphite Substrates

Very limited data are available on the mechanical properties for pyrolytic-carbon-infiltrated graphite substrates. As already mentioned, the main interest in infiltrating graphite substrates with pyrolytic carbon is to seal the internal pores to significantly reduce the gas permeability. Data presented in Table 9 give the flexural strength and other properties for a base electrographite and after impregnation and infiltration [53]. The flexural strength is reported to increase from 4200 psi for the base graphite with a density of 1.71 g/cm^3 to a value of 8000 psi after gas-pyrolysis treatment to a density of 1.85 g/cm^3 [53]. Pyrolytic carbon infiltration from a density of 1.78 to 1.80 g/cm^3 was associated with a flexural strength increase with the grain from 4350 to 4900 psi [61]. Infiltration of fluid-coke particles in the pyrolytic carbon consolidation process [84] gave fracture stresses of up to 10,000 psi and diametral compression strengths on 0.5-in.-diameter by 0.5-in. cylinders of 5900 psi.

Bend and compressive strength data for EY9 graphite before and after infiltration are given in Table 10 [44]. The bend strength increases from 5200 to 9410 psi and the compressive strength from 8410 to 17,800 psi with treatment. After subsequent heating to 2500°C, both the bend and compressive strengths decrease; however, the strengths are still greater than those of the EY9 graphite before treatment. In the same study [44] a series of petroleum-coke bars were prepared containing 39 to 58% deposited carbon. The final density was reported to be about 1.8 g/cm^3. Bend strength was reported to vary from 10,000 to 17,500 psi, with no clear-cut relationship to the amount of carbon deposited. Compressive strengths of 40,000 and 70,000 to 90,000 psi were reported for the bars containing 39 and 93% deposited carbon, respectively [44]. Lampblack was used as the substrate in producing a bar containing 93% deposited carbon. The bend and compressive strengths decreased with heat treatment.

These results illustrate the trends in strength reported for pyrolytic-carbon-infiltrated graphite and coke-particle substrates. The increase in graphite strength with infiltration is open to speculation and is beyond the intent of this section.

TABLE 9

Effect of Various Carbon Impregnants on the Characteristics of an Electrographite after Final Heat Treatment to 2500°C [a]

Property	Base graphite	Pitch impregnated	Resin impregnated	Gas-pyrolysis treated	Improvement in properties over base material valuable for
Bulk density (g/cm^3)	1.71	1.82	1.77	1.85	Moderator graphite, rocket chokes
Open porosity (%)	14	9.5	2.0	1.2	Metallurgical applications
Permeability (cm^2/sec)	3.5×10^{-1}	2×10^{-1}	10^{-5}	10^{-5}	High-temperature gas seals and vanes, metallurgical applications
Flexural strength ($lb/in.^2$)	4200	5000	5500	8000	High-temperature metallurgical applications, spark machining electrodes, transistor jigs, hot-pressing dies
Resistivity (10^{-4} ohm-cm)	8.8	5.4	7.5	5.5	High-temperature furnace electrodes
Thermal conductivity (cal/cm-sec-°C)	0.13	0.20	0.15	0.19	Hot-pressing and continuous-casting dies
Hardness (Shore)	55	56	60	65	
Oxidation rate per hour at 500°C (%)	0.09	0.08	0.06	0.04	High-temperature seals and vanes, transistor jigs

[a] From Ref. 53.

3. Carbon-Fiber Substrates

Mechanical property data have been reported for a large variety of carbon-fiber substrates bonded with pyrolytic carbon [4, 54-58, 70-82, 84, 87-90, 99, 112-115].

Carbon-felt substrates have been investigated in more detail than other carbon-fiber substrates. These substrates are prepared by controlled pyrolysis of rayon and wool felts. The as-carbonized density for rayon-precursor felt substrates falls between 0.1 and 0.2 g/cm^3. Wool-precursor felt substrates are more compacted and have an as-carbonized density of about 0.5 g/cm^3. The effects of composite density and pyrolytic-carbon-deposition temperature on the mechanical properties of infiltrated felt substrates are shown in Figs. 27 through 34 [72, 75, 77, 112]. All of the data given, except in Fig. 28 (wool-precursor felt substrate), are for rayon-precursor felt substrates. At the same infiltrated density, wool-precursor felt substrates are approximately twice as strong as rayon precursors [54, 71, 72, 76]. This is thought to be associated with the higher fiber volume and scaly surface texture of the natural wool fibers. The mechanical strength for the rayon-base infiltrated composites is seen to increase with density and to decrease with increase in deposition and heat-treatment temperatures. The effect of fiber volume on flexural strength and modulus is shown in Figs. 35 and 36 [87]. An increase in strength with increase in fiber volume is seen along with a decrease in strength with heat treatment.

The decrease in strength with increasing deposition temperature has been explained by Pierson, Theis, and Smatana [77] as being due to a change in microstructure from laminar to granular (Fig. 15). Bokros [3] had reported a similar relationship between strength and microstructure for pyrolytic carbon deposited in a fluid bed. The fracture stress for Bokros' pyrolytic carbon deposits, however, is from two to four times higher than that found for infiltrated felt composites.

Table 11 [55] gives data on flexural and shear strength measured on a variety of substrates infiltrated by different techniques. Carbon felt is the only substrate for which sufficient strength data are available for a comparison of the three principal infiltration processes. At comparable densities the flexural strength of the substrate infiltrated by the isothermal process is

TABLE 10

Mechanical Tests on EY9 Graphite before and after Deposition Treatment[a]

	Mean Young's modulus (10^6 psi)	Mean bend strength (psi)	Mean compressive strength (psi)
Before treatment	1.8	5200	8,410
After treatment at 750°C in benzene to give a 6 to 7% weight increase	2.7	9410	17,800
After subsequent heating to 2500°C	2.6	8200	15,000

[a] From Ref. 44.

greater than of the substrates infiltrated by either the thermal- or pressure-gradient process [55]. This difference is also thought to be associated with differences in pyrolytic carbon microstructure.

From the idealized rule-of-mixture plot shown in Fig. 26, a tensile strength of 200,000 psi would be predicted for a 50% fiber-volume, unidirectional, high-modulus single-filament layup composite bonded with pyrolytic carbon. Strengths in this range, however, have not been reported for carbon-carbon composites. As shown in Table 12 [78], a tensile strength of 103,000 psi and a modulus of 20 x 10^6 psi have been measured for Morganite type II filaments bonded with pyrolytic carbon. Flexural strengths of 120,000 to 140,000 psi have been reported for PAN-based unidirectional layup composites bonded with pyrolytic carbon [113]. Ring tensile strengths of typically 60,000 psi have been reported for pyrolytic-carbon-bonded filament-wound cylinders processed with high-modulus Thornel yarns [88, 115]. Estimated tensile strengths of up to 120,000 psi [115] were obtained for the solid bonded composites on correcting for porosity. The difference in measured strengths and predicted strengths based on the idealized rule-of-mixtures relationship shown in Fig. 26 is due to filament orientation, nonideal coupling, and load transfer from fiber to fiber through the matrix,

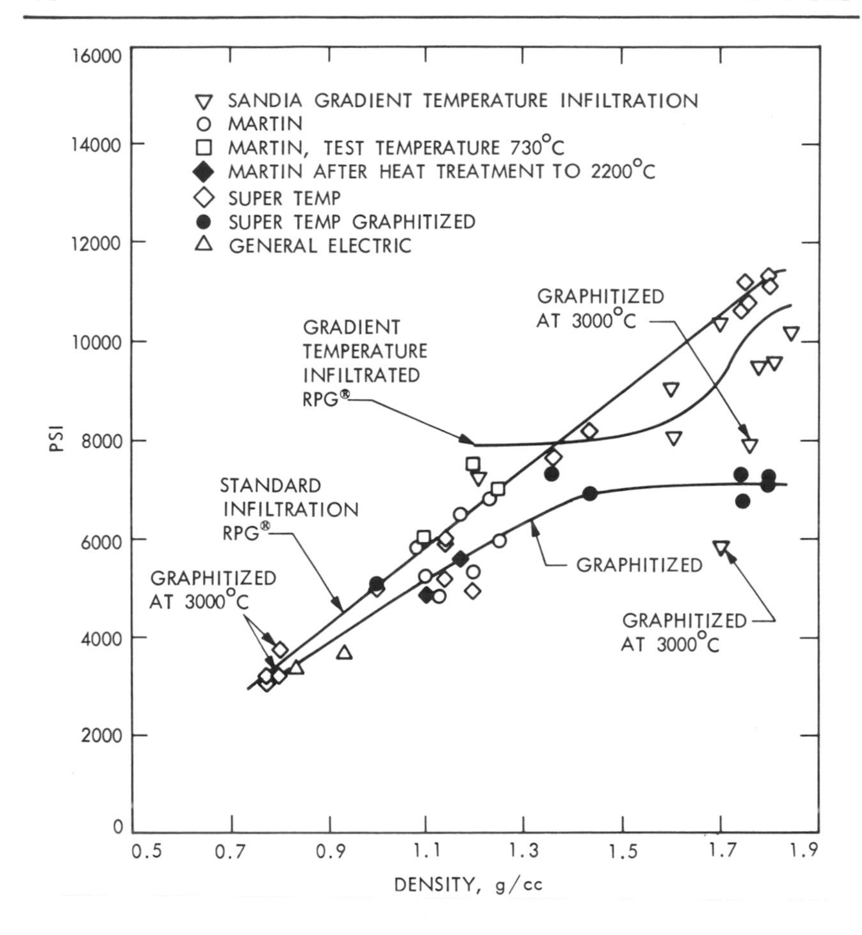

FIG. 27. Flexural strength versus density for infiltrated rayon-precursor carbon felt. From Ref. 72.

and the low matrix failure strain. The effect of filament orientation on strength and modulus is shown in Figs. 37 and 38 [99].

In brittle-fiber, ductile-matrix composites the high plasticity of the matrix facilitates load transfer from fiber to fiber, and high strengths obeying the rule-of-mixtures relationship are obtained [111]. In pyrolytic-carbon-bonded composites failure is initiated in the matrix due to its low strain to failure. Load transfer in the composite from fiber to fiber is therefore limited to the fiber stress level at the composite failure strain.

Plotting the fiber stress at the composite failure strain against percent fiber volume yields a linear rule-of-mixture relationship [88, 116]. This observation is consistent with the definition of the rule-of-mixtures relationship.

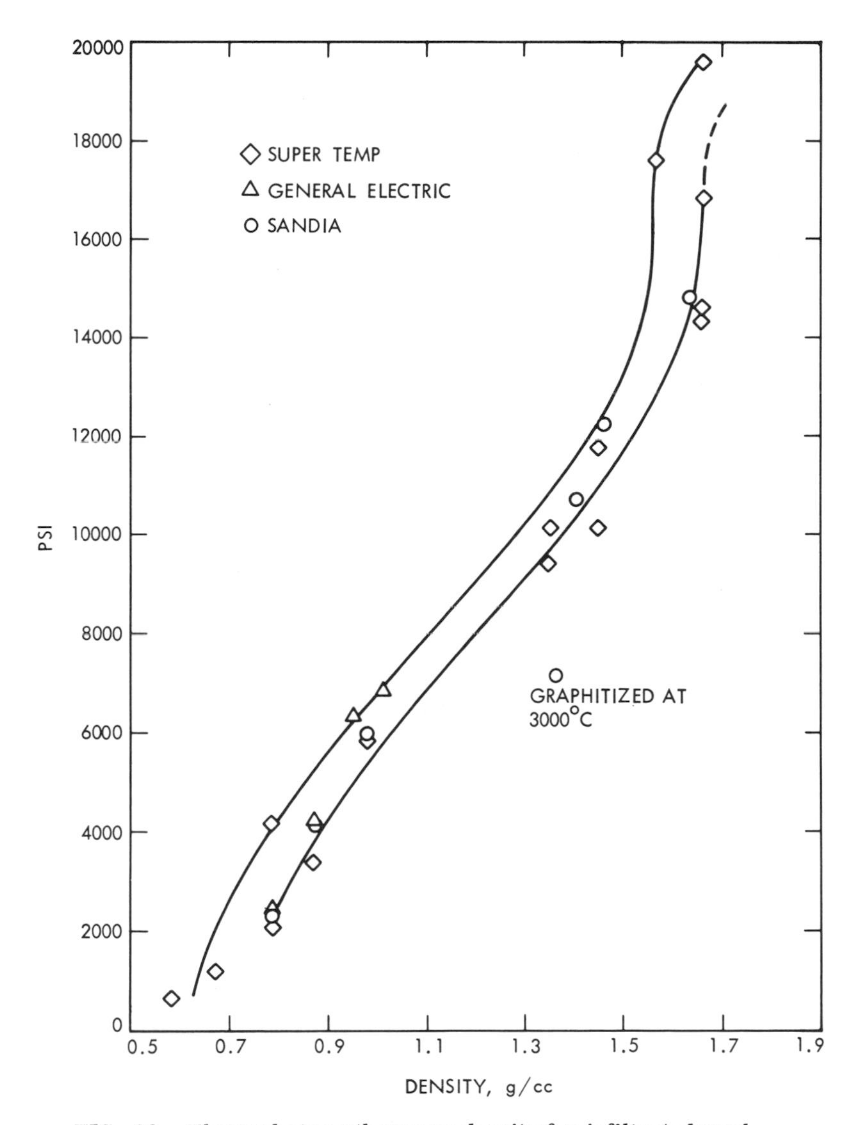

FIG. 28. Flexural strength versus density for infiltrated wool-precursor carbon felt. From Ref. 72.

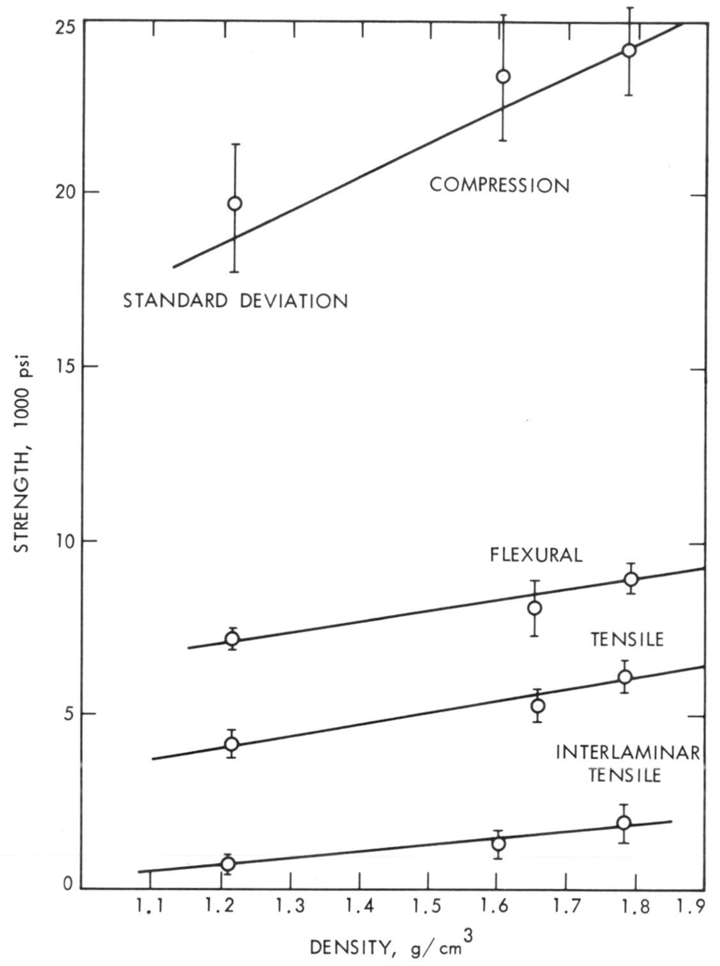

FIG. 29. Mechanical properties versus density for infiltrated carbon felt. From Ref. 75.

C. Thermal Properties

1. Thermal Expansion

The thermal and other properties of pyrolytic-carbon-infiltrated porous substrates have not been characterized to the same extent as the mechanical properties. In thermal properties graphite substrates containing a low volume percentage of pyrolytic carbon would be expected to be relatively the same as the graphite substrates before infiltration. The thermal properties of fibrous composites, on the other hand, would be a weighted average of the phases present and depend on the structure and orientation of the fiber, pyrolytic carbon, and other phases present, as well as internal accommodations. I am unaware of any systematic analysis or qualitative treatment of the thermal properties of infiltrated fibrous composites. Such

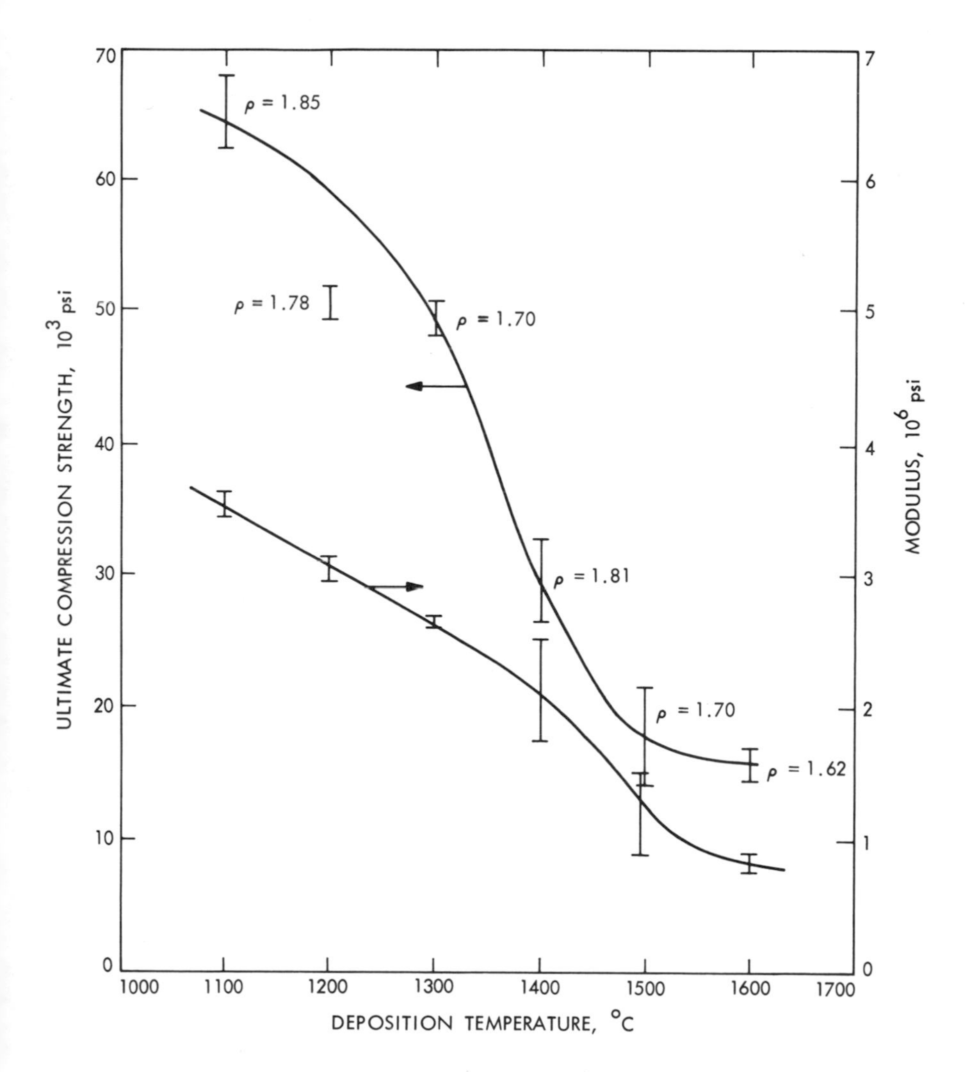

FIG. 30. Compressive strength and modulus versus deposition temperature for carbon felt. From Ref. 77.

materials are in their early stage of development and characterization, and only limited data are available.

Bokros [3] has reviewed the thermal expansion relationship for pyrolytic carbon with preferred orientation and structure. This relationship was summarized as follows [3]: "The thermal expansion of any aggregate, whether it be fully graphitic or turbostratic, therefore depends primarily on

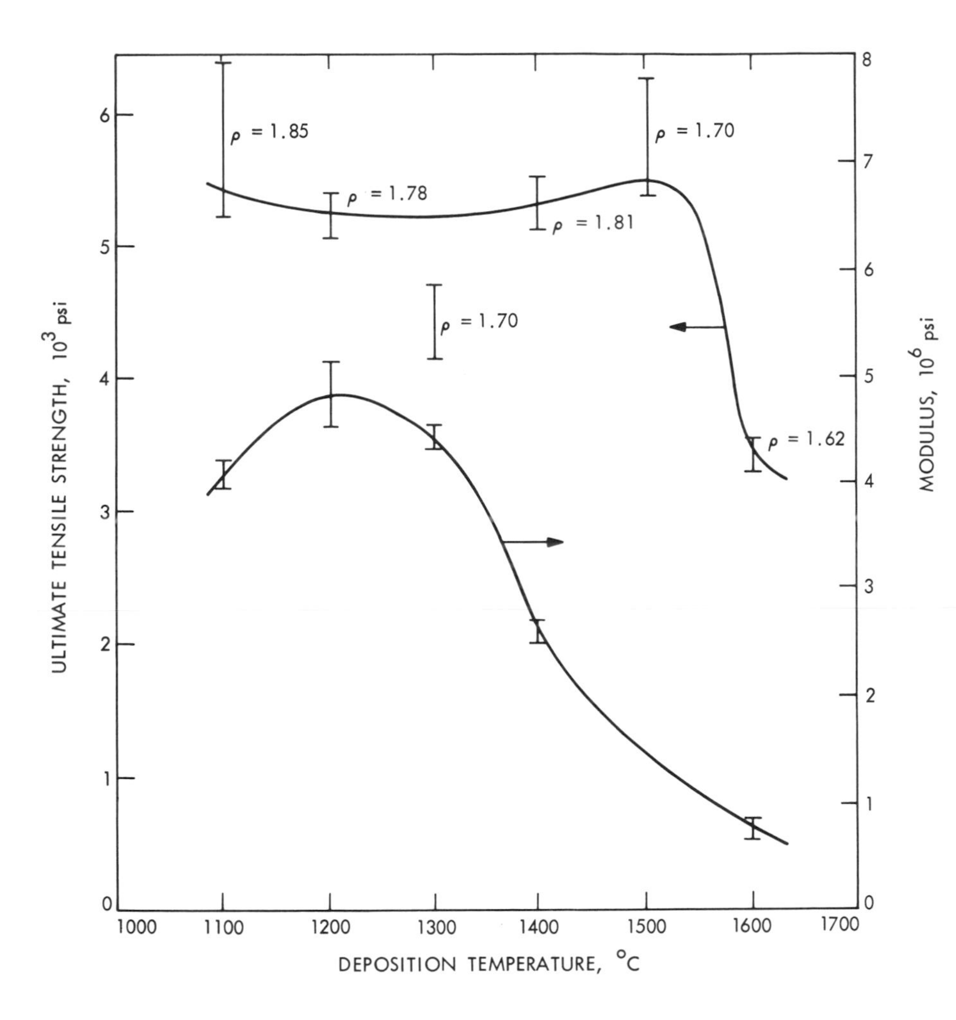

FIG. 31. Tensile strength and modulus versus deposition temperature for carbon felt. From Ref. 77.

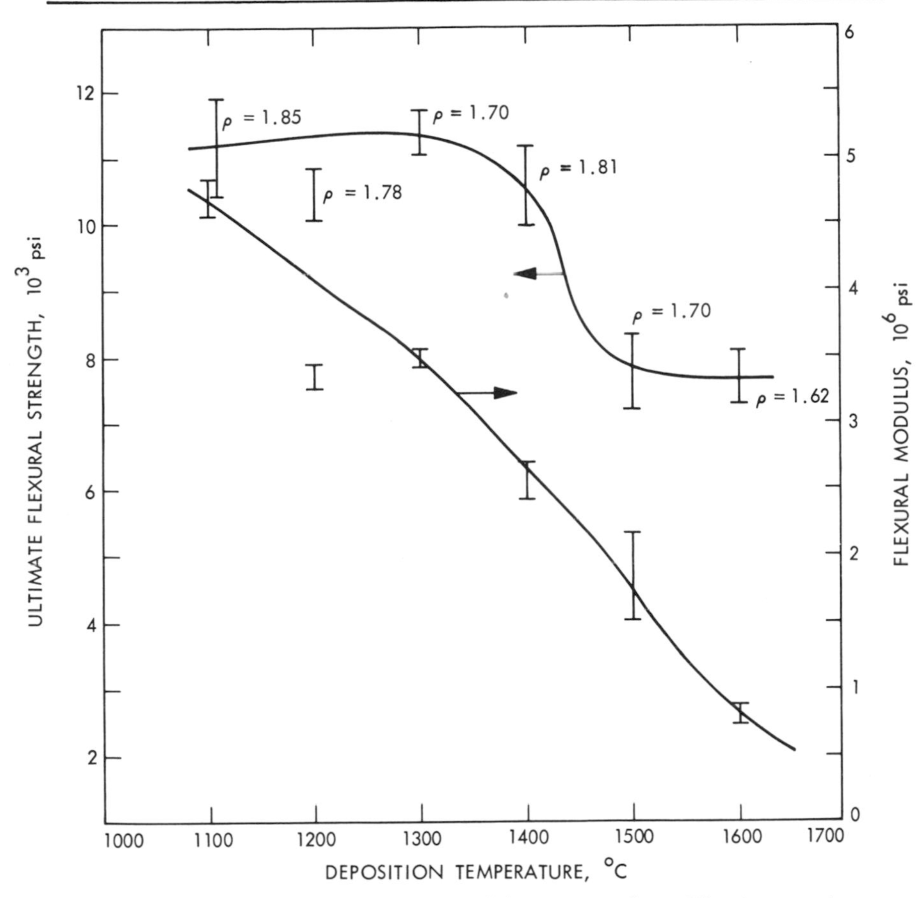

FIG. 32. Flexural strength and modulus versus deposition temperature for carbon felt. From Ref. 77.

the degree of preferred orientation of the crystallites in the aggregate and internal accommodations that result from porosity or intercrystalline elastic restraints." A review of the bulk expansivities at 400° C for pyrolytic carbon showed a primary systematic dependence on preferred orientation and a secondary dependence on density [3].

The effects of density and heat treatment on the linear thermal expansion of rayon- and wool-precursor infiltrated felts are shown in Figs. 39 and 40 [72,76]. The anisotropy in thermal expansion parallel and perpendicular to the surface and the decrease in thermal expansion with heat treatment are illustrated in

TABLE 11

Strength of Infiltrated Carbon-Carbon Composites [a]

Substrate	Infiltration process [b]	Bulk density (g/cm^3)		Flexural strength (psi)	Shear strength (psi)	Ref.
		Initial	After infiltration			
CY2 Omniweave	TGI	0.69	1.60	9,230	—	57
CY2 Omniweave [c]	TGI	0.73	1.56	12,280	—	57
Modmor II omniweave	TGI	0.73	1.64	7,430	—	57
Modmor II omniweave [c]	TGI	1.10	1.52	16,010	—	57
Carbon felt	TGI	0.1	1.60	8,060	—	75
Carbon felt	TGI	0.1	1.80	9,000	—	75
Carbon felt	TGI	0.1	1.85	10,150	1920	76
Carbon felt	PDI	0.1	1.67	8,900	3900	54
Carbon felt	II	0.1	1.60	9,500	—	70
Carbon felt	II	0.1	1.80	11,000	—	72
Pluton B-1/EC-201 [d]	PDI	0.99	1.43	5,700 [e] 11,500 [f]	—	82
Pluton B-1/EC-201 [g]	PDI	1.06	1.51	10,600	3700	54
Thornel 50 orthogonal weave	PDI	0.82	1.53	20,700	7500	54

Wool felt	II	0.4	1.65	18,000	—	72
G-1550 Graphite fabric[h]	II	0.5	1.63	24,430[i] 18,670[j]	4000	114
Carbon fabric[k]	II	0.5	1.65	18,230[i] 19,370[j]	6010	114

[a] Adapted from Ref. 55.

[b] Key: TGI, thermal-gradient infiltration; PDI, pressure-differential infiltration; II, isothermal infiltration.

[c] Compressed to one-half width, 42-h initial treatment at 1100° C, Morganite Ltd. tow.

[d] Tape wrap cylinder, 3M carbon fabric.

[e] Axial.

[f] Circumferential.

[g] Fabric layup plate.

[h] HITCO 8 Harness Satin graphite fabric.

[i] Edgewise orientation.

[j] Flatwise orientation.

[k] Kureha square-weave carbon fabric.

TABLE 12

Tensile Strength and Modulus of Parallel-Filament and Filament-Wound Substrates[a]

Material	Specimen	Tensile strength (psi) and modulus (10^6 psi)[b]		
		20°C	1650°C	2480°C
Parallel-filament CY2-5; phenolic 22323 carbonized 815°C	Round dog bone	30,000 (4.2)	35,500 (3.5)	34,500 (2.3)
45°-0°-45° Filament CY2-5, phenolic 22323 carbonized 815°C	Round dog bone	11,400 (1.0)	13,000 (2.2)	10,000 (1.0)
CVD/felt	Round dog bone	7,000 (2.7)	9,200 (2.1)	11,600 (2.0)
Parallel-filament WYB, pitch graphitized 1400°C	Miniature NOL ring	50,000	—	—
Parallel-filament Thornel 50, pitch graphitized 1400°C	Miniature NOL ring	87,600	—	—
Parallel-filament Morganite II, CVD matrix 1090°C	Flat dog bone	103,000 (20.0)	—	—
Filament-wound CY2-5, CVD matrix 1090°C, graphitized 2980°C	3-in. burst tube, 45°	7,100 (2.8)	—	—
Filament-wound CY2-5, CVD matrix 1090°C, graphitized 2980°C	3-in. burst tube, 70°	11,000 (2.8)	—	—
CVD/felt 1090°C	3-in. burst tube	4,500 (2.5)	—	—
CVD/felt 1090°C, graphitized 2980°C	3-in. burst tube	3,800 (2.0)	—	—

[a] Data from Ref. 78.

[b] Values for modulus are shown in parentheses.

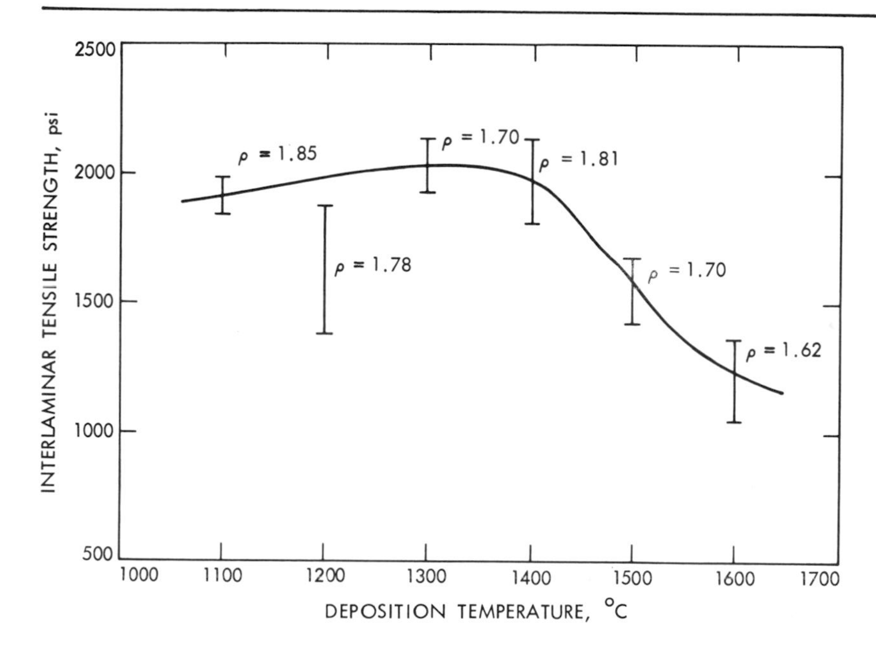

FIG. 33. Shear strength versus deposition temperature for carbon felt. From Ref. 77.

Fig. 40 [76]. Other investigators have reported similar behavior [58, 70, 99, 112]. The thermal expansion of infiltrated wool-precursor substrates has been reported to be much greater than that of rayon-base materials [72, 76, 80].

The effect of filament orientation on the coefficient of thermal expansion (CTE) for filament-wound (FW) cones is shown in Fig. 41 [99]. The average CTE values for CVD/felt are seen to be lower than those for CVD/FW. Expansion coefficients at 815°C of 4.0×10^{-6} °C^{-1} have been reported for composites with densities of 1.68 and 1.76 g/cm^3, respectively [89]. In both cases the core was made from Modmor type II fiber with CXH and C-3000 used as the wrap fiber [89]. Expansion coefficients of 1.8 to 2.0×10^{-6} °C^{-1} were reported for infiltrated unidirectional Modmor type II [89]. Thermal expansion data for a variety of other infiltrated substrates have been reported by Stover [80].

These results illustrate the trends and differences that have been reported for pyrolytic-carbon-infiltrated fibrous substrates. Additional

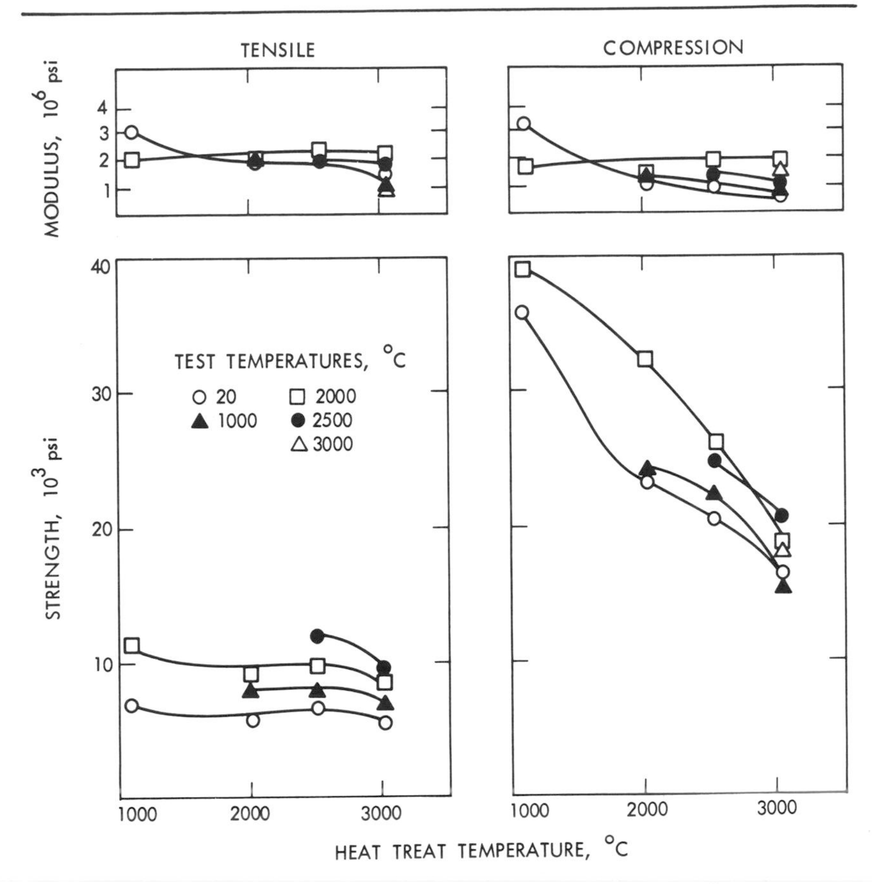

FIG. 34. Strength and modulus of infiltrated felt composites versus heat-treatment temperature. Test temperatures: ○, 20°C; ▲, 1000°C; □, 2000°C; ●, 2500°C; △, 3000°C. From Ref. 112.

studies are needed to define and characterize in more detail the interrelationship of pyrolytic carbon, fiber, and structure with thermal expansion.

2. Thermal Conductivity

The thermal conductivity of infiltrated fibrous composites is even less well defined and characterized than the mechanical properties and thermal expansion. Results for rayon- and wool-base infiltrated composites are summarized in Fig. 42 [72]. The striking feature of this figure is the dependence of thermal conductivity on density and orientation. The reader is referred to other references for thermal conductivity data for a variety of infiltrated substrates [58, 70, 75, 76, 84, 89, 99, 112].

The thermal conductivity of pyrolytic graphite and its temperature dependence have also been fairly well reviewed by Bokros [3]. A similar review for infiltrated fibrous composites is not possible at this time because

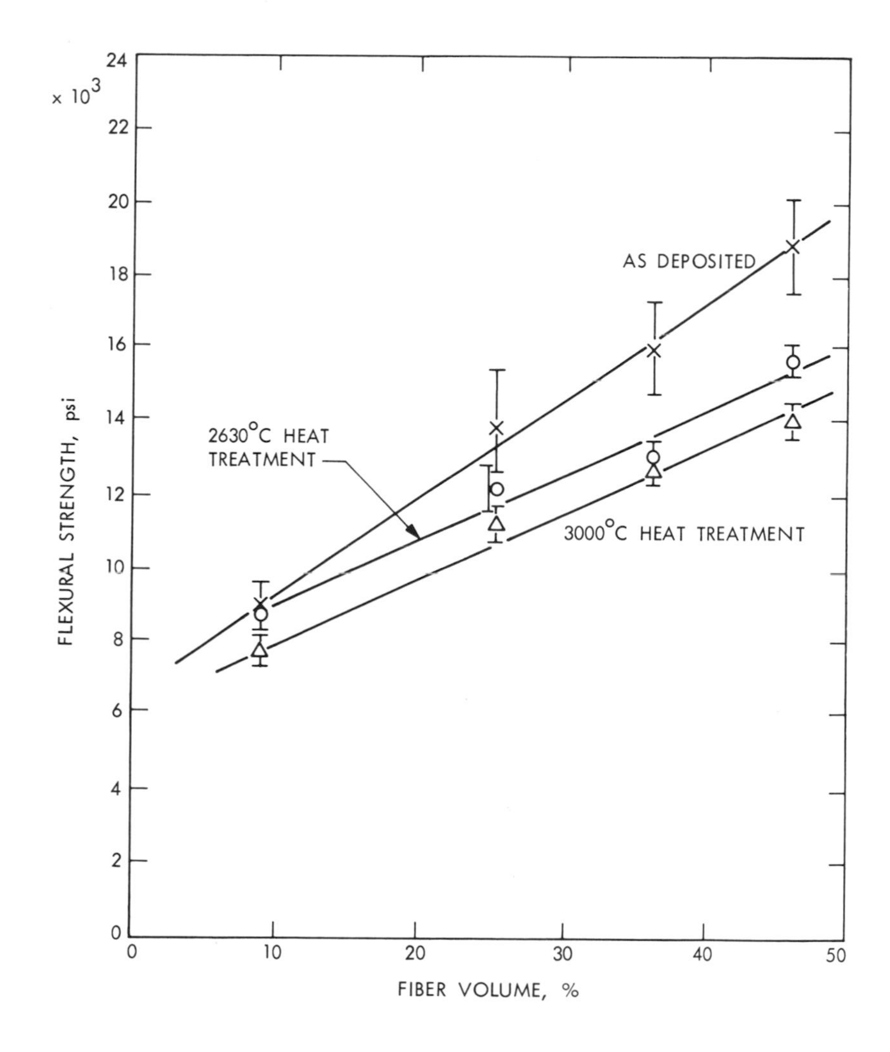

FIG. 35. Flexural strength versus fiber volume for infiltrated rayon-precursor carbon felt. From Ref. 87.

of a lack of thermal conductivity data and a lack of understanding of the relationship between thermal conductivity and the structure of pyrolytic carbon and fiber. Orientation and crystallite-size effects of each component along with their graphitizability and interdependency need to be characterized as a function of fiber type and processing conditions.

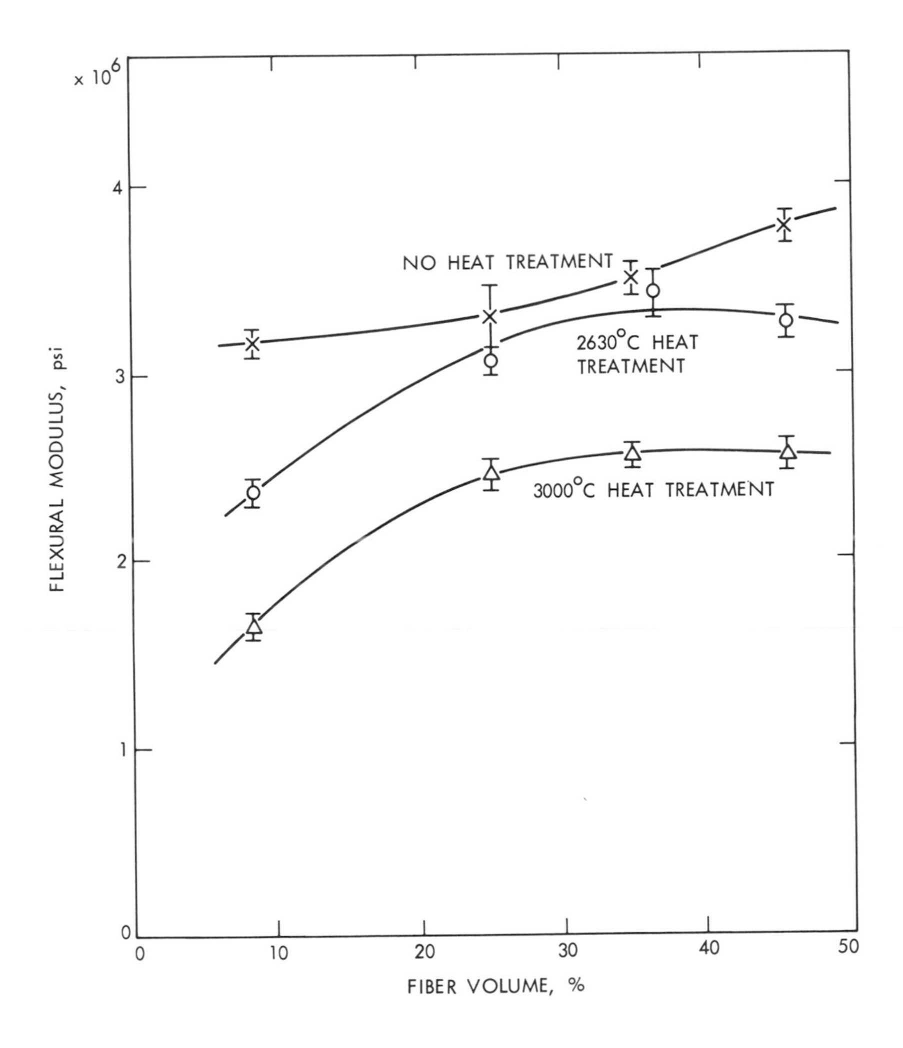

FIG. 36. Flexural modulus versus fiber volume for infiltrated rayon-precursor carbon felt. From Ref. 87.

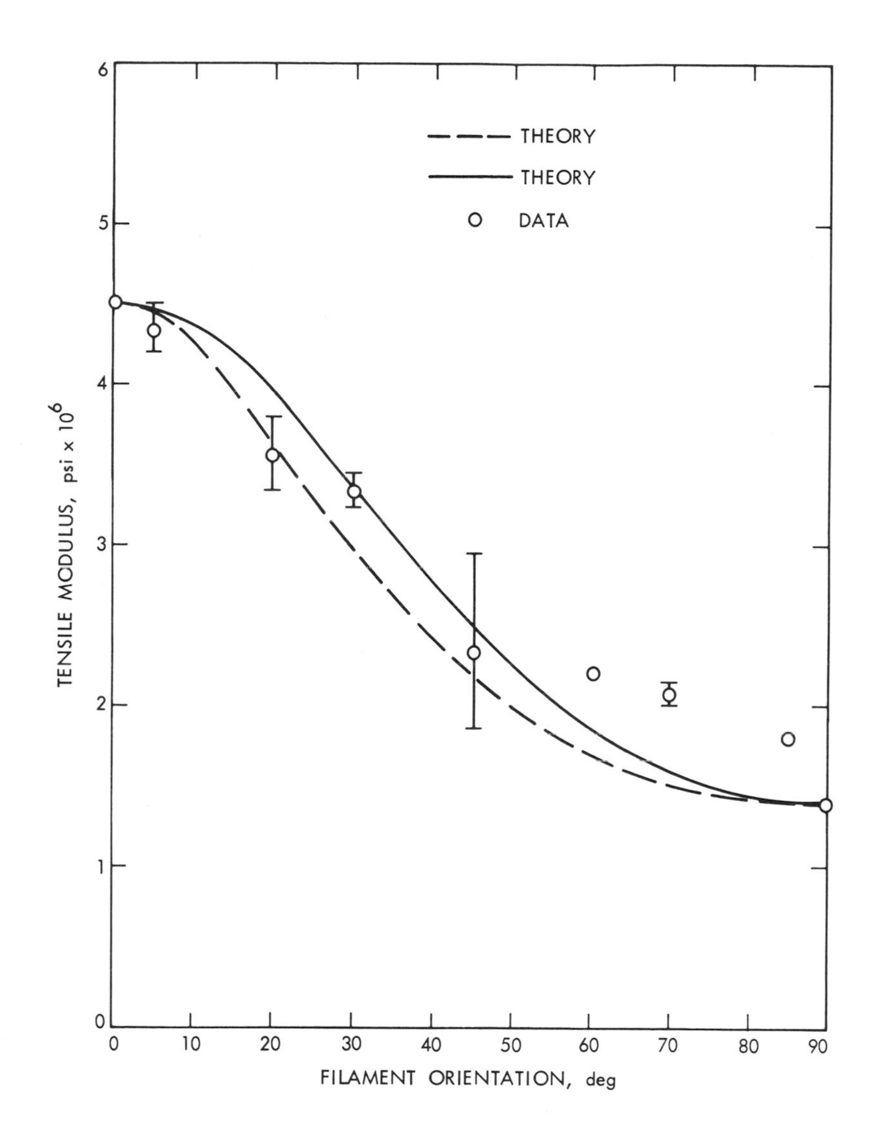

FIG. 37. Tensile modulus versus filament orientation. The curves show values calculated from theory; the circles show experimental values. From Ref. 99.

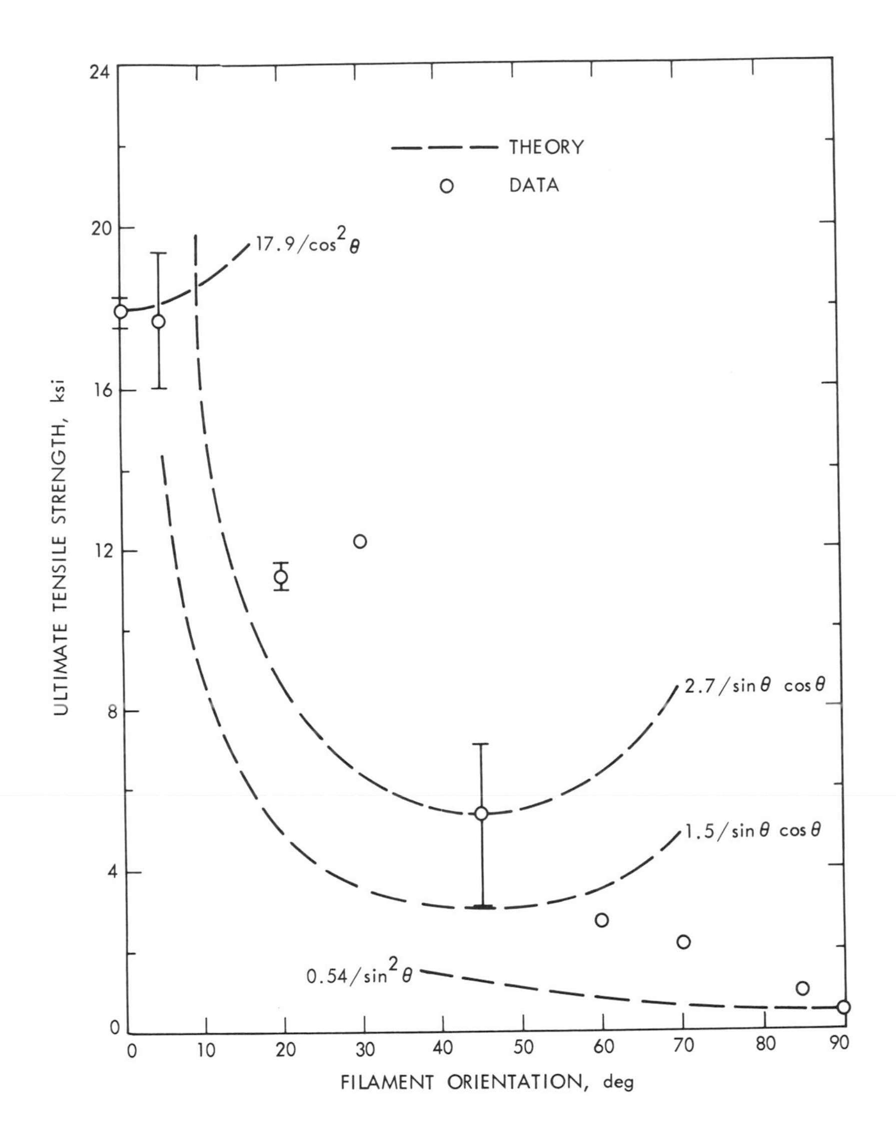

FIG. 38. Tensile strength versus filament orientation. The curves show values calculated from theory; the circles show experimental values. From Ref. 99.

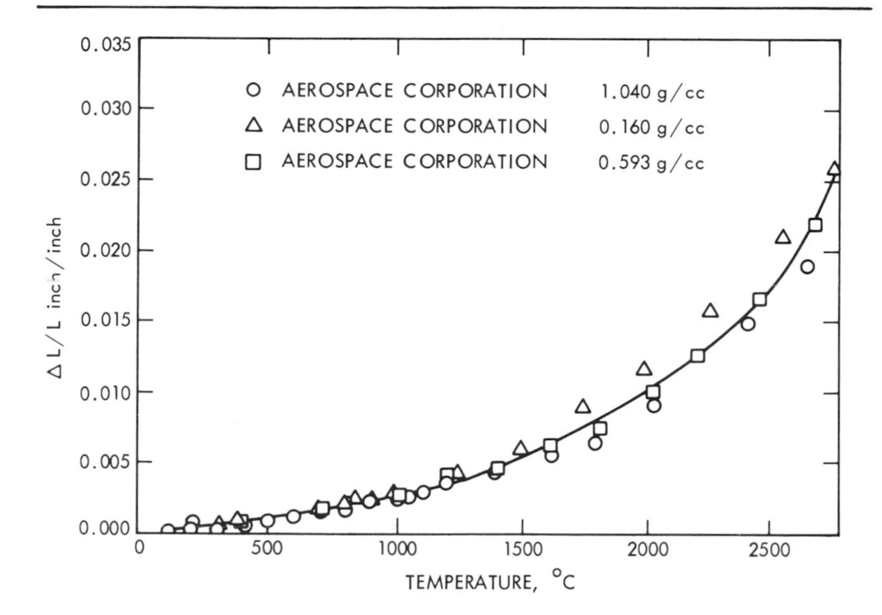

FIG. 39. Thermal expansion versus temperature for rayon-precursor infiltrated carbon felt parallel to the felt layers. Sample density: ○, 1.040 g/cm^3; △, 0.160 g/cm^3; □, 0.593 g/cm^3. From Ref. 72.

VI. APPLICATIONS FOR PYROLYTIC-CARBON-INFILTRATED POROUS SOLIDS

Free-standing pyrolytic graphite as well as pyrolytic-graphite-coated articles have found many uses in the aerospace industry for rocket nozzles, thrust chambers, nose cones, and heat shields. The main characteristic that makes pyrolytic graphite of interest for these applications is its anisotropy in structure and properties. In the direction parallel to the deposition surface pyrolytic graphite is like a metal. It has high strength and is a good conductor of heat and electricity. Perpendicular to the deposition surface pyrolytic graphite is like a ceramic and is a good insulator. The anisotropy of pyrolytic graphite is somewhat of a paradox. It creates interest in this material for certain high-temperature applications and also contributes to the problems associated with producing closed shapes or using this material for certain other applications. The appearance of cracks and delaminations, high residual stresses, low interlaminar shear strength, and problems in joining pyrolytic graphite to support structures tend to restrict

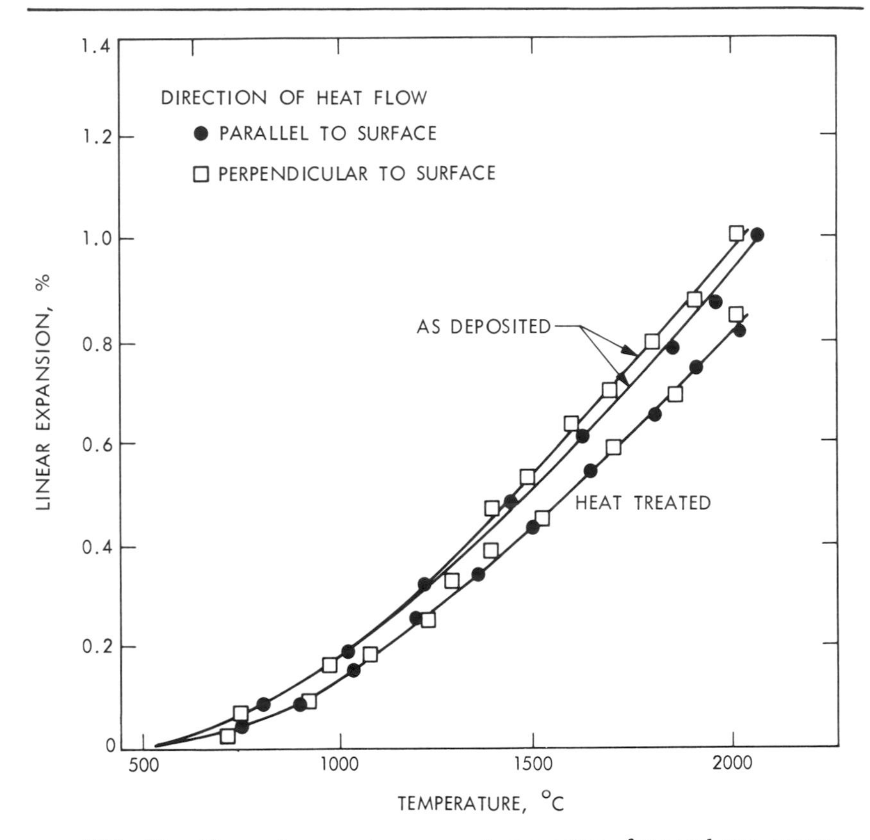

FIG. 40. Thermal expansion versus temperature for wool-precursor infiltrated carbon felt. Direction of heat flow: ●, parallel to surface; □, perpendicular to surface. From Ref. 76.

the applications for this material. However, materials and design engineers have made significant advances in utilizing a brittle, anisotropic material like pyrolytic graphite in the aerospace industry by learning how to design and live with its characteristics.

Combining porous particulate and porous fibrous substrates with pyrolytic carbon has resulted in the manufacturing of a new class of composite materials containing many of the desirable features of pyrolytic graphite as well as minimizing some of the undesirable features of monolithic pyrolytic graphite plates and closed shapes. In this new class of composites, particularly the fibrous composites, the materials engineers and scientists have

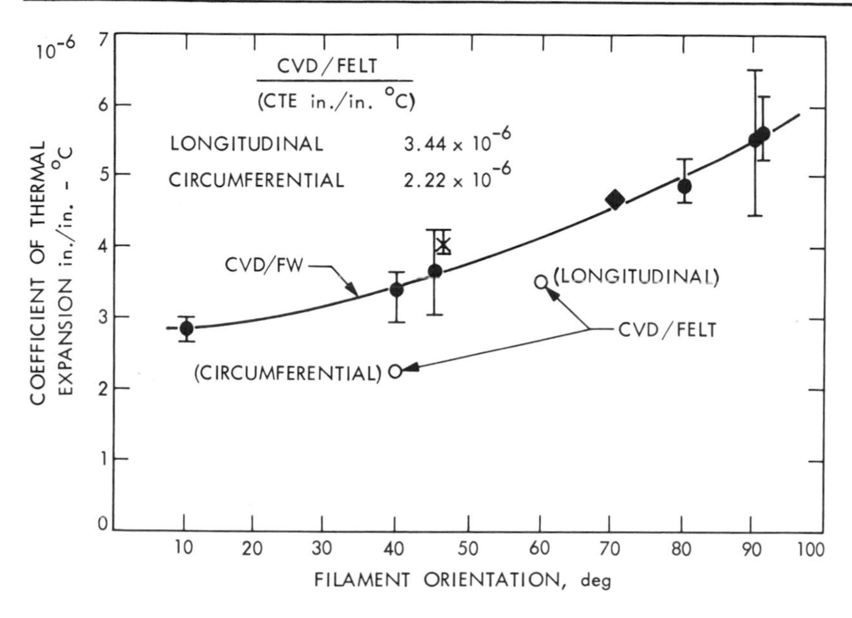

FIG. 41. Coefficient of thermal expansion as a function of filament orientation. Key: ●, average of four full-size heat shields; ◆, one Malta-size cone; X, average of two burst cylinders. From Ref. 99.

traded off some of the anisotropy of monolithic pyrolytic graphite and the problems associated therewith for a more reliable fibrous composite capable of being fabricated in a larger variety of thicknesses and shapes. In reviewing the applications for infiltrated porous solids, particulate substrates will be examined first. The intention here is to cite specific classes of applications for infiltrated particulate substrates. This application section is not intended to be an exhaustive survey.

Initial infiltration work was done in England in the 1950s to produce impervious graphites for the Dragon reactor [32, 33, 44, 64-67, 117]. Considerable success was achieved in producing graphites with low gas permeabilities. Other more economical methods were developed for producing impervious graphites. Similar work was sponsored in the United States for the gas-cooled reactor. Recent work sponsored by the U.S. Atomic Energy Commission [60] using the vacuum-pressure pulsing infiltration technique has been successful in reducing the helium permeabilities of

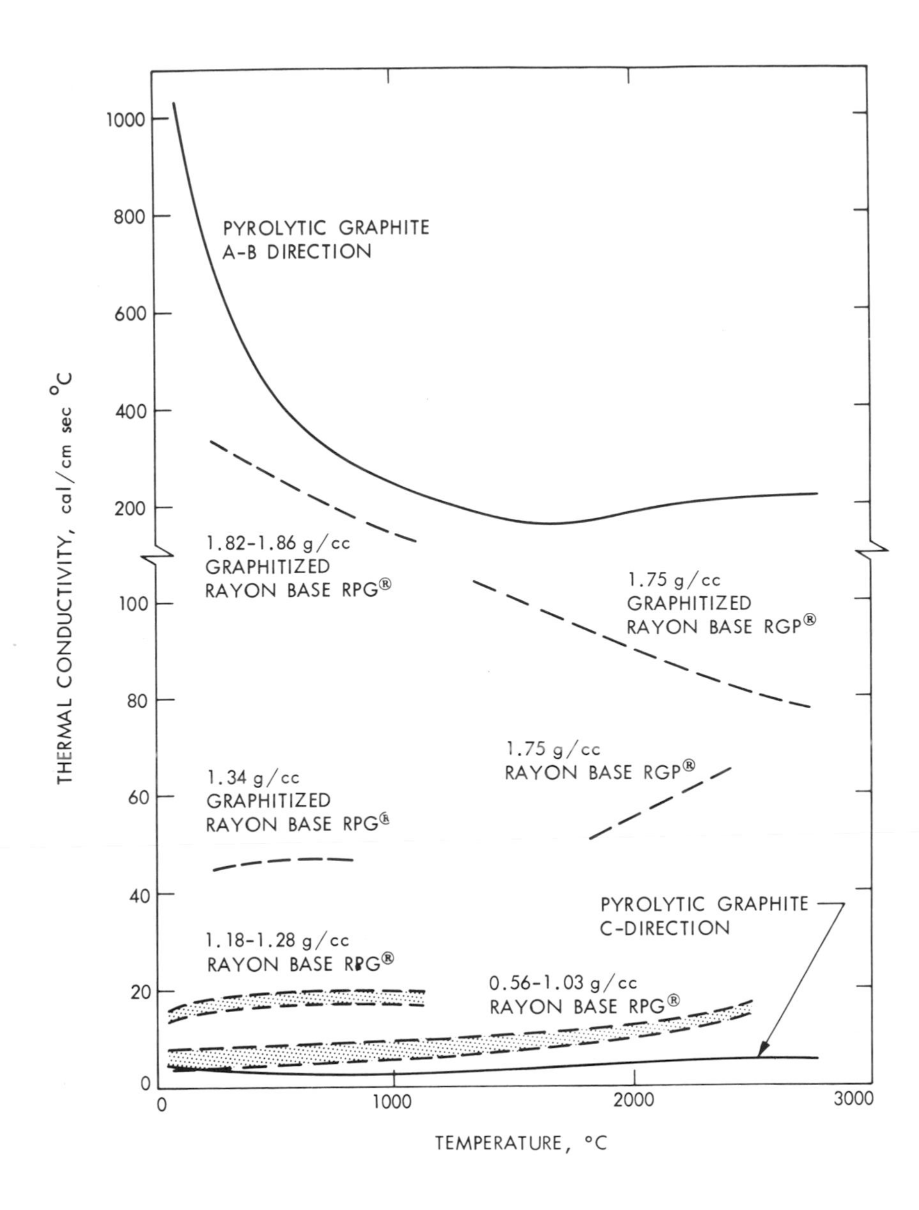

FIG. 42. Thermal conductivity as a function of density and graphitization. From Ref. 72.

graphite from 10^{-2} to 10^{-8} cm^2/sec, as required for the molten-salt breeder reactor. This technique is reported to offer considerable promise.

Other applications have been reported. Pyrolytic-carbon-infiltrated graphite is used in the electronics industry in crucibles for crystal-growing operations [61, 118]. Molded carbon components that are subsequently infiltrated with pyrolytic carbon have several applications in the foundry industry as well as use in spark-erosion techniques [119, 120]. The electrical properties of graphite are particularly important in spark-erosion applications. An additional application for infiltrated porous substrates that takes advantage of the electrical properties of graphite is in electrochemical grinding. Particulate grinding wheels have been made from silicon carbide and aluminum oxide by casting and dry-pressing, followed by suitable drying, carbonization, and infiltration with pyrolytic carbon, which becomes the principal bond [121]. Metal-impregnated and pyrolytic-carbon-infiltrated composites have been reported to show significantly greater oxidation resistance than graphites of comparable densities [122].

There may be other specific applications for pyrolytic-carbon-infiltrated particulate that I am unaware of. Such mechanical applications as vanes for rotating motors, bearings, seals, and pressure plates, are virgin areas where infiltrated particulate solids could find increasing usage. Still other potential application areas could be in heat shields, sintering and hot-pressing components, sliding electrical contacts, and parts in equipment and machinery in contact with corrosive substances [123].

The major interest for infiltrated substrates is in the area of fibrous structures. During the past 4 years research and development in pyrolytic-carbon-infiltrated fibrous substrates has nearly paralleled the interest in pyrolytic graphite during the early 1960s.

In the latter half of the 1960s a family of materials known as RPG was developed by the Super-Temp Company. These materials are produced by infiltrating rayon- and wool-precursor substrates [70-72, 75-78]. Starting with a rayon-precursor felt substrate with a density of 0.1 g/cm^3, virtually complete densification to 1.92 g/cm^3 has been reported [70]. This family of materials is finding potential application in the same aerospace areas that have attracted wide interest for monolithic pyrolytic graphite. Uniform parts up to several feet in cross section have been produced from felt-based

RPG in thicknesses of up to 6 in. [4, 57]. Such alloying elements as niobium, tantalum, zirconium, silicon, boron, and hafnium can be incorporated up to 33 wt% in RPG materials [4, 122]. Other infiltrated substrates that have found application or have potential application in the aerospace industry include RPP (Ling Temco Vought Aerospace Corp.), Carbitex (Carborundum Co.), and Pyro-carb (HITCO Corp.).

Sandia Laboratories has provided the main support for pyrolytic-carbon-infiltrated felt and filament-wound materials for heat-shield applications [58, 78, 124-128]. Two successful heat-shield flight tests conducted during 1970 have been reported [126]. Work at General Electric [89, 90, 129] and AVCO [130, 131] has also demonstrated the potential application of infiltrated fibrous substrates in the aerospace industry for nose tips and heat shields. A number of papers presented at the Air Force Materials Symposium described the potential for infiltrated composites [132-135]. In the rocket-nozzle and thrust-chamber areas infiltrated fabric layup substrates and infiltrated filament-wound structures have shown considerable promise [136-138]. Development of infiltrated carbon-carbon potential structures for the space-shuttle thermal protection system has likewise been reported [139-140]. In oxidizing environments pyrolytic carbon has been reported to show significantly lower oxidation than carbon binders from pitch and resin as well as carbon from rayon-precursor fibers [141-142].

In the field of prosthetics infiltrated carbon-carbon composites are currently being evaluated as load-bearing structures and for compatibility with animal tissue and fluids [143]. The greatest application potential as currently envisioned for infiltrated fibrous substrates is in the area of friction materials for brakes, clutches, and other energy-absorbing devices [144, 145]. Composites are of interest because of their high strength and thermal stability at elevated temperatures, low oxidation, and superior coefficient of friction under applied loads. It is believed that the applications listed here, and many others, for infiltrated particulate and fibrous substrates will come to fruition during the current decade.

ACKNOWLEDGMENTS

I am deeply grateful to many of my colleagues for their kind permission to use and reproduce published work. Special thanks are expressed to the workers at Sandia Laboratories as well as to Dr. Jack Bokros of Gulf General Atomics, whose published figures and data are used extensively throughout this chapter.

REFERENCES

1. H. M. Ezekiel, High Strength, High Modulus Graphite Fibers, Technical Report AFML-TR-70-100, Air Force Materials Laboratory, Wright-Patterson Air Force Base, Ohio, January 1971.

2. H. B. Palmer and C. F. Cullis, in Chemistry and Physics of Carbon, Vol. 1 (P. L. Walker, Jr., ed.), Dekker, New York, 1965, pp. 265-325.

3. J. C. Bokros, in Chemistry and Physics of Carbon, Vol. 5 (P. L. Walker, Jr., ed.), Dekker, New York, 1969, pp. 1-118.

4. W. H. Smith and D. H. Leeds, in Modern Materials, Vol. 7, Academic Press, New York, 1970, pp. 139-221.

5. A. G. Gaydon and H. G. Wolfhard, Flames, 2nd ed., Chapman and Hall, London, 1960, Chapter VIII.

6. R. A. Schwind, Carbon Deposition in Porous Media by the Thermal Decomposition of Methane, Ph.D. thesis, Washington University, St. Louis, 1968.

7. E. C. W. Smith, Proc. Roy. Soc. (London), A174, 110 (1940).

8. R. Cabannes, J. Phys. Radium, 17, 492 (1956).

9. G. Porter, in Fourth Symposium on Combustion, Cambridge, Mass., 1953, pp. 248-252.

10. G. Porter, in Combustion Research and Reviews, Butterworths, London, 1955, pp. 108-114.

11. R. C. Anderson, Literature on the Combustion of Petroleum, American Chemical Society, Washington, D.C., 1959.

12. C. R. Kinney and P. L. Walker, Jr., the Pennsylvania State University, Summary Report on A.E.C. Contract No. AT (30-1)-1710, November 1958.

13. C. R. Kinney and R. S. Slysh, in Proceedings of the Fourth Conference on Carbon, Pergamon Press, New York, 1960, p. 301.

14. R. O. Grisdale, A. C. Pfister, and W. Van Roosbroeck, Bell System Tech. J., 30, 271 (1951).

15. R. O. Grisdale, J. Appl. Phys., 24, 1082 (1953).

16. C. R. Kinney and P. L. Walker, Jr., the Pennsylvania State University, Quarterly Progress Reports NYO-6681, February 1955, and NYO-6682, May 1955.

17. G. R. Johnson and R. C. Anderson, in Proceedings of the Fifth Conference on Carbon, Vol. 1, Pergamon Press, New York, 1962, p. 395.

18. W. E. Sawyer and A. Man, U.S. Pat. 229, 335 (1880).

19. W. L. Voelper, U.S. Pat. 683,085 (1901).

20. R. Gomer and L. Meyer, J. Chem. Phys., 23, 1370 (1955).

21. L. Meyer, J. Chem. Phys., 28, 619 (1958).

22. P. A. Tesner, in Seventh Symposium on Combustion, Academic Press, New York, 1959, p. 546.

23. R. Iley and H. R. Riley, J. Chem. Soc., 1362 (1948).

24. R. J. Diefendorf, J. Phys. Radium, 57 (1960).

25. R. J. Diefendorf, in High Temperature Technology, Butterworths, London, 1964, p. 313.

26. R. J. Diefendorf, in Reactivity of Solids (J.W. Mitchell, ed.), Wiley, New York, 1969, pp. 461-475.

27. R. E. Duff and S. H. Bauer, Los Alamos Technical Report LA-2556, September 1961.

28. R. J. Diefendorf, in Carbon Composite Technology Symposium, University of New Mexico, January 1970, pp. 127-142.

29. K. V. Ingold and F. P. Lassing, Can. J. Chem., 31, 30 (1953).

30. F. O. Rice and M. T. Murphy, J. Amer. Chem. Soc., 66, 765 (1944).

31. R. E. Kinney and D. J. Crowley, Ind. Eng. Chem., 46, 258 (1954).

32. A. R. G. Brown, D. Clark, and J. Eastabrook, J. Less Common Metals, 1, 94 (1959).

33. A. R. G. Brown and W. Watt, in Industrial Carbon and Graphite, Society of Chemical Industry, London, 1958, pp. 86-100.

34. D. B. Murphy, H. B. Palmer, and C. R. Kinney, in Industrial Carbon and Graphite, Society of Chemical Industry, London, pp. 77-85.

35. T. J. Hirt and H. B. Palmer, Carbon, 1, 65 (1963).

36. T. J. Hirt and H. B. Palmer, in Proceedings of the Fifth Conference on Carbon, Vol. 1, Pergamon Press, New York, 1962, pp. 406 and 639.

37. H. B. Palmer, Carbon, 1, 55 (1963).

38. H. B. Palmer and T. J. Hirt, J. Phys. Chem., 67, 709 (1963).

39. B. Eisenberg and H. Bliss, in Symposium on Recent Kinetic Studies, Columbus, Ohio, Preprint 1A, May 1966.

40. G. I. Kozlov and V. G. Knorre, in Combustion and Flame, Butterworths, London, 1959, pp. 98-104; H. S. Glick, in 7th Symposium on Combustion, Butterworths, London, 1959, pp. 98-104.

41. L. Kramer and J. Happel, in The Chemistry of Petroleum Hydrocarbons, Vol. II (B. T. Brooks, ed.), Reinhold, New York, 1955.

42. J. L. Hudson and J. Heicklen, Carbon, 6, 405 (1968).

43. J. Heicklen, J. L. Hudson, and L. Armi, Carbon, 7, 365 (1969).

44. R. L. Bickerdike, A. R. G. Brown, G. Hughes, and H. Ranson, in Proceedings of the Fifth Conference on Carbon, Vol. 1, Pergamon Press, New York, 1962, p. 575.

45. R. W. Thiele, Ind. Eng. Chem., 31, 916 (1939).

46. O. Vohler, P. L. Reiser, and E. Sperk, Carbon, 6, 397 (1968).

47. K. Hedden and E. Wicke, in Proceedings of the Third Conference on Carbon, Pergamon Press, New York, 1957, p. 249.

48. W. T. Barry and W. H. Sutton, Conference on Behavior of Plastics in Advanced Flight Vehicle Environments, Dayton, Ohio, February 17, 1960.

49. W. T. Barry and C. A. Ganlin, Final Report USAF BSD Contract 269-486, General Electric Co., MSD, June 1961.

50. L. McAllister et al., AIChE, Chem. Eng. Progr. Symp. Series, 59 (40), 17 (1963).

51. E. Weger, J. Brew, and R. Schwind, Final Report, USAF BSD TR 66-385, November 1966.

52. E. Weger, J. Brew, and R. A. Servais, Final Report, SAMSO TR 68-123, January 1968.

53. A. R. Ford, Engineer, 224, 444 (1967).

54. W. V. Kotlensky and J. Pappis, in Summary of Papers, Ninth Biennial Conference on Carbon, DCIC, June 1969, p. 77.

55. W. V. Kotlensky, in SAMPE 16th National Symposium and Exhibit, Vol. 16, SAMPE, Azusa, Calif., 1971, p. 257.

56. D. H. Leeds, W. V. Kotlensky, and W. H. Smith, in Proceedings of the 17th Refractory Composites Working Group, April 1971, p. 559.

57. D. M. Forney, Jr. (ed.), Technical Report AFML-TR-70-133, Part III, August 1970.

58. H. M. Stoller and E. R. Frye, Sandia Report SC-DC-71 3653, April 1971.

59. V. K. Pecik, K. I. Makarov, and P. A. Tesner, Khim. Prom. (Moscow), 11, 808 (1964).

60. R. L. Beatty and D. V. Kiplinger, Nuclear Appl. Tech., 8, 488 (1970); also ORNL-4344 VC-80, Oak Ridge National Laboratory Semi Annual Progress Report, August 31, 1968, p. 230.

61. D. A. Schreiber and J. C. Davidson, paper presented to the American Ceramic Society, October 23-25, 1968, Pasadena, Calif.

62. L. Bochirol, M. Ortega, and R. Teoule, in Abstracts of Papers, Eighth Biennial Conference on Carbon, State University of New York, at Buffalo, June 1967, p. 107.

63. T. R. Jenkins, J. B. Morris, and H. J. C. Tulloch, paper presented at the Seventh Biennial Conference on Carbon, Cleveland, Ohio, June 1965.

64. R. L. Bickerdike, H. C. Ranson, C. Vivante, and G. Hughes, Dragon Project Report 139, January 1963.

65. R. L. Bickerdike and A. R. G. Brown, in Nuclear Graphite (O.E.E.C. Dragon Project Symposium 1959), O.E.E.C.-E.N.E.A., 1961, p. 109).

66. R. L. Bickerdike and G. Hughes, ibid., p. 91.

67. K. W. Carley-Macauly and M. Mackenzie, in Special Ceramics (P. Popper, ed.), Academic Press, New York, 1963, p. 151.

68. J. M. Hutcheon, B. Longstaff, and R. K. Warner, in Industrial Carbon and Graphite, Society of Chemical Industry, London, 1958, pp. 259-271.

69. P. L. Walker, Jr., L. G. Austin, and S. P. Nandi, in Chemistry and Physics of Carbon, Vol. 2 (P. L. Walker, Jr., ed.), Dekker, New York, pp. 320-380.

70. R. M. Williams, Advances in Structural Composites, Vol. 12, SAMPE, Azuza, Calif., October 1967.

71. D. H. Leeds and W. H. Smith, in Summary of Papers, Ninth Biennial Conference on Carbon, DCIC, June 1969, p. 201.

72. D. H. Leeds, S. W. Bauer, A. P. Valeriani, and W. H. Smith, SAMPE Journal, 6, 35 (1970).

73. D. W. Bauer, W. V. Kotlensky, J. W. Warren, W. H. Smith, and E. Gray, in Summary of Papers, Tenth Biennial Conference on Carbon, DCIC, June 1971, p. 57.

74. D. W. Bauer and W. V. Kotlensky, ibid., p. 59.

75. H. O. Pierson, Sandia Laboratories Development Report SC-DR-67-2969, December 1967.

76. H. O. Pierson and J. F. Smatana, Sandia Laboratories Development Report SC-DR-69-530, September 1969; also reported in Chemical Vapor Deposition, Second International Conference (J. M. Blocher, Jr., and J. C. Withers, eds.), the Electrochemical Society, New York, 1970, p. 487.

77. H. O. Pierson, J. D. Theis, and J. F. Smatana, Sandia Laboratories Development Report SC-DR-68-264, May 1968.

78. E. R. Frye, Carbon Composite Technology Symposium, University of New Mexico, January 1970, pp. 173-191.

79. J. J. Gebhardt, pp. 171-240 in Ref. 57.

80. E. R. Stover, Technical Report AFML-TR-69-67, Vol. II, May 1969.

81. J. J. Gebhardt, in Summary of Papers, Tenth Biennial Conference on Carbon, DCIC, June 1971, p. 53.

82. W. V. Kotlensky, pp. 241-293 in Ref. 57.

83. E. Gyarmati and H. Nickel, Carbon, 8, 400 (1970).

84. W. H. Pfeifer, W. J. Wilson, N. M. Griesenauer, M. F. Browning, and J. M. Blocher, Jr., in Chemical Vapor Deposition, Second International Conference (J. M. Blocher, Jr., and J. C. Withers, eds.), the Electrochemical Society, New York, 1970, p. 463.

85. F. E. Papalegis and R. G. Bourdeau, AFML Technical Documentary Report ML-TDR-64-201, July 1964.

86. G. M. Jenkins and J. O. Medwell, Carbon, 6, 645 (1968).

87. H. O. Pierson, B. Granoff, D. M. Schuster, and J. F. Smatana, paper presented at the International Conference on Carbon Fibres, Their Composites and Applications, the Plastics Institute, London, 1971.

88. W. V. Kotlensky, D. W. Bauer, J. W. Warren, W. H. Smith, and J. G. Campbell, in Summary of Papers, Tenth Biennial Conference on Carbon, DCIC, June 1971.

89. E. R. Stover, Technical Report AFML-TR-69-67, Vol. III, April 1970; also Vol. IV, May 1971.

90. E. R. Stover, W. C. Marx, L. Markowitz, and W. Mueller, Technical Report AFML-TR-70-283, March 1971.

91. R. L. Bond (ed.), Porous Carbon Solids, Academic Press, New York, 1967, p. 161.

92. E. Shoffner, Lockheed Missile and Space Co., private communication, May 1969.

93. J. S. Conroy, R. S. Slysh, D. B. Murphy, and C. R. Kinney, in Proceedings of the Third Conference on Carbon, Pergamon Press, New York, 1959, p. 395.

94. C. R. Kinney, in Proceedings of the First and Second Conferences on Carbon, Waverly Press, Baltimore, 1956, p. 83.

95. D. B. Murphy and W. V. Kotlensky, cited in Ref. 94, p. 90.

96. D. B. Murphy and W. V. Kotlensky, cited in Ref. 16.

97. R. E. Franklin, Proc. Roy. Soc. (London), A209, 196 (1951).

98. H. L. Riley, Quart. Rev., 1, 59 (1947).

99. H. M. Stoller, J. L. Irwin, B. Granoff, F. F. Wright, Jr., and J. H. Gieske, Sandia Laboratories Report No. SC-DC-71 4046, June 1971.

100. L. R. McCreight et al., Ceramic and Graphite Fibers and Whiskers, Academic Press, New York, 1965, p. 56.

101. R. Bacon, J. Appl. Phys., 31, 284 (1960).

102. F. E. Papaligis and R. G. Bourdeau, Technical Report AFML-TR-65-35, May 1965.

103. F. E. Papaligis and R. G. Bordeau, Technical Report AFML-TDR-64-201, July 1964.

104. R. L. Hough, Technical Report AFML-TR-64-336, December 1964.

105. R. M. Bushong, Technical Report AFML-TDR-64-297, November 1964.

106. W. V. Kotlensky, K. H. Titus, Jr., and H. E. Martens, Nature, 193, 1066 (1962).

107. W. V. Kotlensky and H. E. Martens, in Proceedings of the Fifth Carbon Conference, Vol. II, Pergamon Press, Ltd., 1963, p. 625.

108. Morganite Research and Development Ltd., Bulletin No. R25/10-68/5M.

109. Union Carbide Corp., Bulletin No. 465-203-GG; also Bulletin No. 465-206-B1.

110. H. R. Rauch, Sr., W. H. Sutton, and L. R. McCreight, Ceramic Fibers and Fibrous Composite Materials, Academic Press, New York, 1968, pp. 49-102.

111. A. Kelly and G. J. Davies, Metallurgical Rev., 10 (37) (1965).

112. H. A. MacKay, Sandia Laboratories Development Report SC-DR-70-871, May 1971.

113. J. R. McLaughlin, Nature, 227, 701 (1970).

114. D. W. Bauer and W. V. Kotlensky, "Relationship Between Structure and Strength for CVD Carbon Infiltrated Felt, Fabric and Woven Substrates," paper accepted for publication in three parts in the SAMPE Quarterly Journal, 1972-1973.

115. C. D. Coulbert, J. S. Waugh, and H. Shimizu, Marquardt Corp. Report 6142, October 1968.

116. T. Guess, Sandia Corp., private communication, March 1971.

117. A. R. G. Brown, A. R. Hall, and W. Watt, Nature, 172, 1145 (1953).

118. F. Rusinko, Speer Carbon Co., private communication, June 1971.

119. R. L. Bickerdike, G. Hughes, E. Parks, and F. J. Rolinson, Foundry Trade Journal, No. 2638, June 29, 1967, p. 3.

120. H. C. Ranson, private communication, March 1971.

121. D. H. Leeds, E. Kelly, J. E. Price, G. Nisula, W. E. Sanders, and I. Weber, in SAMPE, 16th National Symposium and Exhibit, Vol. 16, SAMPE, Azusa, Calif., 1971, p. 272.

122. D. H. Leeds, E. Kelly, and J. Heicklen, Ind. Eng. Chem. Prod. Res. Develop., 9, 573 (1970).

123. K. Koziol and C. Conradty, private communications, August 1971.

124. J. E. McDonald, in Carbon Composite Technology Symposium, University of New Mexico, January 1970, p. 13.

125. D. F. McVey, I. Averbach, and D. D. McBride, Paper No. 70-155, AIAA 8th Aerospace Sciences Meeting, New York, January 19-21, 1970.

126. D. J. Ragali and H. W. Schmitt, paper presented at the Conference on Continuum Aspects of Graphite Design, ORNL, Gatlinburg, Tenn., November 1970.

127. J. L. Irwin and O. J. Burchett, in SAMPE 16th National Symposium and Exhibit, Vol. 16, SAMPE, Azusa, Calif., 1971, p. 179.

128. J. L. Irwin, ibid, p. 190.

129. E. R. Stover, in Carbon Composite Technology Symposium, University of New Mexico, January 1970, p. 55.

130. L. E. McAllister, AVCO Corp., private communication, April 1971.

131. L. E. McAllister and A. R. Taverna, in Summary of Papers, Tenth Biennial Conference on Carbon, DCIC, June 1971, p. 66.

132. S. Channon, paper presented at the Air Force Materials Symposium 70, Miami Beach, Fla., May 18-22, 1970.

133. C. Pratt, paper presented at the Air Force Materials Symposium 70, Miami Beach, Fla., May 18-22, 1970.

134. J. Latva, paper presented at the Air Force Materials Symposium 70, Miami Beach, Fla., May 18-22, 1970.

135. P. Pirrung, paper presented at the Air Force Materials Symposium 70, Miami Beach, Fla., May 18-22, 1970.

136. W. H. Armour and R. M. Hale, Technical Report AFML-TR-70-26, March 1970.

137. P. Pirrung, Air Force Materials Laboratory, private communications, September 1971.

138. J. Campbell, Marquardt Corp., private communications, September 1970.

139. W. C. Jones and G. M. Studdert, in SAMPE 16th National Symposium and Exhibit, Vol. 16, SAMPE, Azusa, Calif., 1971, p. 65.

140. K. R. Carnahan and R. W. Kiger, in Summary of Papers, Tenth Biennial Conference on Carbon, DCIC, June 1971, p. 68.

141. J. W. Warren, Super-Temp Co., private communication, August 1971.

142. J. Latva, Air Force Materials Laboratory, private communication, September 1971.

143. D. H. Leeds, Super-Temp Co., private communication, July 1971.

144. D. W. Bauer and W. V. Kotlensky, in Summary of Papers, Tenth Biennial Conference on Carbon, DCIC, June 1971, p. 59.

145. N. Hooton, Bendix Corp., private communication, August 1970.

AUTHOR INDEX

Numbers in brackets are reference numbers. Underlined numbers give the pages on which the complete references are listed.

L

M

N

O

T

SUBJECT INDEX

Other books of interest to you...

Because of your interest in our books, we have included the following catalog of books for your convenience.

Any of these books are available on an approval basis. This section has been reprinted in full from our ***material science*** catalog.

If you wish to receive a complete catalog of MDI books, journals and encyclopedias, please write to us and we will be happy to send you one.

material science

including
Polymers, Plastics, Fibers, and Coatings
Metals and Metallurgy
Ceramics and Glass
Vacuum Science

ALTGELT and SEGAL
Gel Permeation Chromatography

edited by KLAUS H. ALTGELT, *Chevron Research Company, Richmond, California,* and LEON SEGAL, *South Regional Research Laboratory, U.S.D.A., New Orleans, Louisiana*

672 pages, illustrated. 1971

Demonstrates the manifold applications of gel permeation chromatography in the field of polymer chemistry. Directed to all research, quality-control, and analytical chemists working with conventional and unconventional polymers and other large molecules in the fields of polymer, cellulose, and petroleum chemistry.

CONTENTS: The sizes of polymer molecules and the GPC separation, *F. W. Billmeyer, Jr. and K. H. Altgelt.* Gel permeation chromatography column packings – types and uses, *D. J. Harmon.* Chromatographic instrumentation and detection of gel permeation effluents, *E. M. Barrall, II and J. F. Johnson.* Peak resolution and separation power in gel permeation chromatography, *D. J. Harmon.* A review of peak broadening in gel chromatography, *R. N. Kelley and F. W. Billmeyer, Jr.* Mathematical methods of correcting instrumental spreading in GPC, *L. H. Tung.* Comparison of different techniques of correcting for band broadening in GPC, *J. H. Duerksen.* Separation mechanisms in gel permeation chromatography, *W. W. Yau, C. P. Malone, and H. L. Suchan.* Gel permeation chromatography and thermodynamic equilibrium. *E. F. Casassa.* Calibration of GPC columns, *H. Coll.* Data treatment in GPC, *L. H. Tung.* The overload effect in gel permeation chromatography, *J. C. Moore.* Gel permeation chromatography using a bio-glas substrate having a broad pore size distribution, *A. R. Cooper, J. H. Cain, E. M. Barrall, II, and J. F. Johnson.* High resolution gel permeation chromatography – using recycle, *K. J. Bombaugh and R. F. Levangie.* Gel permeation chromatography with high loads, *K. H. Altgelt.* Fast gel permeation chromatography, *J. N. Little, J. L. Waters, K. J. Bombaugh, and W. J. Pauplis.* Extension of GPC techniques, *G. Meyerhoff.* Phase distribution chromatography (PDC) of polystyrene, *R. H. Casper and G. V. Schulz.* Apparent and real distribution in GPC (experiments with PMMA samples), *K. C. Berger and G. V. Schulz.* The instrument spreading correction in GPC. I: The general shape function using a linear calibration curve, *T. Provder and E. M. Rosen.* The instrument spreading correction in GPC. II: The general shape function using the Fourier transform method with a nonlinear calibration curve, *E. M. Rosen and T. Provder.* Behavior of micellar solutions in gel permeation chromatography: A theory based on a simple model, *H. Coll.* Gel permeation analysis of macromolecular association by an equilibrium method, *B. F. Cameron, L. Sklar, V. Greenfield, and A. D. Adler.* Gel filtration chromatography, *B. F. Cameron.* Determination of polymer branching with gel permeation chromatography. Abstract of a review, *E. E. Drott and R. A. Mendelson.* Fractionation of linear polyethylene with gel permeation chromatography. Part III, *N. Nakajima.* Application of GPC in the study of stereospecific block copolymers, *R. D. Mate and M. R. Ambler.* Composition of butadiene-styrene copolymers by gel permeation chromatography, *H. E. Adams.* A direct GPC calibration for low molecular weight polybutadiene, employing dual detectors, *J. R. Runyon.* Quantitative determination of plasticizers in polymeric mixtures by GPC, *D. F. Alliet and J. M. Pacco.* Evaluation of pulps, rayon fibers, and cellulose acetate by GPC and other fractionation methods, *W. J. Alexander and T. E. Muller.* Characterization of the internal pore structures of cotton and chemically modified cottons by gel permeation, *L. F. Martin, F. A. Blouin, and S. P. Rowland.* Application of GPC to studies of the viscose process. I: Evaluation of the method, *L. H. Phifer and J. Dyer.* Application of GPC to studies of the viscose process. II: The effects to steeping and alkali-crumb aging, *J. Dyer and L. H. Phifer.* Gel permeation chromatography calibration. I: Use of calibration curves based on polystyrene in THF and integral distribution curves of elution volume to generate calibration curves for polymers in 2,2,2-trifluoroethanol, *T. Provder, J. C. Woodbrey, and J. H. Clark.* Modification of a gel permeation chromatograph for automatic sam-

(continued)

ALTGELT and SEGAL *(continued)*

ple injection and on-line computer data recording, *A. R. Gregges, B. F. Dowden, E. M. Barral, II, and T. T. Horikawa.* Characterization of crude oils by gel permeation chromatography, *H. H. Oelert, D. R. Latham, and W. E. Haines.* Separation and characterization of high-molecular-weight saturate fractions by gel permeation chromatography, *J. H. Weber and H. H. Oelert.* Fractionation of residuals by gel permeation chromatography, *E. W. Albaugh, P. C. Talarico, B. E. Davis, and R. A. Wirkkala.* Combined gel permeation chromatography–NMR techniques in the characterization of petroleum residuals, *F. E. Dickson, R. A. Wirkkala, and B. E. Davis.* A rapid method of identification and assessment of total crude oils and crude oil fractions by gel permeation chromatography, *J. N. Done and W. K. Reid.* Gel permeation analysis of asphaltenes from steam stimulated oil wells, *C. A. Stout and S. W. Nicksic.* GPC separation and integrated structural analysis of petroleum heavy ends, *K. H. Altgelt and E. Hirsch.*

AMERICAN VACUUM SOCIETY *Experimental Vacuum Science and Technology*

edited by THE AMERICAN VACUUM SOCIETY EDUCATION COMMITTEE

288 pages, illustrated. 1973

A collection of experiments, which are graded from simple procedures to sophisticated vacuum processes, and designed to aid instructors and introduce students to the basic concepts and techniques of the field of vacuum science. Includes an extensive bibliography to stimulate further investigation. Especially useful for all students and teachers in the many fields of the basic sciences and engineering where vacuum methods and techniques are important.

CONTENTS: **Section 1: Procedures in Vacuum Production and Measurement,** *W. Brunner and H. Patton.* **Section 2: Experiments which Illustrate the Characteristics of the Vacuum Environment:** Demonstration of the outgassing of different vacuum materials, *F. Rosebury.* Comparison of gas evolution phenomenon from glass and metal system envelopes during baking, *R. Lawson.* Determination of the net quantity of gas flowing through a cylindrical tube, *K. Busen.* **Section 3: Experiments which Illustrate the Dependence of the Physical Properties of Gases on Gas Density:** Measurement of the pumping action of an ionization gauge, *H. Farber.* Study of the linearity of an ionization gauge, *J. Miller, III.* Calibration of gauges, *C. Morrison.* **Section 4: Experiments which Examine Physical and Chemical Interactions at Surfaces:** Study of the sorption of gases for different gas–sorbent combinations, *K. Wear.* The use of sorbents as traps and pumps, *H. Farber.* Sorption of gases by titanium, *H. Farber.* Investigation of the passage of oxygen across a silver barrier, *K. Busen.* **Section 5: Processes Requiring a Vacuum Environment:** Thin film evaporation, *M. Thomas.* Fabrication of a nichrome resistor, *R. Riegert and G. Breitweiser.* Sputtering, *P. Grosewald.* Ejection patterns in single crystal sputtering, *G. Wehner.* **Section 6: Special Projects:** Study of the sublimation of ice at various pressures, *W. Parker.* Study of friction, *P. McElligott.* Measurement of the mean free path of conduction electrons in silver, *R. Olson and J. Wilson.* Construction and use of a cathode ray tube, *B. Kendall and H. Luther.* Construction of a vacuum triode using solder glass techniques, *J. King and J. Orsula.* Experiments using solder glass techniques, *D. Whitcomb.* **Section 7: Speculations:** Original thought experiments, *M. Carbone.* Provocative ideas and questions, *N. Milleron.*

BEER *Liquid Metals: Chemistry and Physics*

(Monographs and Textbooks in Material Science Series, Volume 4)

edited by SYLVAN Z. BEER, *Converta Enterprises, Inc., Syracuse, New York*

742 pages, illustrated. 1972

Presents a comprehensive review of the research done on the liquid state of metals, bringing together the latest advances, as well as data previously scattered among a wide variety of publications. Of prime importance to chemists, physicists, research metallurgists, metallurgical engineers, and materials scientists working in the areas of liquid-state theory, the theory of metals, process metallurgy involving liquid metals, and high-temperature chemistry.

CONTENTS: On the thermodynamic formalism of metallic solutions, *C. H. P. Lupis.* Kinetics of evaporation of various elements from liquid iron alloys under vacuum, *R. Ohno.* Relation between thermodynamic and electrical properties of liquid alloys, *D. N. Lee and B. D. Lichter.* The surface tension of liquid metals, *B. C. Allen.* Significant structure theory applied to liquid metals, *S. M. Breitling and H. Eyring.* Diffraction analysis of liquid metals and alloys, *C. N. J. Wagner.* The optical properties of liquid metals, *J. N. Hodgson.* Effect of pressure on the properties of liquid metals, *A. Rapoport.* Sound propagation in liquid metals, *R. T. Beyer and E. M. Ring.* The viscosity of liquid metals, *R. T. Beyer and E. M. Ring.* Magnetic properties of liquid metals, *R. Dupree and E. F. W. Seymour.* Diffusion in liquid metals, *N. H. Nachtrieb.* Electromigration in liquid alloys, *S. G. Epstein.* Electronic nature of liquid metals and liquid metal theory, *J. E. Enderby.* Structure and properties of noncrystalline metallic alloys produced by rapid quenching of liquid alloys, *B. C. Giessen and C. N. J. Wagner.*

BLACK and PRESTON ***High-Modulus Wholly Aromatic Fibers***

(Fiber Science Series, Volume 5)

edited by W. Bruce Black, *Monsanto Textiles Company, Pensacola, Florida* and Jack Preston, *Monsanto Textiles Company, Chemstrand Research Center, Durham, North Carolina*

304 pages, illustrated. 1973

Based on a symposium on high-modulus aromatic fibers held by the American Chemical Society in Boston on April 13, 1972. The first formal publication of research which shows the relationship of fiber properties to polymer structure. Extremely significant reading for all fiber scientists and material scientists; plastics scientists and engineers interested in fiber-reinforced plastics; spacecraft and aircraft oriented engineers; and scientists in the industrial fiber, sports equipment, and airframe fields.

CONTENTS: High-modulus wholly aromatic fibers: Introduction to the Symposium and historical perspective, *W. Black*. High-modulus wholly aromatic fibers. I. Wholly ordered polyamide-hydrazines and poly-1,3,4,-oxadizole-amides, *J. Preston, W. Black and W. Hofferbert, Jr.* High-modulus wholly aromatic fibers. II. Partially ordered polyamide-hydrazides, *J. Preston, W. Black, and W. Hofferbert, Jr.* Self-regulating polycondensations. II. A study of the order present in polyamide-hydrazides derived from terephthaloyl chloride and p-aminobenzhydrazide, *R. Morrison, J. Preston, J. Randall, and W. Black.* Self-regulating polycondensations. III. NMR analysis of oligomers derived from terephthaloyl chloride and p-aminobenzhydrazide, *J. Randall, R. Morrison, and J. Preston.* Some physical and mechanical properties of some high-modulus fibers prepared from all-para aromatic polyamide-hydrazides, *W. Black, J. Preston, H. Morgan, G. Raumann, and M. Lilyquist.* Morphology and crystal structure of wholly aromatic all-para polyamide-hydrazide polymers, *V. Holland.* X-ray study of an all-para wholly aromatic polyamide-hydrazide[a,b], *R. Miller.* Molecular weight characterization of wholly para-oriented, aromatic polyamide-hydrazides and wholly aromatic polyamides, *J. Burke.* Construction and properties of fabrics of high-modulus organic fibers useful for composite reinforcing, *M. Lilyquist, R. DeBrunner, and J. Fincke.* Mechanical properties of a high-modulus polyamide-hydrazide fiber in composites and of the polyamide-hydrazide fiber and fabric composites, *D. Zaukelies and B. Daniels.* Tire cord application of high-modulus fibers derived from polyamide-hydrazides, *G. Raumann and J. Brownlee.* The application of high-modulus fibers to ballistic protection, *R. Laible, F. Figucia, and W. Ferguson.* High-modulus wholly aromatic fibers. III. Random copolymers containing hydrazide and/or carbonamide linkages, *J. Preston, H. Morgan, and W. Black.*

BOLKER ***Natural and Synthetic Polymers:*** *An Introduction*

by Henry I. Bolker, *Department of Chemistry, McGill University, Montreal, Quebec*

in preparation. 1973

Presents a unified approach to polymer chemistry, with equal emphasis on natural and synthetic polymers, and is arranged in a logical sequence of topics based on increasing complexity of molecular architecture. Useful as a textbook for a first course in polymer chemistry and as a reference book for workers in the field.

CONTENTS: Introduction • Natural condensation polymers: The linear polysaccharides • Synthetic condensation (step-growth) polymers • Addition (chain-growth) polymers • Stereoregularity in addition polymers • Branched homopolymers: Synthetic and natural • Natural heteropolymers: I. Heteropolysaccharides • Natural heteropolymers: II. Nucleic acids • Copolymers and copolymerization • Cross-linking in synthetic polymers • Natural heteropolymers: III. Polypeptides and proteins • Lignins.

BROWNING ***Analysis of Paper***

by B. L. Browning, *The Institute of Paper Chemistry, Appleton, Wisconsin*

352 pages, illustrated. 1969

Provides comprehensive coverage of methods for chemical analysis of paper. Is of value to manufacturers of paper and paper board, suppliers of components or of additives introduced into paper, converters and printers, purchasers and users, librarians, and others concerned with the properties, behavior, and applications of paper that are related to composition.

CONTENTS: Paper as a commodity • Sampling and preparation of sample • Determination of moisture • Fiber analysis • Fiber quality methods • Lignin • Rosin size • Starch • Proteins • Coatings • Waxes and oils • Fillers and white coating pigments • Dyes and colored pigments • Acidity and alkalinity • Residues and impurities • Biological control agents • General identification of additives in paper • Synthetic resins • Wet-strength agents • Polysaccharides and gums • Miscellaneous additives • Noncellulose fibers • Specks and spots • Permanence of paper • Paper in forensic science.

BUTLER, O'DRISCOLL, and SHEN ***Reviews in Macromolecular Chemistry***

(Book Edition)

edited by George B. Butler, *Department of Chemistry, University of Florida,*

(continued)

BUTLER, O'DRISCOLL, and SHEN *(continued)*

Gainesville, and KENNETH F. O'DRISCOLL, *Department of Chemical Engineering, University of Waterloo, Ontario, Canada* and MITCHEL SHEN, *Department of Chemical Engineering, University of California, Berkeley*

Vol. 1 *out of print*

Vol. 2 388 pages, illustrated. 1968

Vol. 3 430 pages, illustrated. 1969

Vol. 4 428 pages, illustrated. 1970

Vol. 5, Part I see NEUSE and ROSENBERG

Vol. 5, Part II 250 pages, illustrated. 1970

Vol. 6 498 pages, illustrated. 1971

Vol. 7 314 pages, illustrated. 1972

Vol. 8 346 pages, illustrated. 1972

Vol. 9 380 pages, illustrated. 1973

Reviews of the currently published literature for those who wish to keep abreast of the new and rapidly advancing developments in macromolecular chemistry. Of interest to organic and physical chemists, biochemists, engineers, and all students and research workers in polymer chemistry and related fields.

CONTENTS:

Volume 9: Mechanical properties of polymers: The influence of molecular weight and molecular weight distribution, *J. Martin, J. Johnson, and A. Cooper*. On the mathematical modeling of polymerization reactors, *W. Ray*. Anionic cyclopolymerization, *C. McCormick and G. Butler*. Carbon-13 NMR of polymers, *V. Mochel*. Thermodynamics of polymerization. IV. Thermodynamics of equilibrium polymerization, *H. Sawada*.

CARROLL *Physical Methods in Macromolecular Chemistry*

a series edited by BENJAMIN CARROLL, *Rutgers—The State University, Newark, New Jersey*

Vol. 1 400 pages, illustrated. 1969

Vol. 2 384 pages, illustrated. 1972

A series which reviews why and how analytical methods are used in the study of macromolecules. Each method is critically discussed by experts in the field. Directed to researchers in polymer chemistry, biopolymers, and organic and inorganic chemistry.

CONTENTS:

Volume 1: Surface chemistry and polymers, *M. Rosoff*. Internal reflection spectroscopy, *J. K. Barr and P. A. Flournoy*. Electric properties of synthetic polymers, *E. O. Forster*. Assessing radiation effects in polymers, *P. Y. Feng and E. S. Freeman*. Fluorescence techniques for polymer solutions, *D. J. R. Laurence*. Insoluble polymers: Molecular weights and their distributions, *H. C. Cheung*.

Volume 2: Gel permeation chromatography in polymer chemistry, *D. D. Bly*. Interactions of polymers with small ions and molecules, *D. J. R. Laurence*. Electric properties of biopolymers: Proteins, *E. O. Forster and A. P. Minton*. Thermal methods, *E. P. Manche and B. Carroll*.

CARTER *Essential Fiber Chemistry*

(Fiber Science Series, Volume 2)

by MARY E. CARTER, *FMC Corporation, American Viscose Division, Marcus Hook, Pennsylvania*

232 pages, illustrated. 1971

Discusses the chemical and physical structure and properties of ten commercially important fibers. Useful to all chemists interested in the research and development of natural and man-made fibers.

CONTENTS: Cotton • Rayon • Cellulose acetate • Wool • Polyamide • Acrylic fibers • Polyethylene terephthalate • Polyolefins • Spandex • Glass.

CONLEY *Thermal Stability of Polymers*

In 2 Volumes

(Monographs in Macromolecular Chemistry Series)

edited by R. T. CONLEY, *Wright State University, Dayton, Ohio*

Vol. 1 656 pages, illustrated. 1970

Vol. 2 in preparation. 1974

CONTENTS: Introduction, *R. T. Conley*. Molecular structure and stability criteria, *R. T. Conley*. The relationship between the kinetics and mechanism of thermal depolymerization, *R. H. Boyd*. Random scission processes, *A. V. Tobolsky, A. M. Kotliar, and T. C. P. Lee*. Fundamental reactions in oxidation chemistry, *P. M. Norling and A. V. Tobolsky*. Thermal and oxidative degradation of polyethylene, polypropylene, and related olefin polymers, *R. H. Hansen*. Thermal and oxidative degradation of natural rubber and allied substances, *E. M. Bevilacqua*. Vinyl and vinylidene polymers, *R. T. Conley and R. Malloy*. Fluorocarbon polymers, *W. W. Wright*. Thermal and thermo-oxidative degradation of polyamides, polyesters, polyethers, and related polymers, *R. T. Conley and R. A. Gaudiana*. Thermosetting resins, *R. T. Conley*. Thermal and thermo-oxidative degradation of cellulosic polymers, *R. T. Conley*. Heterocyclic polymers, *G. P. Shulman*. Degradation of inorganic polymers, *J. Economy and J. H. Mason*.

D'ALELIO and PARKER *Ablative Plastics*

edited by GAETANO F. D'ALELIO, *Department of Chemistry, University of Notre Dame, Indiana,* and JOHN A. PARKER, *NASA, Ames Research Center, Moffet Field, California*

504 pages, illustrated. 1971

Provides the comprehensive and rational approach required for the design and production of reliable head shields for future space missions. Includes discussions on the various aspects of heat-rejection mode as a function of heating rate; the nature of the heat transfer, both radiative and conductive; and the nature of degrading polymers. A valuable reference for all aerospace scientists, polymer chemists, physicists, and aerodynamic engineers.

CONTENTS: Ablative polymers in aerospace technology, *D. L. Schmidt*. Hypervelocity heat protection—a review of laboratory experiments, *N. S. Vojvodich*. A review of ablative studies of interest to naval applications, *F. J. Koubek*. Structural design and thermal properties of polymers, *G. F. D'Alelio*. Characterization of an epoxy-anhydride ablative system using com-

(continued)

D'ALELIO and PARKER *(continued)*

puter treatment of analytical results, *C. G. Taylor and E. L. Pendleton.* The synthesis and characterization of some potential ablative polymers, *R. Y. Wen, L. F. Sonnabend, and R. Eddy.* Thermal degradation and curing of polyphenylene, *D. N. Vincent.* Thermosetting polyphenylene resin–its synthesis and use in ablative composites, *N. Bilow and L. J. Miller.* Structural ablative plastics, *R. M. Lurie, S. F. D'Urso, and C. K. Mullen.* Prediction of heat shield performance in terms of epoxy resin structure, *G. J. Fleming.* Ablative resins for hyperthermal environments, *B. S. Marks and L. Rubin.* The development of polybenzimidazole composites as ablative heat shields, *R. R. Dickey, J. H. Lundell, and J. A. Parker.* Ablative degradation of a silicon foam, *T. McKeon.* Thermophysical characteristics of high-performance ablative composites, *M. L. Minges.* The design and development of a high-heating rate thermogravimetric analyzer suitable for use with ablative plastics, *A. M. Melnick and E. J. Nolan.* Pyrolysis kinetics of nylon 6-6, phenolic resin and their composites, *H. E. Goldstein.* Pyrolysis-gas chromatography as a tool for studying the degradation of ablative plastics, *R. M. Ross.* Nonequilibrium flow and the kinetics of chemical reactions in the char zone, *G. C. April, R. W. Pike, and E. G. del Valle.* Arc-image testing of ablation materials, *E. M. Liston.* Development and characterization of a radio frequency-transparent ablator, *E. L. Strauss.* Tailoring polymers for entry into the atmosphere of Mars and Venus, *R. G. Nagler.*

DIGGLE *Oxides and Oxide Films*

in multi-volumes

(The Anodic Behavior of Metals and Semiconductors Series)

edited by JOHN W. DIGGLE, *Research School of Chemistry, The Australian National University, Canberra*

Vol. 1 552 pages, illustrated. 1972

Vol. 2 424 pages, illustrated. 1973

Treats the anodic behavior of metals and semiconductors and peripheral areas in an authoritative and interdisciplinary manner. The initial volumes deal with the physics and chemistry of oxides and oxide films. Of great value for all those involved in electrochemistry, materials science, solid state physics, electrical engineering, metallurgy, corrosion science, semiconductors, and electrochemical technology.

CONTENTS:

Volume 1: Passivation and passivity, *V. Brusić.* Mechanisms of ionic transport through oxide films, *M. J. Dignam.* Electronic current flow through ideal dielectric films, *C. A. Mead.* Electrical double layer at metal oxide-solution interfaces, *S. Ahmed.*

Volume 2: Anodic oxide films: Influence of solid-state properties on electrochemical behavior, *A. Vijh.* Dielectric loss mechanism in amorphous oxide films, *D. M. Smyth.* Porous anodic films in aluminum, *G. C. Wood.* Dissolution of oxide phases, *J. W. Diggle.*

FOURT and HOLLIES

Clothing: Comfort and Function

(Fiber Science Series, Volume 1)

by LYMAN FOURT and NORMAN HOLLIES, *Gillette Research Institute, Rockville, Maryland*

272 pages, illustrated. 1970

A unified review of the present state of knowledge in the science of clothing. Of interest to textile scientists, fiber producers and marketers, textile converters, and garment makers.

CONTENTS: The factors involved in the study of clothing • Clothing considered as a system interacting with the body • Clothing considered as a structural assemblage of materials • Heat and moisture relations in clothing • Physiological and field testing of clothing by wearing it • Physical properties of clothing and clothing materials in relation to comfort • Differences between fibers with respect to comfort • Current trends and new developments in the study of clothing.

FRISCH and REEGEN *Ring-Opening Polymerization*

(Kinetics and Mechanisms of Polymerization Series, Volume 2)

edited by KURT C. FRISCH, and SIDNEY L. REEGEN, *Polymer Institute, University of Detroit, Michigan*

544 pages, illustrated. 1969

Covers the polymerization of important classes of cyclic monomers such as ethylene and propylene oxide, alkylenimines, and sulfides, lactones, lactams, cyclic silicone compounds, and cyclic nitrogen containing heterocycles. Of great interest to the industrial, commercial, and academic worlds as it has application to elastomers, coatings, fibers, films, and foams.

CONTENTS: 1,2 Epoxides, *Y. Ishii and S. Sakai.* 1,3 Epoxides and higher epoxides, *P. Dreyfuss and M. P. Dreyfuss.* Cyclic formals, *J. Furukawa and K. Tada.* Cyclic sulfides, *P. Sigwalt.* Alkylenimines, *M. Hauser.* Lactones, *R. D. Lundberg and E. F. Cox.* Lactams, *H. K. Reimschuessel.* Cyclic siloxanes and silazanes, *E. E. Bostick.* Nitrogen-containing heterocyclic compounds, *V. Kargin and V. Kabanov.* N-carboxy-α-amino acid anhydrides, *Y. Shalitin.*

FRISCH and SAUNDERS
Plastic Foams

(Monographs on Plastic Series, Volume 1)

edited by KURT C. FRISCH, *University of Detroit, Michigan,* and JAMES H. SAUNDERS, *Monsanto Company, Pensacola, Florida*

Part I 464 pages, illustrated. 1972

Part II 704 pages, illustrated. 1973

Gives an integrated picture of the fundamental principles, technology, and applications of foams, and offers a thorough treatment of specific types of plastic foams. Emphasis is placed on the newer trends in this science.

Of particular value to chemists and engineers engaged in research and development, and marketing and production personnel in the polymer and plastics industry.

CONTENTS:

GARG, SVALBONAS, and GURTMAN
Analysis of Structural Composite Materials

(Monographs and Textbooks in Material Science Series, Volume 6)

by SABODH GARG, *Systems, Science and Software Company, La Jolla, California,* VYTAS SVALBONAS, *Grumman Aerospace Corporation, Bethpage, New York,* and GERRY GURTMAN, *Systems, Science and Software Company, La Jolla, California*

552 pages, illustrated. 1973

Compares various theories for the static and dynamic analysis of structural composite materials. Deals with the elastic properties of laminated composites and particulate and unidirectional fiber reinforced composites, composite strength, and stress wave propagation. May be used as a textbook for a graduate course in composites and is of interest to researchers and analysts in any industry that uses composites.

CONTENTS:

HAM *Vinyl Polymerization*
In 2 Parts

(Kinetics and Mechanisms of Polymerization Series, Volume 1)

edited by GEORGE E. HAM, *Geigy Chemical Corporation, Ardsley, New York*

Part I 560 pages, illustrated. 1967

Part II 432 pages, illustrated. 1969

"The book is a good introduction to the series. It has provided a sound basis for subsequent volumes and should serve as an important reference text for students and researchers in polymer chemistry."—B. D. Gesner, Bell Telephone Labs., *SPE Journal*

"The book is highly recommended."—Arthur Tobolsky, *The American Scientist*

CONTENTS:

(continued)

HAM *(continued)*

M. Morton. Mechanisms of cationic polymerization, *Z. Zlámal.* Radiation-induced polymerization, *Y. Tabata.*

HENCH and DOVE *Physics of Electronic Ceramics*

In 2 Parts

(Ceramics and Glass: Science and Technology Series, Volume 2)

edited by LARRY L. HENCH and DEREK B. DOVE, *College of Engineering, University of Florida, Gainesville.*

Part A 584 pages, illustrated. 1971

Part B 576 pages, illustrated. 1972

A highly useful treatise which deals with the physical basis for the behavior of electronic ceramics.

Fundamental physical theories describing each type of electronic ceramics are presented, with discussions included that relate the theories to the applications of the materials. Of special value to graduate students who have had a course in modern physics and also of interest to materials scientists and engineers in electronics, communications, and ceramics.

CONTENTS:

HENCH and GOULD *Characterization of Ceramics*

(Ceramics and Glass: Science and Technology Series, Volume 3)

edited by LARRY L. HENCH and ROBERT W. GOULD, *University of Florida, Gainesville*

672 pages, illustrated. 1971

Focuses on the two major directions which comprise the distinct discipline of ceramic characterization: the exploration of the factors that control the properties of the final product and the rapid development of high resolution analytical technisues used for ceramic materials. A particularly timely textbook for an advanced undergraduate or graduate materials science curriculum. Valuable for all materials scientists and ceramic and materials engineers working on the development of an effective and economical materials characterization program.

CONTENTS:

KATON *Organic Semiconducting Polymers*

(Monographs in Macromolecular Chemistry Series)

edited by J. E. KATON, *Miami University, Oxford, Ohio*

328 pages, illustrated. 1968

CONTENTS: Basic physics of semiconductors, *D. E. Hill*. Theoretical aspects of the electronic behavior of organic macromolecular solids, *H. A. Pohl*. Recent experimental aspects of the electronic behavior of organic macromolecular solids, *S. Kanda and H. A. Pohl*. Semiconducting organic polymers containing metal groups, *B. A. Bolto*. Semiconducting biological polymers, *D. D. Eley*.

KETLEY *The Stereochemistry of Macromolecules*

In 3 Volumes

edited by A. D. KETLEY, *W. R. Grace & Co., Clarksville, Maryland*

Vol. 1 424 pages, illustrated. 1967

Vol. 2 400 pages, illustrated. 1967

Vol. 3 476 pages, illustrated. 1968

CONTENTS:

Volume 1: Ziegler-Natta polymerization: Catalysts, monomers, and polymerization procedures, *D. O. Jordan*. The mechanism of Ziegler-Natta catalysis. I. Experimental foundations, *D. F. Hoeg*. Mechanism of Ziegler-Natta polymerization. II. Quantum-chemical and crystal-chemical aspects, *P. Cossee*. Copolymerization of olefins by Ziegler-Natta catalysts, *I. Pasquon, A. Valvassori, and G. Sartori*. Polymerization of dienes by Ziegler-Natta catalysts, *W. Marconi*. Manufacture and commercial applications of stereoregular polymers, *M. Compostella*.

Volume 2: Stereospecific polymerization of vinyl-type monomers and dienes by alkali-metal-based catalysts, *D. Braun*. Stereospecific polymerization of vinyl ethers, *A. D. Ketley*. Ionic polymerization of aldehydes, ketones, and ketenes, *G. F. Pregaglia and M. Binaghi*. Stereospecific polymerization of epoxides, *T. Tsuruta*. Stereochemistry of free-radical polymerizations, *W. Cooper*. Conformational effects induced in polymers by rigid matrices, *N. Marans*. Simple stereoregular polymers in biological systems, *J. N. Baptist*.

Volume 3: Chain conformation and crystallinity, *P. Corradini*. High-resolution nuclear magnetic resonance of synthetic polymers, *J. C. Woodbrey*. Vibrational analyses of the infrared spectra of stereoregular polypropylene, *T. Miyazawa*. Optically active stereoregular polymers, *M. Farina and G. Bressan*. Physical properties of stereoregular polymers in solid state, *J. F. Johnson and R. S. Porter*. Properties of synthetic linear stereoregular polymers in solution, *V. Crescenzi*. Macromolecules as information storage systems, *A. M. Liquori*. Automata theories of hereditary tactic copolymerization, *H. H. Pattee*. Effect of microtacticity on reactions of polymers, *M. M. van Beylen*. Degradation of stereoregular polymers, *H. H. G. Jellinek*.

KURYLA and PAPA *Flame Retardancy of Polymeric Materials*

a series edited by WILLIAM C. KURYLA and ANTHONY J. PAPA, *Union Carbide Corporation, South Charleston, West Virginia*

Vol. 1 352 pages, illustrated. 1973

Vol. 2 296 pages, illustrated. 1973

A series concerned with the various modes of rendering polymeric materials fire resistant, which emphasizes specific reagents and techniques in use today. Treats each class of polymer separately to aid the fabricator in gaining an understanding of the specific problems associated with its flammability characteristics, and to review the science and practical solution to its flame retardancy. Of special benefit to industrial polymer chemists, plastics engineers, and fabricators of polymeric materials.

CONTENTS:

Volume 1: Available flame retardants, *W. Kuryla*. Inorganic flame retardants and their mode of action, *J. Pitts*. Fire retardation of polyvinyl chloride and related polymers, *M. O'Mara, W. Ward, D. Knechtges, and R. Meyer*. Fire retardation of wool, nylon, and other natural and synthetic polyamides, *G. Crawshaw, A. Delman, and P. Mehta*.

Volume 2: Fire retardation of polystyrene and related thermoplastics, *R. Lindemann*. Fire retardation of polyethylene and polypropylene, *R. Schwarz*. Flame retardation of natural and synthetic rubbers, *H. Fabris and J. Sommer*. Flame retardancy of phenolic resins and urea- and melamine-formaldehyde resins, *N. Sunshine*.

LEFEVER *Aspects of Crystal Growth*

(Preparation and Properties of Solid State Materials Series, Volume 1)

edited by R. A. LEFEVER, *Sandia Laboratories, Albuquerque, New Mexico*

Vol. 1 296 pages, illustrated. 1971

Concerns certain aspects of the growth and properties of single crystals. Directed to both beginning and experienced crystal growers, material scientists, and solid state physicists.

CONTENTS: A review of the preparation of single crystals by fused melt electrolysis and some general properties, *W. Kunnmann*. The role of mass transfer in crystallization processes, *W. Wilcox*. Exploratory flux crystal growth, *A. Chase*.

LOWRY *Markov Chains and Monte Carlo Calculations in Polymer Sciences*

(Monographs in Macromolecular Chemistry Series)

edited by GEORGE G. LOWRY, *Western Michigan University, Kalamazoo*

344 pages, illustrated. 1970

Written for the polymer chemist who, although not primarily concerned with mathematical theories, desires a working knowledge of the topics treated. Begins with an introduction to the principles involved and later exemplifies some significant applications of Markov chain theory and Monte Carlo methods.

CONTENTS: Introduction: Deterministic and stochastic approaches, *G. G. Lowry*. Markov chains, *J. Myhre*. Monte Carlo methods, *M. Fluendy*. Polymer conformation as a Markov chain problem, *J. Kinsinger*. Polymer conformation and the excluded-volume problem, *S. Windwer*. Higher order Markov chains and statistical thermodynamics of linear polymers, *J. Mazur*. Copolymer composition and tacticity, *F. P. Price*. Molecular-weight distributions, *G. G. Lowry*.

McCULLOUGH *Concepts of Fiber-Resin Composites*

(Monographs and Textbooks in Material Science Series, Volume 2)

by R. L. McCULLOUGH, *Boeing Scientific Research Laboratories, Seattle, Washington*

128 pages, illustrated. 1971

Presents basic concepts of composite material systems. Introduces the study of composite systems by discussing where and how composite materials are used and how their components are selected. A valuable reference for materials scientists, and students and research management engaged in the exploration of composite materials.

CONTENTS: Materials • Composite structures • Composite properties • The interphase region • Synopsis.

MAY and TANAKA *Epoxy Resins: Chemistry and Technology*

edited by CLAYTON MAY, *Shell Development Company, Emeryville, California* and YOSHIO TANAKA, *Research Institute for Polymers and Textiles, Yokohama, Japan*

704 pages, illustrated. 1973

Brings together the contributions of a number of outstanding researchers in the field of epoxy resins. Not only emphasizes the chemistry and technology of epoxy resins, but also deals with many industrial applications. Of great value for polymer chemists and technicians, and a wide variety of engineers.

CONTENTS: Introduction to epoxy resins, *C. May*. Synthesis and characteristics of epoxides, *Y. Tanaka, A. Okada, and I. Tomizuka*. Epoxide-curing reactions, *Y. Tanaka and T. Mika*. Curing agents and modifiers, *T. Mika*. Properties of cured resins, *D. Kaelble*. Epoxy-resin adhesives, *A. Lewis and R. Saxon*. Epoxy-resin coatings, *G. Somerville and I. Smith*. Electrical and electronic applications, *A. Breslau*. Epoxy laminates, *J. DelMonte*. Polymer stabilizers and plasticizers, *W. Port*. Analysis of epoxides and epoxy resins, *H. Jahn and P. Goetzky*. Toxicity, hazards, and safe handling, *H. Borgstedt and C. Hine*.

MILLICH and CARREHER *Interfacial Synthesis*

edited by FRANK MILLICH, *Department of Chemistry, University of Missouri, Kansas City*, and CHARLES E. CARREHER, JR., *Department of Chemistry, University of South Dakota, Vermillion*

in preparation. 1973

Summarizes the accomplishments in interfacial synthesis to date. Speculates on mechanism, discusses the complex matter of synthetic control, and points out the beneficial aspects of interfacial synthesis in comparison to alternative methods of synthesis. Of fundamental concern to graduate students and teachers of organic chemistry, physical chemists, macromolecular biochemists, and polymer chemists who engage in interfacial synthesis.

CONTENTS: Stirring in organic chemical synthesis, *J. Rushton*. High-speed stirring in interfacial synthesis, *J. Rushton*. Problems and solutions in kinetics and mechanisms, *J. Bradbury and P. Crawford*. Interface effects on chemical reaction rate, *F. MacRitchie*. Liquid-vapor interfacial polycondensations, *L. Sokolov*. Copolycondensation and macroscopic kinetics, *L. Sokolov and V. Nikonov*. The role of the particle-water interface in polymerization, *J. Gardon*. Biochemical reactions at an interface, *R. Baier and D. Cadenhead*. Commercial application of interfacial synthesis, *E. Oliver and Y. Yen*. Polycarboxylic esters, *S. Temin*. Polycarbonates, *H. Vernaleken*. Polycondensations with carbon suboxide, *I. Daniewska*. Polyamides, *V. Nikonov and V. Savinov*. Polyesteramides, *I. Panayotov*. Polyurethanes, *T. Tanaka and T. Yokoyama*. Polyureas, *K. Stueben and A. Barnabeo*. Polyphosphonates, polyphosphates, and polyphosphites, *F. Millich, J. Teague, L. Lambing, and D. Hackathorn*. Other phosphorus containing polymers, *C. Carreher, Jr*. Organometallic polymers, *C. Carreher, Jr.* Modification of natural polymers by interfacial methods, *M. Horio.* Interfacial modifications of poly(vinyl alcohol) and related polymers, *M. Tsuda.* High temperature resistant polymers made by interfacial polymerization, *H. Mark and S. Atlas.*

MYERS and LONG *Characterization of Coatings: Physical Techniques*

In 2 Parts

(Treatise on Coatings Series, Volume 2)

edited by RAYMOND R. MYERS, *Paint Research Institute, Kent State University, Ohio,* and J. S. LONG, *Department of Chemistry, University of Southern Mississippi, Hattiesburg*

Part I 696 pages, illustrated. 1969

Part II in preparation. 1973

Explores the scientific frontier that has developed since the appearance of Mattiello's treatise on coatings. Emphasizes the urgent need of the paint industry to master new technological concepts and instrumental techniques to match the rapid pace of development of its products. Written for the working paint scientist, the laboratory assistant, technician, and superintendent. Also a valuable reference for the formulator and personnel engaged in the production of raw materials for the paint industry.

CONTENTS:

Part I: The intrinsic properties of polymers, *A. Tawn.* Surface areas, *D. Gans.* Adhesion of coatings, *A. Lewis and L. Forrestal.* Mechanical properties of coatings, *P. Pierce.* The ultimate tensile properties of paint films, *R. Evans.* Gas chromatography, *J. Haken.* Thermoanalytical techniques, *P. Garn.* Microscopy in coatings and coating ingredients, *W. Lind.* Radioactive isotopes, *G. Coe.* Infrared spectroscopy *C. Smith.* Ultraviolet and visible spectroscopy, *F. Spagnolo and E. Scheffer.* Color of polymers and pigmented systems, *G. Ingle.* Photoelastic coatings, *A. Blumstein.*

Part II: Dielectric properties, *S. Negami.* Gel permeation chromatography, *K. Boni.* Infrared Fourier transform spectroscopy, *M. Low.* Interfacial energetics, *D. Gans.* Nuclear magnetic resonance, *M. Levy.* Particle sizing, *B. DeWitt.* Scanning electron microscopy, *L. Princen.* Solubility, *J. Gordon and J. Teas.* Transport properties, *G. Park.* Viscometry, *K. Oesterle.* X-ray analysis, diffraction, and emission, *R. Scott.*

MYERS and LONG *Film-Forming Compositions*

In 3 Parts

(Treatise on Coatings Series, Volume 1)

edited by RAYMOND R. MYERS, *Paint Research Institute, Kent State University, Ohio,* and J. S. LONG, *Department of Chemistry, University of Southern Mississippi, Hattiesburg*

Part I 584 pages, illustrated. 1967

Part II 448 pages, illustrated. 1968

Part III 608 pages, illustrated. 1972

Devoted to materials which form, or aid the formation of, continuous films. Discusses vehicles and resins, placing considerable emphasis on procedures for developing a suitable vehicle for conveying dissolved or suspended solids to a substrate and imparting to the surface those protective and decorative properties for which the coating was designed. Of inestimable value to chemists, formulators, laboratory assistants, technicians, and production superintendents of the coatings industry. Also valuable for laboratory personnel of raw material suppliers, chemists in the plastics and similar industries, and as an excellent reference treatise for libraries.

Part I: Acrylic ester emulsions and water-soluble resins, *G. Allyn.* Acrylic ester resins, *G. Allyn.* Alkyd resins, *W. M. Kraft, E. G. Janusz, and D. Sughrue.* Asphalt and asphalt coatings, *S. H. Alexander.* Chlorinated rubber, *H. E. Parker.* Driers, *W. J. Stewart.* Epoxy resin coatings, *G. R. Somerville.* Hydrocarbon resins and polymers, *D. F. Koenecke.* Hydrocarbon solvents, *W. W. Reynolds.* Natural resins, *C. L. Mantell.* Polyethers and polyesters, *A. C. Filson.* Urethane coatings, *A. Damusis and K. C. Frisch.* Vehicle manufacturing equipment, *A. F. Steioff.*

Part II: Styrene-butadiene latexes in protective and decorative coatings, *F. A. Miller.* Starch polymers and their use in paper coating, *T. F. Protzman and R. M. Powers.* Cellulose esters and ethers, *J. B. G. Lewin.* Drying oils – modifications and use, *A. E. Rheineck and R. O. Austin.* Paint and painting in art, *S. Rees Jones.* Rosin and modified rosins and resins, *C. L. Mantell.* Urea and melamine resins, *H. P. Wohnsiedler and W. L. Hensley.* *Vinyl resins W. H. McKnight and G. S. Peacock.* Vinyl emulsions, *H. D. Cogan and A. L. Mantz.*

Part III: Dimer acids in surface coatings, *J. Boylan.* Emulsion technology, *L. Princen.* Phenolic resins for coatings, *S. Richardson and W. Wertz.* Plasticization of coatings, *F. Ball.* Fatty polyamides and their applications in protective coatings, *D. Wheller and D. Peerman.* Polycarbonate resins, *D. Fox and K. Goldblum.* Reactive polyesters, *F. Ball.* Varnishes, *L. Montague.* Reactive silanes as adhesion promoters to hydrophilic surfaces, *E. Plueddemann.* Surface-active agents, *T. Ginsberg.* Shellac, *J. Martin.* Tall oil in surface coatings, *R. Perez.* Silicones in protective coatings, *L. Brown.*

MYERS and LONG *Formulations*

(Treatise on Coatings Series, Volume 4)

edited by RAYMOND R. MYERS, *Department of Chemistry, Kent State University, Ohio,* and J. S. LONG, *University of Southern Mississippi, Hattiesburg*

Part I in preparation. 1973

NEUSE and ROSENBERG *Metallocene Polymers*

(Reviews in Macromolecular Chemistry Series, Volume 5, Part I)

by EBERHARD NEUSE, *F. J. Weck Company, City of Industry, California,* and HAROLD ROSENBERG, *Air Force Materials Laboratory, Wright-Patterson Air Force Base, Ohio*

170 pages, illustrated. 1970

Presents a comprehensive and critical account of the progress made in the synthesis and characterization of metallocene polymers.

CONTENTS: Introduction • Macromolecular compounds with pendent metallocenyl groups • Macromolecular compounds with intrachain metallocenylene groups • Conclusions.

O'CONNOR *Instrumental Analysis of Cotton Cellulose and Modified Cotton Cellulose*

(Fiber Science Series, Volume 3)

edited by ROBERT T. O'CONNOR, *Agricultural Research Service, U.S.D.A., New Orleans, Louisiana*

512 pages, illustrated. 1972

Describes the applications of instrumental procedures specifically developed by the textile chemist to meet today's demands. Particularly geared to textile chemists and others in the textile industry, and to paper and wood manufacturers. Also of interest to polymer chemists, analytical chemists, and other researchers involved with the processes by which fibers are blended and modified.

CONTENTS: Elemental analysis: Detection, identification, and quantitative determination of metals and nonmetallic elements, *R. T. O'Connor.* Infrared spectroscopy and physical properties of cellulose, *C. Y. Liang.* Light microscopy in the study of cellulose, *M. L. Rollins and I. V. de Gruy.* Electron microscopy of cellulose and cellulose derivatives, *M. L. Rollins, A. M. Cannizzaro, and W. R. Goynes.* Instrumental methods in the study of oxidation, degradation, and pyrolysis of cellulose, *P. K. Chatterjee and R. F. Schwenker, Jr.* X-ray diffraction, *V. W. Tripp and C. M. Conrad.* Wide-line nuclear magnetic resonance spectroscopy, *R. A. Pittman and V. W. Tripp.* The infrared spectra of chemically modified cotton cellulose, *R. T. O'Connor.*

PEARL *The Chemistry of Lignin*

by IRWIN A. PEARL, *The Institute of Paper Chemistry, Appleton, Wisconsin*

360 pages, illustrated. 1967

CONTENTS: Nebulous concept of lignin • Isolation of lignin • Chemical structure of lignin • Biosynthesis and formation of lignin • Reactions of lignin in major pulping and bleaching processes • Chemical reactions of lignin • Physical properties of lignin and its preparations • Biological decomposition of lignin • Thermal decomposition of lignin • Linkage of lignin in the plant • Utilization of lignin and its preparations.

PETERLIN *Plastic Deformation of Polymers*

edited by A. PETERLIN, *Research Triangle Institute, Research Triangle Park, North Carolina*

318 pages, illustrated. 1971

Explores the actual mechanism of deformation that occurs during the formation of fibers and films. Focuses primarily on three topics: what happens on a molecular, crystalline, and supercrystalline level; how the original structure influences the deformation; and what determines the useful mechanical properties of fibers and films. A valuable tool for the staff of research and development laboratories working with plastics, rubbers, fibers, and films, and for students and faculty involved in material science and polymer chemistry.

CONTENTS: Infrared studies of the role of monoclinic structure in the deformation of polyethylene, *Y. Kikuchi and S. Krimm.* Infrared studies of drawn polyethylene. Part I. Changes in orientation and conformation of highly drawn linear polyethylene, *W. Glenz and A. Peterlin.* Structure of oriented polyacrylonitrile films, *J. L. Koenig, L. E. Wolfram, and J. G. Grasselli.* Morphology and deformation behavior of "row-nucleated" polyoxymethylene film, *C. A. Garber and E. S. Clark.* Plastic deformation of polypropylene. VI. Mechanism and properties, *F. J. Baltá-Calleja and A. Peterlin.* Polyethylene crystallized under the orientation and pressure of a pressure capillary viscometer. Part I., *J. H. Southern and R. S. Porter.* Heat relaxation of drawn polyoxymethylene, *A. Siegmann and P. H. Geil.*

Retraction of cold-drawn polyethylene and polypropylene, *D. Hansen, W. F. Kracke, and J. R. Falender.* Electron spin resonance studies of free radicals in mechanically loaded nylon 66, *G. S. P. Verma and A. Peterlin.* Transition from linear to nonlinear viscoelastic behavior. Part I. Creep of polycarbonate, *I. V. Yannas and A. C. Lunn.* Yielding behavior of glassy polymers. III. Relative influences of free volume and kinetic energy, *K. C. Rusch and R. H. Beck, Jr.* Yielding of quenched and annealed polymethyl methacrylate, *D. H. Ender.* Yield phenomenon in oriented polyethylene terephthalate, *M. Parrish and N. Brown.* Electron paramagnetic resonance investigation of molecular bond rupture due to ozone in deformed rubber, *K. I. DeVries, E. R. Simonson, and M. L. Williams.* Factors affecting the depth of draw in a cold-forming operation, *H. L. Li, P. J. Koch, D. C. Prevorsek, and H. J. Oswald.* Quantitative structural characterization of the mechanical properties of isotactic polypropylene, *R. J. Samuels.*

RAVVE *Organic Chemistry of Macromolecules: An Introductory Textbook*

by A. Ravve, *Continental Can Company, Chicago, Illinois*

512 pages, illustrated. 1967

CONTENTS: **Part I: Introduction:** Historical introduction and definitions • Physical properties of macromolecules • Molecular weights of polymers • **Part II: Polymerization Reactions-Mechanisms:** Addition polymerization: Mechanism of free-radical polymerization • Ionic polymerization • Polymerization with the aid of complex catalysts • Stereospecific polymerization • Bulk, solution, suspension, and emulsion polymerization. **Part III: Common Addition Polymers:** Macroalkanes • Polymers and copolymers from dienes and polyenes • Styrene and styrene-like polymers and polyacrylics • Halogen-bearing addition polymers and vinyl esters and ethers. **Part IV: Condensation Polymers:** Mechanism of polycondensation reactions • polyesters • Polyamides • Polycarbamates, polyureas, and polycarbodiimides • Phenoplasts • Aminoplasts • Ladder and semiladder polymers. **Part V: Naturally Occurring Polymers:** Polysaccharides • Proteins • Polynucleotides. **Part VI: Reactions of Polymers:** Graft and block copolymers • Reactions of polymers • Degradation of polymers.

REICH and STIVALA *Autoxidation of Hydrocarbons and Polyolefins: Kinetics and Mechanisms*

by Leo Reich, *Picatinny Arsenal, Dover, New Jersey,* and Salvatore S. Stivala, *Department of Chemistry, Stevens Institute of Technology, Hoboken, New Jersey*

544 pages, illustrated. 1969

CONTENTS: Introduction • Oxidation of simple hydrocarbons in absence of inhibitors and accelerators • Oxidation of simple hydrocarbons in presence of antioxidants • Oxidation of simple hydrocarbons in presence of metal catalysts • Weak chemiluminescence during hydrocarbon autoxidation • Qualitative aspects of autoxidation of saturated polyolefins • Quantitative aspects of autoxidation of saturated polyolefins • Investigation of polyolefin oxidation by various techniques.

REMBAUM and SHEN *Biomedical Polymers*

edited by Alan Rembaum, *California Institute of Technology, Pasadena* and Mitchel Shen, *University of California, Berkeley*

304 pages, illustrated. 1971

A collection of papers given at the Symposium on Biomedical Polymers held in Pasadena in 1969. Of interest to scientists in polymer chemistry, polymer physics, biochemistry, bioengineering, materials science, pharmacology, and surgery.

CONTENTS: Problems in blood-tissue reactions to polymeric materials, *B. Zweifach.* Past, present, and future of artificial kidney treatment, *B. Barbour.* The chemistry and properties of the medical-grade silicones, *S. Braley.* Correlation of the surface charge characteristics of polymers with their antithrombogenic characteristics, *S. Srinivasan and P. Sawyer.* Persistent polarization in polymers and blood compatibility, *P. Murphy, A. Lacroix, S. Merchant, and W. Bernhard.* Selection, characterization, and biodegradation of surgical epoxies, *A. Cupples and R. Schubert.* Foreign body reactions to plastic implants, *D. Ocumpaugh and H. Lee.* Rapid *in vitro* screening of polymers for bio-compatibility, *C. Homsy, K. Ansevin, W. O'Bannon, S. Thompson, R. Hodge, and M. Estrella.* Improved membranes for hemodialysis, *F. Martin, H. Shuey, and C. Saltonstall, Jr.* Control of polymer morphology for biomedical applications, 1. Hydrophilic polycarbonate membranes for dialysis, *R. Kesting.* Surgical adhesives in ophthalmology, *M. Refojo.* Medical uses for polyelectrolyte complexes, *M. Vogel, R. Cross, H. Bixler, and R. Guzman.* Potentialities of a new class of anticlotting and antihemorrhagic polymers, *T. Yen, M. Daver, and A. Rembaum.* Synthesis and properties of a new class of potential biomedical polymers, *A. Rembaum, S. Yen, R. Landel, and M. Shen.* Recognition polymers, *D. Bradley.* The challenge for high polymers in medicine, surgery, and artificial internal organs, *H. Lee and K. Neville.*

RICHARDSON *Optical Microscopy for the Materials Sciences*

(Monographs and Textbooks in Material Science Series, Volume 3)

by James H. Richardson, *Aerospace Corporation, El Segundo, California*

(continued)

RICHARDSON *(continued)*

704 pages, illustrated. 1971

Provides in one volume a comprehensive survey of the techniques for preparation and optical examination of specimens in the broad area of materials sciences. A highly useful text for the university or vocational school student studying the microstructure of materials or metallography and also a valuable practical reference for all researchers using microscopy, as well as for engineers and industrial metallographers.

CONTENTS: The Brightfield optical microscope • The microscopy of phase structures • Photomicrography • Photomacrography • Specimen preparation • Reagents and techniques for specimen preparation • Laboratory safety • Examination of the specimen • Accessories • Laboratory design.

ROGERS *Permselective Membranes*

edited by CHARLES E. ROGERS, *Case Western Reserve University, Cleveland, Ohio*

224 pages, illustrated. 1971

Encompasses a broad range of topics pertaining to the expanding field of permselective membranes, including theoretical aspects of transport behavior, new and unusual methods for the preparation or modification of membrane materials, and the effects of experimental conditions on the permselectivity of membranes to both ionic and nonionic penetrants. Of value to biophysicists, polymer chemists, physicists, biochemists, and chemical engineers, as well as biologists and physiologists.

CONTENTS: Transport of dissolved oxygen through silicone rubber membrane, *S. Hwang, T. Tang, and K. Kammermeyer*. Gas transport in segmented block copolymers, *K. Ziegel*. Transport of noble gases in poly(methyl acrylate), *W. Burgess, H. Hopfenberg, and V. Stannett*. Permeation of gases at high pressures, *S. Stern, S. Fang, and R. Jobbins*. Permeation of gases through modified polymer films III. Gas permeability and separation characteristics of gamma-irradiated Teflon FEP copolymer films, *R. Huang and P. Kanitz*. Theoretical interpretation of the effect of mixture composition on separation of liquids in polymers, *M. Fels and R. Huang*. Permselectivity of solutes in homogeneous water-swollen polymer membranes, *H. Yasuda and C. Lamaze*. Ion-exchange selectivity coefficients in the exchange of calcium, strontium, cobalt, nickel, zinc, and cadmium ions with hydrogen ion in variously cross-linked polystyrene sulfonate cation exchangers at 25°C, *M. Reddy and J. Marinsky*. Ion and water transport through permselective membranes, *N. Lakshminarayanaiah*. Permeability of cellulose acetate membranes to selected solutes, *H. Lonsdale, B. Cross, F. Graber, and C. Milstead*. Transport through permselective membranes, *C. Rogers and S. Sternberg*.

SCHEY *Metal Deformation Processes: Friction and Lubrication*

(Monographs and Textbooks in Material Science Series, Volume 1)

edited by JOHN A. SCHEY, *University of Illinois at Chicago Circle*

822 pages, illustrated. 1970

A comprehensive treatment of all aspects of friction and lubrication in metal deformation processes. Of aid to metallurgists, mechanical engineers, chemists, and physicists.

CONTENTS: Background and system of approach, *J. Schey*. Friction effects in metalworking processes, *J. Schey*. Friction, lubrication, and wear mechanisms, *C. Riesz*. Lubricants, *C. Riesz*. Lubricant properties and their measurements, *J. Schey*. Rolling lubrication, *J. Schey*. Wire drawing lubrication, *J. Newnham*. Hot extrusion lubrication, *S. Kalpakjian*. Forging lubrication, *S. Kalpakjian*. Cold forging and cold extrusion lubrication, *J. Newnham*. Sheet metal working lubrication, *J. Newnham*.

SEGAL *High-Temperature Polymers*

edited by CHARLES L. SEGAL, *North American Aviation, Inc., Canoga Park, California*

208 pages, illustrated. 1967

CONTENTS: Introduction, *C. L. Segal*. Thermally stable polymers with aromatic recurring units, *C. S. Marvel*. Inorganic polymer chemistry, *J. R. Van Wazer*. Kinetics and gaseous products of thermal decomposition of polymers, *H. L. Friedman*. Studies of stability of condensation polymers in oxygen-containing atmospheres, *R. T. Conley*. Thermal degradation of polymers. III: Mass spectrometric thermal analysis of condensation polymers, *G. P. Shulman*. Viscoelastic relaxation mechanism of inorganic polymers. V: Counterion effects in bulk polyelectrolytes, *A. Eisenberg, S. Saito, and T. Sasada*. Thermal stability of carborane-containing polymers, *J. Green and N. Mayes*. Synthesis and thermal stability of structurally related aromatic Schiff bases and acid amides, *A. D. Delman, A. A. Stein, and B. B. Simms*. New high-temperature polymers. II: Ordered aromatic copolyamides containing fused and multiple ring systems, *F. Dobinson and J. Preston*. Synthesis of fusible branched polyphenylenes, *N. Bilow and L. J. Miller*.

SEGAL, SHEN, and KELLEY
Polymers in Space Research

edited by CHARLES L. SEGAL, *Whittaker Corporation, San Diego,* MITCHEL SHEN, *University of California, Berkeley,* and FRANK N. KELLEY, *Air Force Rocket Propulsion Laboratory, Edwards, California.* Associate Editors: GEORGE F. PEZDIRTZ, *NASA Langley Research Center, Hampton, Virginia,* and W. DAVID ENGLISH, *Astropower Laboratory, McDonnell Douglas Aeronautics Company, Huntington Beach, California*

480 pages, illustrated. 1970

CONTENTS:

Part I: Recent Developments in the Synthesis, Characterization, and Evaluation of Thermally Stable Polymers

Introduction, *C. Segal and G. Pezdirtz.* Aromatic polymers: Single- and double-stranded chains, *J. Stille.* Thermally stable spiropolymers, *J. Hodgkin and J. Heller.* Isomeric and substituent effects in some dibenzoylbenzene-diamine polymers, *A. Volpe, L. Kaufman, and R. Dondero.* Arylsulfimide polymers. III. The syntheses of some monomeric aryl-1,2-disulfonic acids and derivatives, *G. D'Alelio, Y. Giza, and D. Feigl.* Properties of heterocyclic condensation polymers, *G. Berry and T. Fox.* Relative thermophysical properties of some polyimidazopyrrolones, *R. Jewell.* Thermal decomposition of polyimides in vacuum, *T. Johnston and C. Gaulin.* Thermomechanical behavior of an aromatic polysulfone, *J. Gillham, G. Pezdirtz, and L. Epps.* Panel discussion on thermally stable polymers, *C. Segal, J. Stille, G. Pezdirtz, G. D'Alelio, H. Levine, and W. Gibbs.*

Part II: Properties of Polymers at Low Temperatures

Introduction, *M. Shen and W. English.* Relaxation behavior of polymers at low temperatures, *J. Sauer and R. Saba.* Thermal properties of polymers at low temperatures, *W. Reese.* Multiple transitions in polyvinyl alkyl ethers at low temperatures, *W. Schell, R. Simha, and J. Aklonis.* Internal friction study of diluent effect on polymers at cryogenic temperatures, *M. Shen, J. Strong, and H. Schlein.* Stress-strain behavior of adhesives in a lap joint configuration at ambient and cryogenic temperatures, *G. Tiezzi and H. Doyle.* Some properties of nitroso rubbers in fluorine at ambient and cryogenic temperatures, *S. Toy, W. English, W. Crane, and M. Toy.* Cryogenic properties of a polyurethane adhesive, *R. Robbins.* Some effects of structure on a polymer's performance as a cryogenic adhesive, *R. Gosnall and H. Levine.*

Part III: Solid Propellents

Introduction, *F. Kelley.* Recent developments in solid-propellent binders, *H. Marsh, Jr.* Saturated hydrocarbon polymers for solid rocket propellents, *A. Di Milo and D. Johnson.* Preparation and curing of poly (perfluoroalkylene oxides), *J. Zollinger, J. Throckmorton, S. Ting, R. Mitsch, and D. Elrick.* Functionality and functionality distribution measurements of binder prepolymers, *A. Muenker and B. Hudson, Jr.*

SERAFINI and KOENIG
Cryogenic Properties of Polymers

edited by TITO T. SERAFINI, *NASA-Lewis Research Center, Cleveland,* and JACK L. KOENIG, *Case Western Reserve University, Cleveland, Ohio*

312 pages, illustrated. 1968

CONTENTS: Cryogenic positive expulsion bladders, *R. F. Lark.* Adhesives for cryogenic applications, *L. M. Roseland.* Glass-, boron-, and graphite-filament-wound resin composites and liners for cryogenic pressure vessels, *M. P. Hanson.* Mechanical behavior of poly(ethylene terephthalate), *I. M. Ward.* Effect of film processing on cryogenic properties of poly(ethylene terephthalate), *R. E. Eckert and T. T. Serafini.* Mechanical properties of epoxy resins and glass/epoxy composites at cryogenic temperatures, *L. M. Soffer and R. Molho.* Mechanical relaxation of poly-4-methyl-pentene-1 at cryogenic temperatures, *M. Takayanagi and N. Kawasaki.* Transitions in glasses at low temperatures, *R. A. Haldon, W. J. Schell, and R. Simha.* Mechanical behavior of poly(ethylene terephthalate) at cryogenic temperatures, *C. D. Armeniades, I. Kuriyama, J. M. Roe, and E. Baer.* Infrared studies of chain folding in polymers II. Poly(ethylene terephthalate), *J. L. Koenig and M. J. Hannon.* Crystallization of poly(ethylene terephthalate) from the glassy amorphous state, *G. S. Y. Yeh and P. H. Geil.* Strain-induced crystallization of poly(ethylene terephthalate), *G. S. Y. Yeh and P. H. Geil.* Molecular motion in polytetrafluoroethylene at cryogenic temperatures, *E. S. Clark.* Synthesis of ultrahigh molecular weight poly(ethylene terephthalate), *L.-C. Hsu.* Development of vulcanizable elastomers suitable for use in contact with liquid oxygen, *P. D. Schuman, E. C. Stump, and G. Westmoreland.* Synthesis of fluorinated polyurethanes, *R. Gosnell and J. Hollander.*

SKEIST *Reviews in Polymer Technology*

edited by IRVING SKEIST, *Skeist Laboratories, Inc., Livingston, New Jersey*

260 pages, illustrated. 1972

Consists of intensive, up-to-date reviews on various aspects of polymer technology. Directed to chemists, engineers, technicians, commercial planners, and others working in the polymers and plastics fields.

CONTENTS: Coupling agents as adhesion promoters, *P. Cassidy and W. Yager.* Processing powdered polyethylene, *A. Zimmerman.* Plastics and other polymers in building, *I. Skeist and J. Miron.* Fire retardance of polymeric ma-

(continued)

SKEIST *(continued)*

terials, *I. Einhorn.* Recent advances in photo-cross-linkable polymers, *G. Delzenne.* Organic colorants for polymers, *T. Reeve.*

SOLOMON *Step-Growth Polymerizations*

(Kinetics and Mechanisms of Polymerization Series, Volume 3)

edited by DAVID H. SOLOMON, *C.S.I.R.O., Melbourne, Australia*

416 pages, illustrated. 1972

Presents a critical and constructive assessment of developments in step-growth polymerization and considers the application of theoretical concepts to commercial systems. Highly recommended to students of polymer science and researchers working in the area of step-growth polymerization, including polymer chemists and other scientists in the paint, coatings, and plastics industries.

CONTENTS: Polyesterification, *D. H. Solomon.* Polyamides, *D. C. Jones and T. R. White.* Polyurethanes: The chemistry of the diisocyanate-diol reaction, *D. J. Lyman.* Cyclopolycondensation, *P. M. Hergenrother.* The reactions of formaldehyde with phenols, melamine, aniline, and urea, *M. F. Drum and J. R. LeBlanc.* Diels-Alder polymerization, *W. J. Bailey.* Inorganic polymers, *J. R. MacCallum.*

STEWART *Infrared Spectroscopy: Experimental Methods and Techniques*

by JAMES E. STEWART, *Durrum Instrument Corporation, Palo Alto, California*

656 pages, illustrated. 1970

A guide to instrumentation and experimental methods and techniques for the infrared spectroscopist. Primarily for those involved with spectroscopy research and for graduate students in chemistry and physics interested in spectroscopy.

CONTENTS: Infrared spectroscopy • The infrared spectrophotometer • Elements of geometric optics • Elements of physical optics • Optical components of infrared spectrophotometers • Optical systems of infrared spectrophotometers • Slit functions and spectral modulation transfer functions of monochromators • Interference spectroscopy • Mechanics of infrared spectrophotometers • Elements of electronics • Infrared detectors • Electronic systems of infrared spectrophotometers • Electromechanical transfer functions of infrared spectrophotometers • Photometric accuracy of infrared spectrophotometers • Experimental methods of infrared transmission spectroscopy • Experimental methods of infrared reflection spectroscopy • Experimental methods of infrared emission spectroscopy • Advanced methods of infrared spectroscopy.

SZEKELY *Blast Furnace Technology: Science and Practice*

edited by JULIAN SZEKELY, *State University of New York, Buffalo*

414 pages, illustrated. 1972

Represents the efforts of academic and industrial researchers, metallurgists, plant operators, and designers concerned with the most up-to-date aspects of ironmaking technology. Valuable reading for all production engineers, plant operators, and designers concerned with ironmaking technology, and also of importance to metallurgical engineers, research scientists—chemists, physicists, engineers—and students in this field.

CONTENTS: Single particle studies applied to direct reduction and blast furnace operations, *R. Bleifuss.* Structural effects in gas–solid reactions, *J. Szekely and J. Evans.* The use of catalysts to enhance the rate of Boudouard's reaction in direct reduction metallurgical processes, *Y. Rao and B. Jalan.* Thirty psi high top–gas pressure operation at NSC Nagoya works, *T. Yatsuzuka, Y. Yamada, and A. Tayama.* Practical application of mathematical models in ironmaking, *D. Christie, C. Kearton, and R. Thomas.* A mass–transport model of erosion of the carbon hearth of the iron blast furnace, *J. Elliott and J. Popper.* Contribution to the study of the reaction mechanism occurring in high temperature zone of the blast furnace, *R. Vidal and A. Poos.* The place of direct reduction in a modern blast furnace–BOF plant, *J. Peart and D. George.* The blast furnace control problem, *J. A. Laslo.* Modern blast furnace design in Germany, *F. Lenger.* The blast furnace–a transition, *F. Berczynski.* Projected performance of a blast furnace with prereduced burdens, *J. Agarwal.*

SZEKELY *The Steel Industry and the Environment*

edited by JULIAN SZEKELY, *Center for Process Metallurgy, State University of New York, Buffalo*

312 pages, illustrated. 1973

Contains the proceedings of the Second C. C. Furnas Memorial Conference on The Steel Industry and The Environment held at the State University of New York at Buffalo in November, 1971. Brings together authors representing different viewpoints on the questions raised in examining the interaction of the steel industry and the environment. Of utmost importance to plant operators, designers, metallurgists, environmental scientists, and all others concerned with the impact of the steel industry on the environment.

CONTENTS: The role of the federal government in environmental pollution control, *K. Johnson*. Control of air pollution in the British iron and steel industry, *F. Ireland*. Health and the steel industry environment, *K. Spring*. The economic impact of the installation and operation of pollution abatement devices, *J. Barker*. Desulfurization of coke oven gas: Technology, economics, and regulatory activity, *R. Dunlap, W. Gorr, and M. Massey*. Treatment of cold-mill wastewaters by ultrahigh-rate filtration, *C. Symons*. Experience with pollution abatement, *C. Black and W. Sebesta*. The interaction of the socioeconomic and ecological environment in American steel mill towns, *L. Thaxton and R. Genton*. Plant availability versus clean air: An economic dilemma that can be solved, *R. Heller*. A survey of wastewater treatment techniques for steel mill effluents, *T. Centi*. Emission of sulfurous gases from blast-furnace slags, *R. Kaplan and G. Ringstorff*. Treatment of waste gases from the basic oxygen furnace in West Germany, *E. Weber*. On the oxidation of cyanides in the stack region of the blast furnace, *H. Sohn and J. Szekely*. Reclaimed scrap and solid metallics for steelmaking, *J. Elliott*.

TALLAN *Electrical Conductivity in Ceramics and Glass*

(Ceramics and Glass: Science and Technology Series, Volume 4)

edited by NORMAN M. TALLAN, *Aerospace Research Laboratories, Wright-Patterson Air Force Base, Ohio*

in preparation. 1973

A text which thoroughly describes several aspects of the electrical conductivity of ceramic materials, and discusses their conductivity, physical dependence on their electronic and ionic defect structures, and the transport mechanisms by which charge and mass move through ceramic materials. Additionally stressed is the use of conductivity measurements to characterize the defect structure and transport properties of ceramics. A great aid to advanced students in materials science, ceramics, and glass.

CONTENTS: General concepts of electrical transport, *D. Adler*. Experimental techniques, *R. Blumenthal and M. Seitz*. Defect structure of ceramic materials, *R. Brook*. Electronic conduction mechanisms, *I. Bransky and J. Wimmer*. Controlled valency effects in electronic conductors, *J. Wagner*. Highly conducting ceramics and the conductor-insulator transition, *J. Honig and R. Vest*. Ionic conductivity and electrochemistry of crystalline ceramics, *J. Patterson*. Conductivity of glass and other amorphous materials, *J. Mackenzie*. Microstructural and polyphase effects, *J. Wimmer and H. Graham*.

TSURUTA and O'DRISCOLL *Structure and Mechanism in Vinyl Polymerization*

edited by TEIJI TSURUTA, *Department of Synthetic Chemistry, Faculty of Engineering, University of Tokyo,* and KENNETH F. O'DRISCOLL, *Department of Chemical Engineering, State University of New York, Buffalo*

552 pages, illustrated. 1969

Presents a general survey of studies on this subject in terms of physical organic chemistry. Topics are organized to focus on the most important chemical features of vinyl compounds and their response to variations in chemical and physical circumstances.

CONTENTS: Historical development of the theory of the reactivity of vinyl monomers, *M. Imoto*. Structure and reactivity of vinyl monomers, *T. Tsuruta*. Initiation in free radical polymerization, *K. F. O'Driscoll and P. Ghosh*. Termination mechanism in radical polymerization, *A. M. North and D. Postlethwaite*. Organometallic compounds as radical-type initiators for vinyl polymerization, *S. Inoue*. Heterogeneous metal peroxides, *T. Otsu*. Polymerization of α, β-disubstituted olefins, *Y. Minoura*. Polymerization of α, β-unsaturated carbonyl compounds, *D. M. Wiles*. Cationic polymerization of vinyl monomers by metal alkyl catalysts, *T. Saegusa*. Rate constants of elementary reactions in cationic polymerization, *T. Higashimura*. Elementary steps in anionic vinyl polymerization, *J. Smid*. Molecular rearrangements in polymerization of vinyl monomers, *A. D. Ketley and L. P. Fisher*.

VOGL *Polyaldehydes*

edited by OTTO VOGL, *Central Research Division, E. I. duPont de Nemours & Company, Wilmington, Delaware*

152 pages, illustrated. 1967

CONTENTS: Preface, *O. Vogl*. Polyaldehydes: introduction and brief history, *O. Vogl*. Polymerization of formaldehyde, *N. Brown*. Polymerization and copolymerization of trioxane, *M. B. Price and F. B. McAndrew*. Polymerization of aliphatic aldehydes, *O. Vogl*. Polymers of haloaldehydes, *I. Rosen*. NMR studies of polyaldehydes, *E. G. Brame, Jr. and O. Vogl*. Polymerization of fluorothiocarbonyl compounds, *W. H. Sharkey*. Crystal structure of polyaldehydes, *P. Corradini*. Morphology of polyoxymethylene, *P. H. Geil*.

VOGL and FURUKAWA *Polymerization of Heterocyclics*

edited by OTTO VOGL, *Department of Polymer Science and Engineering, Univer-*

(continued)

VOGL and FURUKAWA *(continued)*

sity of Massachusetts, Amherst, and JUNJI FURUKAWA, *Kyoto University, Japan*

216 pages, illustrated. 1973

Reviews the polymerization of cyclic ethers and thio ethers, lactones, and lactams. Also covers preparation, polymerization, and properties of perfluoro epoxides, kinetics of cyclic ether polymerization, and the influence of ring strain on the rate of polymerization and living polymers based on cationic ring opening polymerization. Valuable to scientists interested in polymer science, heterocyclic chemistry, polymer engineering, and materials science.

CONTENTS: Introduction, *J. Furukawa.* Polymerization of cyclic ethers, *T. Saegusa.* Polymerization of perfluoro epoxides, *H. Eleuterio.* Specific nature of the polymerization of heterocyclics, *N. Enikolpoyan.* New trioxane copolymers, *H. Cherdron.* Alkylene sulfide polymerizations, *F. Lautenschlaeger.* Lactone polymerization and polymer properties, *G. Brode and J. Koleske.* Lactam polymerization, *J. Sebenda.*

WALKER and THROWER ***Chemistry and Physics of Carbon:*** *A Series of Advances*

a series edited by PHILIP L. WALKER and PETER A. THROWER, *Department of Material Sciences, Pennsylvania State University, University Park*

Vol. 1 400 pages, illustrated. 1965

Vol. 2 400 pages, illustrated. 1966

Vol. 3 464 pages, illustrated. 1968

Vol. 4 416 pages, illustrated. 1968

Vol. 5 400 pages, illustrated. 1969

Vol. 6 368 pages, illustrated. 1970

Vol. 7 424 pages, illustrated. 1970

Vol. 8 480 pages, illustrated. 1973

Vol. 9 272 pages, illustrated. 1973

Vol. 10 288 pages, illustrated. 1973

Vol. 11 in preparation. 1974

CONTENTS:

Volume 1: Dislocations and stacking faults in graphite, *S. Amelinckx, P. Delavignette, and M. Heerschap.* Gaseous mass transport within graphite, *G. F. Hewitt.* Microscopic studies of graphite oxidation, *J. M. Thomas.* Reactions of carbon with carbon dioxide and steam, *S. Ergun and M. Mentser.* Formation of carbon from gases, *H. B. Palmer and C. F. Cullis.* Oxygen chemisorption effects on graphite thermoelectric power, *P. L. Walker, Jr., L. G. Austin, and J. J. Tietjen.*

Volume 2: Electron microscopy of reactivity changes near lattice defects in graphite, *G. R. Hennig.* Porous structure and adsorption properties of active carbons, *M. M. Dubinin.* Radiation damage in graphite, *W. N. Reynolds.* Adsorption from solution by graphite surfaces, *A. C. Zettlemoyer and K. S. Narayan.* Electronic transport in pyrolytic graphite and boron alloys of pyrolytic graphite, *C. A. Klein.* Activated diffusion of gases in molecular-sieve materials, *P. L. Walker, Jr., L. G. Austin and S. P. Nandi.*

Volume 3: Nonbasal dislocations in graphite, *J. M. Thomas and C. Roscoe.* Optical studies of carbon, *S. Ergun.* Action of oxygen and carbon dioxide above 100 millibars on "pure" carbon, *F. M. Lang and P. Magnier.* X-ray studies of carbon, *S. Ergun.* Carbon transport studies for helium-cooled high-temperature nuclear reactors, *M. R. Everett, D. V. Kinsey, and E. Römberg.*

Volume 4: X-ray diffraction studies on carbon and graphite, *W. Ruland.* Vaporization of carbon, *H. B. Palmer and M. Shelef.* Growth of graphite crystals from solution, *S. B. Austerman.* Internal friction studies on graphite, *T. Tsuzuku and M. H. Saito.* Formation of some graphitizing carbons, *J. D. Brooks and G. H. Taylor.* Catalysis of carbon gasification, *P. L. Walker, Jr., M. Shelef, and R. A. Anderson.*

Volume 5: Deposition, structure and properties of pyrolytic carbon, *J. C. Bokros.* The thermal conductivity of graphite, *B. T. Kelly.* The study of defects in graphite by transmission electron microscopy, *P. A. Thrower.* Intercalation isotherms on natural and pyrolytic graphite, *J. G. Hooley.*

Volume 6: Physical adsorption of gases and vapors of graphitized carbon blacks, *N. N. Avgul and A. V. Kiseleyv.* Graphitization of soft carbons, *J. Maire and J. Méring.* Surface complexes on carbons, *B. R. Puri.* Effects of reactor irradiation on the dynamic mechanical behavior of graphites and carbons, *R. E. Taylor and D. E. Kline.*

Volume 7: The kinetics and mechanism of graphitization, *D. B. Fischbach.* The kinetics of graphitization, *A. Pacault.* Electronic properties of doped carbons, *A. M. Marchand.* Positive and negative magnetoresistances in carbons, *P. Delhaes.* The chemistry of the pyrolytic conversion of organic compounds to carbon, *E. Fitzer, K. Mueller and W. Schaefer.*

Volume 8: The electronic properties of graphite, *I. Spain.* Surface properties of carbon fibers, *D. McKee and V. Mimeault.* The behavior of fission products captured in graphite by nuclear recoil, *S. Yajima.*

Volume 9: Carbon fibers from rayon presursors, *R. Bacon.* Control of structure of carbon for use in bioengineering, *J. Bokros, L. LaGrange, and F. Schoen.* Deposition of pyrolytic carbon in porous solids, *W. Kotlensky.*

Volume 10: The thermal properties of graphite, *B. Kelly and R. Taylor.* Lamellar reactions in graphitizable carbons, *M. Robert, M. Oberlin, and J. Mering.* Methods and mechanisms of growth of synthetic diamond, *F. Bundy, H. Strong, and R. Wentorff, Jr.*

† *Volume edited by Philip L. Walker*

Volume 11: Structure and physical properties of carbon fibers, *W. Reynolds*. Highly oriented pyrolytic graphite, *A. Moore*. Evaporated carbon films, *I. McLintock and J. Orr*. Deformation mechanisms in carbons, *G. Jenkins*.

WARD *Chemical Modification of Papermaking Fibers*

(Fiber Science Series, Volume 4)
by KYLE WARD, JR., *Institute of Paper Chemistry, Appleton, Wisconsin*
256 pages, illustrated. 1973

Bridges the gap between research and industrial applications in the field of chemical modification of papermaking fibers. Deals with the chemical changes which produce new or improved properties in paper products. Of particular importance to researchers and technologists in the paper, textile, and related industries, and students of polymer and organic chemistry.

CONTENTS: Introduction • Esterification • Etherification • Oxidation • Crosslinking • Graft polymerization onto cellulose.

WASLEY *Stress Wave Propagation in Solids: An Introduction*

(Monographs and Textbooks in Material Science Series, Volume 5)
by RICHARD J. WASLEY, *Department of Chemistry, University of California, Livermore*
344 pages, illustrated. 1973

Provides the fundamentals necessary for the study of the propagation of short duration, high-intensity, nonelastic, mechanical stress disturbances in solids. The first part of the book treats some of the dynamic analyses of elastic solid media which obey Hooke's law. The last section discusses some of the theoretical and experimental aspects of one-dimensional stress waves and shock loading and response. For scientists and engineers desiring further knowledge in the field of stress wave propagation. Also of interest to advanced undergraduate and graduate students of engineering and physics.

CONTENTS: Elasticity: Quasistatic and dynamic response • Wave propagation in extended media • Wave propagation in semi-extended media: Reflection and refraction • Wave propagation in circular cylindrical rods • Selected applications of concepts of elasticity • Nonelastic material behavior • One-dimensional stress wave investigations • Nonelastic (shock) one-dimensional strain wave investigations.

WEINBERG *Tools and Techniques in Physical Metallurgy*

In 2 Volumes
edited by FRED WEINBERG, *University of British Columbia, Vancouver*
Vol. 1 416 pages, illustrated. 1970
Vol. 2 376 pages, illustrated. 1970

Aids the non-specialist in understanding and making use of the new instruments and techniques of physical metallurgy.

CONTENTS:

Volume 1: Temperature measurement, *R. Bedford, T. Dauphinee, and H. Preston-Thomas*. X-ray diffraction, *C. M. Mitchell*. Crystal growth and alloy preparation, *F. Weinberg and J. T. Jubb*. Quantitative metallography, *J. R. Blank and T. Gladman*. Metallography, *H. E. Knechtel, W. F. Kindle, J. L. McCall, and R. D. Buchheit*.

Volume 2: Electron microscopy, *E. Smith*. Scanning electron microscopy, *O. Schaaber*. Field-ion microscopy, *B. Ralph*. Thermionic-emission microscopy, *W. L. Grube and S. R. Rouze*. Electron-probe microanalysis, *L. C. Brown and H. Thresh*. Emission spectrography and atomic absorption spectrophotometry, *G. L. Mason*.

WILSON *Radiation Chemistry of Monomers-Polymers-Plastics*

by JOSEPH E. WILSON, *Department of Chemistry, Bishop College, Dallas, Texas*
in preparation. 1974

Provides an up-to-date survey of the radiation chemistry of monomers, polymers, and plastics. Gives essential information on radiation properties, measurement, and detection, and the primary chemical results of the interaction of radiation with matter. Of particular interest to polymer and radiation chemists.

CONTENTS: (tentative): Types and sources of radiation • Fundamental effects of the irradiation of matter • Short-term chemical effects of radiation absorption • Radiation chemistry of small molecules • Radiolytic polymerization in homogeneous systems • Radiolytic polymerization in the solid state • Radiation-induced polymerization in thermosetting, polyester, and emulsion systems • Irradiation of polymers: Crosslinking versus scission • Radiolytic grafting of monomers on polymeric films • Radiolytic grafting on fibers.

YOCUM and NYQUIST *Functional Monomers: Their Preparation, Polymerization, and Applications*

In 2 Volumes
edited by RONALD H. YOCUM, *The Dow Chemical Company, Freeport, Texas,* and

(continued)

YOCUM and NYQUIST *(continued)*

EDWIN B. NYQUIST, *The Dow Chemical Company, Midland, Michigan*

Vol. 1 712 pages, illustrated. 1973

Vol. 2 321 pages, illustrated. 1973

A practical reference work which deals with functional monomers. Presents a broad technical background on the preparation and polymerization of individual functional monomers and their applications to various areas of industry. Of special interest to both academic and industrial chemists, particularly those working in the paint, coatings, and textile industry.

CONTENTS:

Volume 1: Acrylamide and other alpha, beta, and unsaturated acids, *D. C. MacWilliams.* Reactive halogenated monomers, *C. F. Raley and R. J. Dólinski.* Hydroxy monomers, *E. B. Nyquist.* Sulfonic acids and sulfonate monomers, *D. A. Kangas.*

Volume 2: Reactive heterocyclic monomers, *D. Tomalia.* Acidic monomers, *L. Luskin.* Basic monomers, vinylpyridines and aminoalkyl (METH) acrylates, *L. Luskin.*

ZIEF *Purification of Inorganic and Organic Materials: Techniques of Fractional Solidification*

edited by MORRIS ZIEF, *J. T. Baker Chemical Co., Phillipsburg, New Jersey*

340 pages, illustrated. 1969

Of interest to the chemist, chemical engineer, and metallurgist.

CONTENTS: Analysis of ultrapure materials, *C. L. Grant.* Optical-emission spectrochemical analysis—arc, spark, and flame, *C. L. Grant.* Spark-source mass spectrography, *P. R. Kennicott.* Atomic-absorption spectroscopy, *J. W. Robinson.* Infrared spectrophotometry, *K. E. Stine and W. F. Ulrich.* Gas-liquid chromatography, *R. A. Keller.* Differential thermal analysis and differential scanning calorimetry, *E. M. Barrall, II, and J. F. Johnson.* Electrical resistance-ratio measurement, *G. T. Murray.* Reduction of cyclohexane content of benzene under steady flow conditions, *J. D. Henry, Jr., M. D. Danyi, and J. E. Powers.* Purification of aromatic amines, *B. Pouyet.* The freezing staircase method, *C. P. Saylor.* Purification of aluminum, *J. L. Dewey.* Concentration of humic acids in natural waters, *J. Shapiro.* Fractionation of polystyrene, *J. D. Loconti.* Purification and growth of large anthracene crystals, *J. N. Sherwood.* Purification of indium antimonide, *A. R. Murray.* Purification of alkaline iodides (KI, RbI, CsI), *D. Ecklin.* Zone melting of metal chelate systems, *K. Ueno, H. Kobayashi, and H. Kaneko.* Purification of dienes, *R. Kieffer.* Purification of kilogram quantities of an organic compound, *J. C. Maire and M. Delmas.* Rapid purification of organic substances, *M. J. van Essen, P. F. J. van der Most, and W. M. Smit.* Investigation of zone-melting purification of gallium trichloride by a radiotracer method, *W. Kern.* Purification of potassium chloride by radio-frequency heating, *R. Warren.* Purification of a metal by electron-beam heating, *R. E. Reed and J. C. Wilson.* Heating by hollow-cathode gas discharge, *W. Class.* Continuous zone refining of benzoic acid, *J. K. Kennedy and G. H. Moates.* Purification of naphthalene in a centrifugal field, *E. L. Anderson.* Zone-melting chromatography of organic mixtures, *H. Plancher, T. E. Cogswell and D. R. Latham.* The concentration of flavors at low temperature, *M. T. Huckle.* Containers for pure substances, *E. C. Kuehner and D. H. Freeman.*

ZIEF and SPEIGHTS *Ultrapurity: Methods and Techniques*

edited by MORRIS ZIEF, *J. T. Baker Chemical Co., Philipsburg, New Jersey,* and ROBERT M. SPEIGHTS, *American Metal Climax, Inc., Golden, Colorado*

720 pages, illustrated. 1972

Brings together for the first time the four essential and interrelated parameters of ultrapurity: preparation, handling, containment, and analysis. Reflects the continuing progress in the preparation of ultrapure chemicals, the explosive growth in developments pertaining to the handling and containment of these materials, as well as the necessity for complete analysis.

Directed to all those working in research, development, or analysis of ultrapure products.

CONTENTS: Purification of alkali halides, *F. Rosenberger.* Purification of organic solvents by frontal-analysis chromatography, *H. Engelhardt.* The preparation of pure sodium and potassium, *R. L. McKisson.* Sublimation of phosphorus pentoxide, *R. D. Mounts.* Purification of proteins by membrane ultrafiltration, *G. J. Fallick.* The purification of p-xylene by partial freezing, *J. R. Gruden and M. Zief.* Purification of isopropylbenzene by preparative gas-liquid chromatography, *J. R. Gruden and M. Zief.* The preparation of ultrapure chemicals by fractional distillation, *H. Plancher and W. E. Haines.* Purification by dry-column chromatography, *F. M. Rabel.* Preparation of ultrapure water, *V. C. Smith.* Preparation and characterization of cholesterol, *I. L. Shapiro.* Contamination problems in trace-element analysis and ultrapurification, *D. E. Robertson.* Airborne contamination, *J. A. Paulhamus.* Glass containers for ultrapure solutions, *P. B. Adams.* Vitreous silica, *G. Hetherington and L. W. Bell.* Ceramics, *C. Garnsworthy.* High-purity chemicals—a challenge to practical analysis, *A. J. Barnard, Jr.* Emission spectroscopy, *E. C. Snooks.* Flame spectrophotometric trace analysis, *D. C. Burrell.* Neutron-activation analysis, *J. J. Kelly.* Visible spectrophotometry, *R. H.*

Weiss. Coulometric titration, *G. W. Higgins*. Information sources for ultrapurification and characterization, *T. E. Connolly*.

ZIEF and WILCOX ***Fractional Solidification***

edited by MORRIS ZIEF, *J. T. Baker Chemical Company, Phillipsburg, New Jersey*, and WILLIAM R. WILCOX, *Aerospace Corporation, Los Angeles*

736 pages, illustrated. 1967

CONTENTS: Introduction, *W. R. Wilcox*. **Part I: Basic Principles:** Phase diagrams, *G. M. Wolten and W. R. Wilcox*. Mass transfer in fractional solidification, *W. R. Wilcox*. Constitutional supercooling and microsegregation, *G. A. Chadwick*. Polyphase solidification, *G. A. Chadwick*. Heat transfer in fractional solidification, *W. R. Wilcox*. **Part II: Laboratory Scale Apparatus:** Laboratory scale apparatus, *E. A. Wynne and M. Zief*. Batch zone melting, *E. A. Wynne*. Progressive freezing, *D. Richman, E. A. Wynne, and F. D. Rosi*. Continuous-zone melting, *J. K. Kennedy and G. H. Moates*. Column crystallization, *R. Albertins, W. C. Gates, and J. E. Powers*. Zone precipitation and allied techniques, *I. A. Eldib*. **Part III: Industrial Scale Equipment:** Proabd refiner, *J. G. D. Molinari*. Newton Chambers' process, *J. G. D. Molinari*. Rotary-drum techniques, *J. C. Chaty*. Phillips fractional-solidification process, *D. L. McKay*. Desalination by freezing, *J. C. Orcutt*. **Part IV: Applications:** Ultrapurification, *P. Jannke, J. K. Kennedy, and G. H. Moates*. Ultrapurity in pharamaceuticals, *P. Jannke*. Ultrapurity in electronic materials, *J. K. Kennedy and G. H. Moates*. Ultrapurity in materials research, *J. K. Kennedy, G. H. Moates, and W. R. Wilcox*. Ultrapurity in crystal growth, *G. H. Moates and J. K. Kennedy*. Bulk purification, *J. D. Loconti*. Analytical applications of fractional solidification, *A. S. Yue*. Materials preparation, *D. Richman and F. D. Rosi*. **Part V: Economics:** Economics of fractional solidification, *J. C. Chaty and W. R. Wilcox*. **Part VI: Appendix:** Introduction, *M. Zief and C. E. Shoemaker*. Survey of inorganic materials, *C. E. Shoemaker and R. L. Smith*. Survey of organic materials, *M. Zief*.

OTHER BOOKS OF INTEREST

CUTLER and DAVIS ***Detergency:*** *Theory and Test Methods*

In 2 Parts

(Surfactant Science Series, Volume 5)

edited by W. G. CUTLER, and R. DAVIS, *Whirlpool Corporation, Benton Harbor, Michigan*

Part 1 464 pages, illustrated. 1972

Part 2 in preparation. 1973

JUNGERMANN ***Cationic Surfactants***

(Surfactant Science Series, Volume 4)

edited by ERIC JUNGERMANN, *Armour-Dial, Inc., Chicago*

672 pages, illustrated. 1970

LINFIELD ***Anionic Surfactants***

(Surfactant Science Series)

edited by WARNER M. LINFIELD, *U.S. Department of Agriculture, Philadelphia, Pennsylvania*

in preparation. 1974

MATTSON and MARK

Activated Carbon: *Surface Chemistry and Adsorption from Solution*

by JAMES S. MATTSON, *Rosenstiel School of Marine and Atmospheric Sciences, University of Miami, Florida*, and HARRY B. MARK, JR., *Department of Chemistry, University of Cincinnati, Ohio*

248 pages, illustrated. 1971

PATRICK ***Treatise on Adhesion and Adhesives***

edited by ROBERT L. PATRICN, *Alpha Research and Development, Inc., Blue Island, Illinois*

Vol. 1 ***Theory***
496 pages, illustrated. 1967

Vol. 2 ***Materials***
568 pages, illustrated. 1969

Vol. 3 ***Special Topics***
264 pages, illustrated. 1973

SCHICK ***Nonionic Surfactants***

(Surfactant Science Series, Volume 1)

edited by MARTIN J. SCHICK, *Central Research Laboratories, Interchemical Corporation, Clifton, New Jersey*

1,120 pages, illustrated. 1967

SHINODA ***Solvent Properties of Surfactant Solutions***

(Surfactant Science Series. Volume 2)

edited by KOZO SHINODA, *Department of Chemistry, Yokohama National University, Japan*

376 pages, illustrated. 1967

SLADE and JENKINS
Thermal Analysis

(Techniques and Methods of Polymer Evaluation Series, Volume 1)

edited by PHILIP E. SLADE, JR., *Monsanto Company, Pensacola, Florida,* and LLOYD T. JENKINS, *Chemstrand Research Center, Durham, North Carolina*

264 pages, illustrated. 1966

SLADE and JENKINS
Thermal Characterization Techniques

(Techniques and Methods of Polymer Evaluation Series, Volume 2)

edited by PHILIP E. SLADE, JR., *Monsanto Company, Pensacola, Florida* and LLOYD T. JENKINS, *Chemstrand Research Center, Durham, North Carolina*

384 pages, illustrated. 1970

STEVENS ***Characterization and Analysis of Polymers by Gas Chromatography***

(Techniques and Methods of Polymer Evaluation Series, Volume 3)

by MALCOLM P. STEVENS, *American University of Beirut, Lebanon*

216 pages, illustrated. 1969

SWISHER ***Surfactant Biodegradation***

(Surfactant Science Series, Volume 3)

by R. D. SWISHER, *Monsanto Company, St. Louis, Missouri*

520 pages, illustrated. 1970

WALTON ***Radome Engineering Handbook:*** *Design and Principles*

(Ceramics and Glass: Science and Technology Series, Volume 1)

edited by JESSE D. WALTON, JR., *Georgia Institute of Technology, Atlanta*

616 pages, illustrated. 1970

JOURNALS OF INTEREST

BIOMATERIALS, MEDICAL DEVICES, AND ARTIFICIAL ORGANS
An International Journal

editor: T. F. YEN, *University of Southern California, Los Angeles*

The aim of this new international journal is to bridge the gap between the theoretical aspects and practical applications of artificial organs and other medical devices, and implantation materials. The basic principles responsible for the success of artificial organs are stressed in order to encourage new research in this field.

4 issues per volume

JOURNAL OF MACROMOLECULAR SCIENCE—*Chemistry*

editor: GEORGE E. HAM, *White Plains, New York*

This international journal provides scientists with a cross-section of the outstanding contributions from laboratories around the world—published four and one-half months from publisher's receipt of last manuscript. The fields covered include anionic, cationic, and free-radical addition polymerization and copolymerization, the manifold forms of condensation polymerization, polymer reactions, molecular weight studies, temperature-dependent properties, rheology, effects of radiation of all forms, polymer degradation, and many others.

8 issues per volume

JOURNAL OF MACROMOLECULAR SCIENCE—*Physics*

editor: PHILLIP H. GEIL, *Case Western Reserve University*

A periodical devoted to the publication of significant fundamental contributions concerning the physics of macromolecular solids and liquids. Papers deal with research in transition mechanisms and structure property relationships, the physics of polymer solutions and melts, glassy and rubbery amorphous solids, and individual polymer molecules and natural polymers, as well as all the areas generally contained in polymer state physics.

4 issues per volume

JOURNAL OF MACROMOLECULAR SCIENCE—*Reviews in Macromolecular Chemistry*

editors: GEORGE B. BUTLER, *University of Florida, Gainesville,* and KENNETH F. O'DRISCOLL, *State University of New York, Buffalo,* and MITCHELL SHEN, *University of California, Berkeley*

Topics in this journal are reviews of certain recent chronological periods, and also have the advantage of reflecting the authors' knowledge, interpretation, and concise summary of the state of knowledge in the given area. Because of the nature of the journal, and the short time between completion of a manuscript and its publication, reviews of this nature more closely approximate a current review than could otherwise be accomplished.

2 issues per volume

POLYMER-PLASTICS TECHNOLOGY AND ENGINEERING

editor: LOUIS NATURMAN, *Stamford, Connecticut*

This journal reflects the increasing importance that polymer applications, processing developments, and mass production of new polymer products will have in the coming years. The emphasis of the articles that comprise the journal will also consider plastics technology and engineering as an important new feature of this publication.

2 issues per volume

Examination On-Approval Policy

Our policy allows instructors to examine a particular book for a period of two months without charge. In the event that the book is definitely adopted for a course as a class text, the instructor may retain the copy as *his desk copy,* provided he advises us of the adoption and the number of students enrolled in the class. If, however, the book will not be used as a class text, the instructor may return it or send us his remittance, less the educational discount.

JOURNAL SUBSCRIPTION INFORMATION

Subscriptions are entered on a calendar year basis. When a subscription is entered, it entitles the subscriber to all issues in the particular volume.

All journal subscriptions are processed after payment has been received. Only prepaid orders will receive service.

Cancellations requested for other than publisher's error will be accepted only prior to the publication of the first issue of each current volume, and will be subject to a handling charge.

Please add **foreign postage** for delivery to all countries outside the U.S. and Canada.

Air mail postage is available upon request.

Indexes to journals are bound into the last issue of each volume with the exception of review journals which are not indexed.

Back volume information and prices are available upon request.

A complimentary copy of any journal is available upon request.

Date____________________

MARCEL DEKKER, INC.
95 Madison Avenue
New York, New York 10016

Please send me a complimentary copy of the following journal(s):

____

____

____

Name____________________________________

Position__________________________________

Company_________________________________

Address__________________________________

City__________________State____________Zip________